Fuels and Combustion

3 EDITION

Samir Sarkar
Retired Professor, Department of Chemical Engineering,
Indian Institute of Technology Bombay, Mumbai

Universities Press

CRC Press
Taylor & Francis Group
Boca Raton London New York

CRC is an imprint of the Taylor & Francis Group,
an informa business

Universities Press (India) Private Limited

Registered Office
3-6-747/1/A & 3-6-754/1, Himayatnagar, Hyderabad 500 029 (A.P.) India
info@universitiespress.com; www.universitiespress.com

Distributed in India, China, Pakistan, Bangladesh, Sri Lanka, Nepal, Bhutan, Indonesia Malaysia, Singapore and Hong Kong by
Orient Blackswan Private Limited
Registered Office
3-6-752 Himayatnagar, Hyderabad 500 029 (A.P.) India

Other Offices
Bangalore / Bhopal / Bhubaneshwar / Chennai
Ernakulam / Guwahati / Hyderabad / Jaipur / Kolkata
Lucknow / Mumbai / New Delhi / Patna

Distributed in the rest of the world by
CRC Press LLC, Taylor and Francis Group,
6000 Broken Sound Parkway, NW, Suite 300, Boca Raton, FL 33487, USA

First published in India by
Universities Press (India) Private Limited 2009

ISBN 978 1 4398 2541 9

Set in Times New Roman 11/13
Triviata Publishing Limited, Chennai 600 083, India

Printed in India by
Graphica Printers, Hyderabad, India

Published by
Universities Press (India) Private Limited
3-6-747/1/A & 3-6-754/1, Himayatnagar, Hyderabad 500 029 (A.P.) India

Preface

Third Edition

The first edition of the book was published in 1974. It was so well accepted by students and professional engineers that it necessitated the publication of an improved edition within a few years. Therefore the second edition of the book was published in 1990.

I retired from the Indian Institute of Technology, Bombay in December 1993. At the end of 2007 I was asked by Universities Press, an associate of Orient Longman Private Limited (now renamed Orient Blackswan Private Limited) to write the third edition of the book. Being physically incapacitated I initially declined. However, the persuasive efforts of Mr Madhu Reddy, who met me in Kolkata, emboldened me to undertake the revision. The first chapter has once again been revised to include the most recent data on global reserves and consumption of fuels. The data in other chapters have also been updated wherever available. The layout of the book has been redesigned to give visibility to all sub-topics and improve readability.

In the process of compilation of the third edition I acknowledge the help received from the library of Development Consultants Private Limited (DCPL), Kolkata, Coal India Limited and my son Prof. Sutanu Sarkar, for his compilation of recent international data about the conversion of coal to oil throughout the world and other fuel related information

I pray that this edition should be useful to students and teachers in India and abroad.

May 2009 *Samir Sarkar*

Preface

Second Edition

Primarily meant for students as a textbook, *Fuels and Combustion* has been well accepted by professional engineers as well.

The book was out of print for a long time. Repeated requests have been received for bringing out its second edition.

The first edition has been critically read several times. The bulk of the text has remained topical after fourteen years of its publication. However, chapter one has been thoroughly revised. Some of the topics in other chapters have been updated. The numerical problems in chapter six and seven have been augmented. All the problems for practising have been provided with numerical answers.

The statistical data needed close scrutiny. The addition of newer data has been possible by the help received from the staff of the Central Fuel Research Institute, Dhanbad, and the Tata Energy Research Institute, New Delhi. I am indeed thankful to them.

Acknowledgement is due to the Curriculum Development Centre of the Indian Institute of Technology, Bombay, for financial support in the preparation of the manuscript.

December 1988

Samir Sarkar
Department of Chemical Engineering
Indian Institute of Technology, Bombay

Preface

First Edition

'Fuels and Combustion' forms an integral part of the undergraduate degree courses in Chemical, Mechanical, Metallurgical and Aeronautical Engineering. While in the vast majority of courses the subject retains a distinct identity, there are instances when it is treated as a minor adjunct to other topics. Apparently this is the result of hasty distribution of available hours, where the first victim is the students' understanding of the subject, leading to difficulties in the application of the knowledge in later years. Since there can be no two opinions on the basic need of the subject for the engineering courses concerned, a definite place has to be found for it in the academic curriculum.

The subject has two aspects, namely, 'Fuels' and 'Combustion'. Their relative importance in a given course is a matter of judgement. However, one should bear in mind that a good treatment of 'combustion' can only follow a sound understanding of different 'fuels'. On the other hand a rigid balance between the different aspects of the subject cannot constitute a set pattern for all the courses. Enough flexibility in this regard should be available to the teacher and the taught. To a chemical engineer, fuels not only supply energy but also provide basic raw materials for a host of vital industries. It is, therefore, understandable why much greater stress should be laid on the processing of fuels for the benefit of chemical engineering students as compared to those of other branches of engineering. On the contrary, an Aeronautical Engineering student may not require more than a cursory treatment of 'solid fuels', and may prefer to raise the 'combustion' content of the subject.

I have planned the present textbook in such a way that all those requiring a course on Fuels and Combustion will find the required material in it. In consonance with the modern methodology of teaching I have laid stress on the systematic development of the subject. The maximum attention has been given to elucidating the fundamental principles, while the analytic approach has been exploited to the fullest possible extent. Indian raw materials and Indian practices have been given due coverage. Numerical problems have been worked out in many sections of the book and additional problems have been furnished for further practices. 'Furnaces and Boilers' has been intentionally excluded from the present textbook. A mere descriptive treatment of this subject is considered uninteresting and undesirable for the engineering students. In fact 'Furnace Design' has received a separate identity in the curriculum of some courses, even at the undergraduate level. This requires a separate textbook on its own.

In preparing the book I have received ample encouragement and support both from the Director of the Indian Institute of Technology, Bombay, and from the Head of the Department of Chemical Engineering, and I am thankful to them. I have benefited from innumerable books and articles and also by discussions with many teachers and students. It will not be fair on my part to single them out. However I must acknowledge the assistance received from TB Ray, J Eqbal and N Murugesan in the various stages of the preparation of the book.

I have derived great pleasure in writing this book. If the students for whom it is primarily meant find it useful, my efforts will have been more than rewarded. Year after year I have been facing the questions of my students on the availability of a suitable textbook. My colleagues have also had a similar experience. I have myself attempted to provide an answer to this question. Any suggestions for the improvement of this work will be gratefully welcomed and duly acknowledged.

ACKNOWLEDGEMENT

I thank my family, especially my wife, who not only inspired me but also supervised and managed the entire process. My thanks also go out to all those who have been associated with bringing out this edition.

Contents

LIST OF TABLES

LIST OF FIGURES

1 Introduction

1.1 ENERGY AND ENERGY RESOURCES

A *fuel* is basically a source of heat. The usual method of producing heat from fuel is by the process of combustion, which is a chemical reaction between the fuel and the oxidant. In the vast majority of cases, air is the oxidant, although oxygen, oxygen-enriched air and other chemicals are also used as oxidants. With the advent of *nuclear fuels* which generate heat by nuclear reactions, the common fuels may be termed *chemical fuels*. *Rocket fuels* are different from conventional fuels; the oxidant for combustion of the fuel is carried by the rocket itself, which therefore does not depend on its surroundings for propulsion. It should be noted that explosives are not regarded as fuels.

Depending upon the state of development, energy resources may be divided into three groups.

- The first group comprises those which are currently important in energy supply—fossil fuels, hydel energy and nuclear energy from fission.
- The second group consists of the resources that are currently used, but only play a limited part from the point of view of the general economy. The quantities used are small on a global scale, though the role of some of these resources is capable of becoming more important later. These include geothermal energy, aeolian energy (wind), chemical energy (batteries), photosynthetic fuels (vegetation and fuel wood), and garbage, municipal wastes and agriresidues.
- The third group covers resources which are currently the subject of research and development at various stages. Some of these are likely to play an important role in the future, such as solar energy, nuclear energy from fusion and energy from the sea (tidal, wave and thermal).

Energy is primarily used by four different sectors.

1. *Domestic*: Energy is used for cooking, lighting, running appliances, cooling and heating.
2. *Transportation*: Energy is used for moving cars, trucks, aircraft, trains and ships.
3. *Industrial*: Energy is used in factories, industrial complexes, research and development and trade.
4. *Health, education and welfare*: Energy is used for supply and distribution of electricity and water, sewage and trash collection and disposal, and administration.

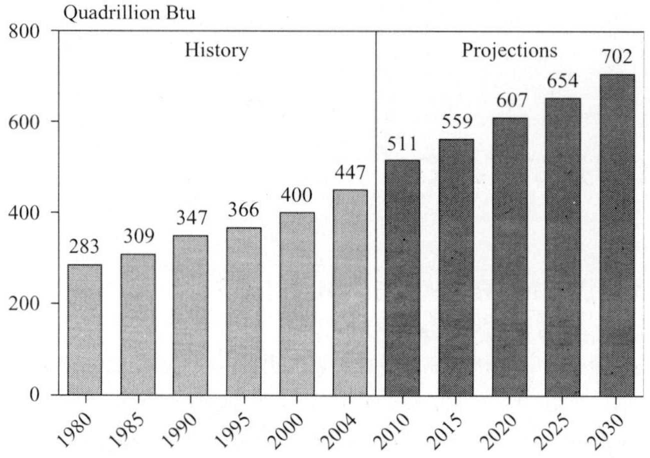

Figure 1.1 World marketed energy consumption, 1980–2030
Sources: History: Energy Information Administration (EIA). *International Energy Annual 2004* (May–July 2006), web site www.eia.doe.gov/iea.

The energy requirements of the advanced countries are almost equally divided among these four heads.

Fuels also have other uses. It is well known that petroleum is also a storehouse of hydrocarbons that are required for the production of petrochemicals. The by-products of carbonisation of coal are of great value in manufacturing chemicals, dyestuffs, pharmaceuticals and perfumery materials. The coke in blast furnaces not only supplies heat but also takes part in the chemical reaction of iron and steel production. Both coal and petroleum are important raw materials in the production of fertilisers, and play a significant role in the world's drive for augmenting food output. Carbon electrodes ultimately depend upon coal and petroleum as the initial raw material. Numerous other examples may be cited for the non-fuel uses of substances classified as fuels. In spite of the great importance of these uses, fuels are primarily recognised as an energy resource.

Currently the energy resources of the world are of great concern because of two factors: the accelerated rate of growth in population, and the growing energy requirement per head. The earth now has about 6,000 million people. The population was about 400 million in AD 1550, 1,100 million in AD 1850 and may reach about 8,203 million in 2030 as per the projection. The annual rate of growth in the world demand per capita for energy utilisation is projected to be about 2% during 2003–2030. The world marketed energy consumption was about 283 quadrillion Btu in 1980, 400 in 2000 and is projected to be 702 in 2030 (Fig. 1.1). The total energy consumption was 29.8 quadrillion kcal in 2003 in the Organisation for Economic Co-operation and Development (OECD), North America.

Table 1.1 gives data about regional energy consumption from 1990 to 2003 and also projections for 2010 to 2030 (see Appendix 1.1 for regional definitions). A complete picture of the energy requirement of the world is available from it. The last column of the table gives the average annual per cent change from 2003 to 2030.

Table 1.1 World total energy consumption by region and fuel, Reference case 1990–2030 (quadrillion kcal)

Region/ Country	History			Projections					Average annual per cent change
	1990	2002	2003	2010	2015	2020	2025	2030	2003–2030
OECD									
OECD North America									
Oil	10.2	11.7	11.9	13.2	14.0	14.7	15.5	16.4	1.2
Natural gas	5.8	7.2	7.1	7.7	8.5	9.0	9.3	9.6	1.1
Coal	5.2	6.0	6.1	6.9	7.2	7.8	8.7	9.6	1.7
Nuclear	1.7	2.3	2.2	2.4	2.5	2.6	2.6	2.6	0.6
Other	2.4	2.5	2.4	2.9	3.0	3.2	3.5	3.7	1.6
Total	**25.3**	**29.7**	**29.8**	**33.1**	**35.2**	**37.3**	**39.6**	**41.9**	**1.3**
OECD Europe									
Oil	7.2	8.0	8.0	8.2	8.3	8.2	8.3	8.4	0.2
Natural gas	2.8	4.5	4.6	5.5	6.2	6.7	7.3	8.0	2.0
Coal	4.4	3.2	3.2	3.2	3.2	3.3	3.3	3.4	0.2
Nuclear	2.0	2.4	2.5	2.4	2.3	2.1	2.0	1.9	1.0
Other	1.2	1.5	1.5	2.0	2.0	2.1	2.1	2.1	1.3
Total	**17.6**	**19.6**	**19.8**	**21.3**	**22.0**	**22.4**	**23.0**	**23.8**	**0.7**
OECD Asia									
Oil	3.7	4.4	4.5	4.7	4.8	4.9	5.1	5.2	0.5
Natural gas	0.7	1.3	1.3	1.4	1.6	1.7	1.7	1.8	1.2
Coal	1.3	2.1	2.1	2.4	2.6	2.6	2.7	2.9	1.2
Nuclear	0.6	1.0	0.9	1.1	1.2	1.3	1.5	1.6	2.0
Other	0.4	0.4	0.5	0.6	0.6	0.6	0.6	0.6	0.9
Total	**6.7**	**9.2**	**9.3**	**10.2**	**10.8**	**11.1**	**11.6**	**12.1**	**1.0**
Total OECD									
Oil	21.1	24.1	24.4	26.1	27.1	27.8	28.9	30.0	0.8
Natural gas	9.3	13.0	13.0	14.6	16.3	17.4	18.3	19.4	1.5
Coal	10.9	11.3	11.4	12.5	13.0	13.7	14.7	15.9	1.2
Nuclear	4.3	5.7	5.6	5.9	6.0	6.0	6.1	6.1	0.3
Other	4.0	4.4	4.5	5.5	5.6	5.9	6.2	6.4	1.5
Total	**49.6**	**58.5**	**58.9**	**64.6**	**68.0**	**70.8**	**74.2**	**77.8**	**1.0**
Non-OECD Europe and Eurasia									
Oil	4.9	2.5	2.5	2.9	3.1	3.3	3.5	3.7	1.4
Natural gas	6.9	5.8	6.1	7.4	8.2	9.0	9.7	10.5	2.0
Coal	3.8	2.1	2.1	2.2	2.5	2.8	2.9	3.2	1.5
Nuclear	0.6	0.7	0.7	0.8	0.9	1.1	1.2	1.2	2.0
Other	0.7	0.8	0.8	1.0	1.2	1.2	1.2	1.3	1.9
Total	**16.9**	**11.9**	**12.2**	**14.3**	**15.9**	**17.4**	**18.5**	**19.9**	**1.8**

Table 1.1 *Continued*

Region/ Country	History			Projections					Average annual per cent change 2003–2030
	1990	2002	2003	2010	2015	2020	2025	2030	
Non-OECD Asia									
Oil	3.5	6.7	7.0	9.7	11.0	12.4	13.9	15.5	3.0
Natural gas	0.8	1.8	2.0	3.1	4.2	5.2	6.4	7.7	5.2
Coal	6.9	9.8	10.4	16.1	19.1	21.9	24.8	28.0	3.8
Nuclear	0.1	0.2	0.2	0.5	0.7	1.0	1.2	1.3	6.3
Other	0.8	1.2	1.3	2.5	2.7	3.0	3.3	3.7	4.0
Total	**12.1**	**19.7**	**20.9**	**31.9**	**37.7**	**43.5**	**49.6**	**56.2**	**3.7**
Non-OECD Middle East									
Oil	1.8	2.6	2.7	3.1	3.4	3.6	3.8	4.0	1.5
Natural gas	1.0	2.0	2.1	2.9	3.5	4.0	4.5	5.2	3.4
Coal	0.03	0.1	0.1	0.1	0.1	0.1	0.1	0.1	0.8
Nuclear	0.0	0.0	0.0	0.03	0.03	0.03	0.03	0.03	
Other	0.03	0.1	0.1	0.2	0.2	0.2	0.2	0.2	4.5
Total	**2.9**	**4.8**	**5.0**	**6.3**	**7.2**	**7.9**	**8.6**	**9.5**	**2.4**
Africa									
Oil	1.1	1.4	1.4	1.9	2.1	2.2	2.3	2.5	2.3
Natural gas	0.4	0.7	0.7	0.9	1.2	1.5	1.9	2.2	4.4
Coal	0.8	0.9	1.0	1.3	1.5	1.5	1.6	1.6	1.7
Nuclear	0.03	0.03	0.03	0.03	0.05	0.05	0.05	0.05	0.7
Other	0.2	0.2	0.2	0.3	0.3	0.3	0.3	0.4	1.9
Total	**2.5**	**3.2**	**3.3**	**4.4**	**5.1**	**5.5**	**6.1**	**6.7**	**2.6**
Central and South America									
Oil	1.9	2.7	2.7	3.2	3.5	3.8	4.1	4.4	1.8
Natural gas	0.6	1.0	1.0	1.6	1.9	2.2	2.6	2.9	3.9
Coal	0.2	0.2	0.2	0.3	0.3	0.4	0.4	0.4	2.9
Nuclear	0.0	0.1	0.1	0.1	0.1	0.1	0.1	0.1	0.5
Other	1.0	1.4	1.5	2.0	2.4	2.8	3.3	3.7	3.3
Total	**3.7**	**5.4**	**5.5**	**7.2**	**8.2**	**9.3**	**10.5**	**11.5**	**2.8**
Total Non-OECD									
Oil	13.2	15.9	16.3	20.8	22.9	25.3	27.6	30.1	2.3
Natural gas	9.7	11.3	11.9	15.9	19.0	21.9	25.1	28.5	3.3
Coal	11.7	13.1	14.6	19.7	23.5	26.7	32.0	33.3	3.3
Nuclear	0.7	1.0	1.0	1.4	1.7	2.3	2.6	2.7	3.5
Other	2.7	3.7	3.9	6.0	6.8	7.5	8.3	9.3	3.3
Total	**38.0**	**45.0**	**47.7**	**63.8**	**69.9**	**83.7**	**95.6**	**103.9**	**3.0**

Table 1.1 *Continued*

Region/ Country	History			Projections					Average annual per cent change 2003–2030
	1990	2002	2003	2010	2015	2020	2025	2030	
Total World									
Oil	34.3	40.0	40.7	46.9	50.0	53.1	56.5	60.1	1.4
Natural gas	19.0	20.5	24.9	30.5	35.3	39.3	43.4	47.9	2.4
Coal	22.6	24.4	26.0	32.2	36.5	40.4	46.7	49.2	2.5
Nuclear	5.0	6.7	6.6	7.3	7.7	8.2	8.7	9.1	1.0
Other	6.7	8.1	8.4	11.5	12.4	13.4	14.5	15.7	2.4
Total	**87.6**	**99.7**	**106.6**	**128.4**	**141.9**	**154.4**	**169.8**	**182.0**	**2.0**

Source: International Energy Outlook 2006, Energy Information Administration, USA.

1.2 CONVENTIONAL ENERGY RESOURCES

1.2.1 Oil

Table 1.2 shows the estimated world oil resources by country as of 1 January 2007. Out of the total resources of 1,317 billion barrels (Table 1.2; Fig 1.2), OPEC countries account for 57.3%; out of the remaining 42.7% for non-OPEC economies, the Middle East contributes the major share; India's share is only 1.2%. Historically, estimates of world oil reserves have generally shown an upward trend. As of 1 January 2007 the proved oil reserves were estimated at 209 billion kilolitres (1,317 billion barrels) which are 2.4 billion kilolitres (about 1%) higher than the estimate for 2005.

Oil remains the dominant energy source but its consumption will decline from 38% in 2003 to 33% in 2030, owing to the rise in oil price (Table 1.1). World oil use is projected to grow from 12.9 gigalitres per day in 2003 to 15.4 gigalitres per day in 2015, and 18.7 gigalitres per day in 2030 (Table 1.3). The higher price of oil has raised the demand for natural gas.

Table 1.3 shows world oil consumption by region until the year 2030 and then projections for 2010 to 2030. Out of the total world oil consumption of 12.7 gigalitres per day in 2003, OECD countries accounted for 7.8 gigalitres per day and non-OECD economies for the rest. India's share was only 0.4 gigalitre per day in 2003, the shortfall being met by extensive imports. The situation practically remains the same in the projection for 2010 to 2030. The average annual per cent change is 2.4 for India, compared to 0.8 for total OECD, 2.3 for total non-OECD and 1.4 for the total world figure.

1.2.2 Natural Gas

The world's natural gas reserves are estimated to be 6,183 trillion cubic feet and the top 20 countries account for 5,602 trillion cubic feet. Almost three-quarters of the world's natural

Table 1.2 World oil reserves by country as of 1 January 2007 (billion barrels)

Country	Oil reserves
Saudi Arabia	262.3
Canada	179.2
Iran	136.3
Iraq	115.0
Kuwait	101.5
United Arab Emirates	97.8
Venezuela	80.0
Russia	60.0
Libya	41.5
Nigeria	36.2
Kazakhstan	30.0
United States	21.8
China	16.0
Qatar	15.2
Mexico	12.4
Algeria	12.3
Brazil	11.8
Angola	8.0
Norway	7.8
Azerbaijan	7.0
Rest of World	65.5
World Total	**1.317.4**

Source: International Energy Outlook 2007. Energy Information Administration, USA.

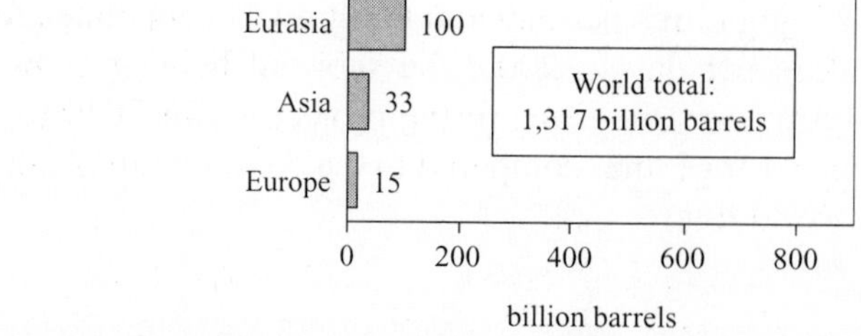

Figure 1.2 World proved oil reserves by geographic region as of 1 January 2007
Source: International Energy Outlook 2007. Energy Information Administration, USA.

Table 1.3 World oil consumption by region, Reference case, 1990–2030 (gigalitres per day)

Region/Country	History			Projections					Average annual per cent change
	1990	2002	2003	2010	2015	2020	2025	2030	2003–2030
OECD									
North America	**3.3**	**3.8**	**3.9**	**4.3**	**4.5**	**4.8**	**5.0**	**5.3**	**1.2**
United States	2.7	3.1	3.2	3.5	3.7	3.9	4.1	4.4	1.2
Canada	0.3	0.3	0.3	0.4	0.4	0.4	0.4	0.4	0.6
Mexico	0.3	0.3	0.3	0.3	0.4	0.4	0.5	0.5	1.7
Europe	**2.2**	**2.4**	**2.5**	**2.5**	**2.5**	**2.5**	**2.5**	**2.6**	**0.2**
Asia	**1.1**	**1.4**	**1.4**	**1.4**	**1.5**	**1.5**	**1.6**	**1.6**	**0.5**
Japan	0.8	0.9	0.9	0.9	0.9	0.9	0.9	0.9	0.1
South Korea	0.2	0.3	0.3	0.4	0.5	0.5	0.5	0.6	1.7
Australia/ New Zealand	0.1	0.2	0.2	0.2	0.2	0.2	0.2	0.2	0.6
Total OECD	**6.6**	**7.6**	**7.8**	**8.2**	**8.5**	**8.8**	**9.1**	**9.5**	**0.8**
Non-OECD									
Europe & Eurasia	**1.5**	**0.8**	**0.8**	**0.9**	**1.0**	**1.0**	**1.1**	**1.1**	**1.4**
Russia	0.9	0.4	0.4	0.4	0.5	0.5	0.5	0.5	0.9
Other	0.6	0.3	0.3	0.4	0.5	0.5	0.6	0.6	2.0
Asia	**1.0**	**2.1**	**2.1**	**2.9**	**3.3**	**3.8**	**4.2**	**4.7**	**3.0**
China	0.4	0.8	0.9	1.4	1.6	1.9	2.1	2.4	3.8
India	0.2	0.4	0.4	0.5	0.5	0.6	0.7	0.7	2.4
Other Non-OECD Asia	0.5	0.9	0.9	1.1	1.2	1.4	1.5	1.6	2.3
Other Non-OECD									
Middle East	**0.6**	**0.8**	**0.9**	**1.0**	**1.0**	**1.1**	**1.2**	**1.2**	**1.5**
Africa	**0.3**	**0.4**	**0.4**	**0.6**	**0.6**	**0.7**	**0.7**	**0.8**	**2.3**
Central & South America	**0.6**	**0.8**	**0.8**	**1.0**	**1.1**	**1.2**	**1.3**	**1.4**	**1.8**
Brazil	0.2	0.3	0.3	0.4	0.4	0.4	0.5	0.5	1.7
Other Central & South America	0.4	0.5	0.5	0.6	0.7	0.7	0.8	0.8	1.9
Total Non-OECD	**4.0**	**4.9**	**4.9**	**6.4**	**7.0**	**8.2**	**8.5**	**9.2**	**2.3**
Total World	**10.6**	**12.5**	**12.7**	**14.6**	**15.5**	**17.0**	**17.6**	**18.7**	**1.4**

Source: International Energy Outlook 2006. Energy Information Administration, USA.

gas reserves are located in the Middle East and Eurasia. India contributes a negligible part (Table 1.4).

Consumption of natural gas is projected to increase on an average by 2.4% per year from 2003 to 2030 (Table 1.5). The higher natural gas prices also make coal more cost competitive, especially in the electric power sector. For many years, natural gas was projected as the fastest growing energy source in the mid-term; however, higher natural gas prices have made coal more cost competitive, especially in the electric power sector. As a result, natural gas use and coal use are projected to increase at similar rates. Demand for natural gas is projected

Table 1.4 World natural gas reserves by country as of 1 January 2007

Country	Reserves (trillion cubic feet)	Per cent of world total
World	6,183	100.0
Top 20 countries	5,602	90.6
Russia	1,680	27.2
Iran	974	15.8
Qatar	911	14.7
Saudi Arabia	240	3.9
United Arab Emirates	214	3.5
United States	204	3.3
Nigeria	182	2.9
Algeria	162	2.6
Venezuela	152	2.5
Iraq	112	1.8
Turkmenistan	100	1.6
Kazakhstan	100	1.6
Indonesia	98	1.6
Norway	82	1.3
China	80	1.3
Malaysia	75	1.2
Uzbekistan	65	1.1
Egypt	59	0.9
Canada	58	0.9
Kuwait	55	0.9
Rest of world	581	9.4

Source: International Energy Outlook 2007. Energy Information Administration, USA.

to rise by an average of 2.4% per year over the period from 2003 to 2030, and the demand for coal by an average of 2.5% per year. The consumption of natural gas in 2003 in billion cubic metres was as follows: world total 2,716.8, OECD total 1,431.9 and non-OECD total 1,284.9, while India's share was 28.3 only (Table 1.5). Total world natural gas consumption is projected to rise from 2,716.8 billion cubic metres in 2003 to 3,796.0 billion cubic metres in 2015 and to 5,150.4 billion cubic metres in 2030 (Table 1.5). The average annual per cent change projected for 2003–2030 is 2.4 for the world, the same for total OECD countries and 3.3 for total non-OECD economies. India's rate of increase in consumption is projected as high as 5.9% per annum, obviously through large-scale imports. This also shows an improvement in the economy of the country.

Despite high rates of increase in natural gas consumption, particularly over the past decade, most regional reserves-to-production ratios have remained high: 66.7 years worldwide, 81.5 years for Russia and more than 100 years for the Middle East. According to the most recent US Geological Survey estimates, a significant volume of natural gas remains to be discovered. Russia is already the world's single largest exporter of natural gas, with net exports

Table 1.5 World natural gas consumption by region, Reference case, 1990–2030 (billion cubic metres)

Region/Country	History			Projections					Average annual per cent change
	1990	2002	2003	2010	2015	2020	2025	2030	2003–2030
OECD									
OECD North America	**636.7**	**781.1**	**775.4**	**837.7**	**925.4**	**982.0**	**1,010.3**	**1,035.8**	**1.1**
United States	543.4	650.9	625.7	660.2	733.0	761.3	762.1	759.3	0.7
Canada	67.9	87.7	95.6	118.4	118.8	133.0	141.0	149.5	1.9
Mexico	25.4	42.5	54.1	59.1	73.6	87.7	107.2	127.0	3.4
OECD Europe	**328.3**	**486.8**	**503.7**	**597.1**	**676.4**	**730.1**	**798.1**	**871.6**	**2.0**
OECD Asia	**79.2**	**135.8**	**141.5**	**152.8**	**167.0**	**178.3**	**184.0**	**192.4**	**1.2**
Japan	53.8	82.1	86.7	92.6	99.0	100.4	101.9	107.5	0.8
South Korea	2.8	22.6	24.6	28.7	33.0	36.1	36.8	36.8	1.7
Australia/ New Zealand	22.6	31.1	30.2	31.5	35.0	41.8	45.3	48.1	1.7
Total-OECD	**1,044.2**	**1,403.7**	**1,431.9**	**1,587.6**	**1,766.8**	**1,890.4**	**1,992.4**	**2,009.8**	**1.5**
Non-OECD									
Non-OECD Europe & Eurasia	**755.6**	**631.1**	**667.9**	**809.4**	**891.5**	**984.8**	**1,061.3**	**1,146.1**	**2.0**
Russia	490.0	413.2	490.0	512.2	553.9	600.0	642.9	677.3	1.6
Other	265.6	217.9	177.9	297.2	337.6	384.8	418.4	468.8	2.7
Non-OECD Asia	**83.2**	**195.3**	**212.3**	**331.1**	**441.5**	**554.7**	**688.8**	**815.0**	**5.1**
China	14.2	31.1	34.0	84.9	110.4	144.3	175.5	198.1	6.8
India	11.3	25.5	28.3	42.5	48.1	62.3	87.7	127.4	5.9
Other Non-OECD Asia	57.7	138.7	150.0	203.7	283.0	348.1	425.6	489.5	4.4
Middle East	**101.9**	**215.1**	**223.6**	**311.3**	**367.9**	**424.5**	**483.9**	**554.7**	**3.4**
Africa	**39.6**	**67.9**	**73.6**	**93.4**	**130.2**	**158.5**	**192.4**	**229.2**	**4.4**
Central & South America	**56.6**	**101.9**	**107.5**	**164.1**	**198.1**	**229.2**	**268.9**	**305.6**	**3.9**
Brazil	2.8	14.2	14.2	28.3	31.1	34.0	39.6	48.1	4.6
Other Central & South America	53.8	87.7	93.3	135.8	167.0	195.2	229.3	257.5	3.8
Total Non-OECD	**1,036.9**	**1,211.3**	**1,284.9**	**1,709.3**	**2,029.2**	**2,351.7**	**2,695.3**	**3,050.6**	**3.3**
Total World	**2,081.1**	**2,615.0**	**2,716.8**	**3,296.9**	**3,796.0**	**4,252.1**	**4,687.7**	**5,150.4**	**2.4**

Source: International Energy Outlook 2006. Energy Information Administration, USA.

of 7.1 trillion cubicfeet, all of it by pipeline. An alternative in many countries is to import or export liquefied natural gas (LNG). Both China and India have limited natural gas reserves and are projected to rely on imports to meet more than 40% of the natural gas demand in 2030, both by pipeline and in the form of LNG.

Table 1.6 World recoverable coal reserves as of 1 January 2003 (billion short tons)

Region/Country	Bituminous and anthracite	Sub-bituminous	Lignite	Total
United States	123.7	110.3	33.5	267.5
Russia	54.1	107.4	11.5	173.0
China	68.6	37.1	20.5	126.2
India	99.3	0.0	2.6	101.9
Other Non-OECD Europe and Eurasia	50.1	18.7	31.3	100.1
Australia & New Zealand	42.6	2.7	41.9	87.2
Africa	55.3	0.2	0.0	55.5
OECD Euripe	19.5	5.0	18.8	43.3
Other Non-OECD Asia	1.4	2.0	8.1	11.5
Brazil	0.0	11.1	0.0	11.1
Other Central & South America	8.5	2.2	0.1	10.8
Canada	3.8	1.0	2.5	7.3
Other	1.8	0.4	0.1	2.3
World total	**528.7**	**298.1**	**170.9**	**997.7**

Source: International Energy Outlook 2007. Energy Information Administration, USA.

Worldwide, the industrial sector remains the most important end-user for natural gas, accounting for 52% of the total growth in natural gas use in the projections; however, natural gas remains an important energy source in the electric power sector, particularly as a fuel for new generating capacity. The electric power sector accounts for 39% of the increase in global natural gas demand over the 2003 to 2030 period.

1.2.3 Coal

The recoverable coal reserves in the world are given in Table 1.6 in terms of the total quantity of all ranks of coal; India stands fourth with 101.9 billion tons out of the world total of 1,000.8 and 270.7 for the USA. India is fortunate to have practically the entire quantity in the form of bituminous coal and anthracite. The total recoverable reserves of coal around the world are estimated to last approximately 180 years at the current levels of consumption (Tables 1.6 and 1.7). Historically, estimates of world recoverable coal reserves, although relatively stable, have declined gradually from 1,174 billion tones at the beginning of 1990 to 1,083 billion tons in 2000 and 1000 billion tons in 2003. The most recent assessment of world coal reserves includes a substantial downward adjustment for Germany: from 73 billion tons to 7 billion tons.

Although coal deposits are widely distributed, 67% of the world's recoverable reserves are located in four countries: the United States (27%), Russia (17%), China (13%) and India (10%). In 2003, these four countries, taken together, accounted for 63% of the total world coal production. By rank, anthracite and bituminous coal account for 53%, sub-bituminous

Table 1.7 World coal consumption by region, Reference case, 1990–2030 (million short tons)

Region/Country	History			Projections					Average annual per cent change
	1990	2002	2003	2010	2015	2020	2025	2030	2003–2030
OECD									
OECD North America	**972**	**1,151**	**1,184**	**1,331**	**1,402**	**1,536**	**1,748**	**1,947**	**1.9**
United States	904	1,066	1,095	1,233	1,276	1,390	1,592	1,784	1.8
Canada	59	68	69	73	97	110	117	123	2.1
Mexico	9	17	20	25	29	36	39	40	2.5
OECD Europe	**1,298**	**892**	**887**	**880**	**878**	**893**	**896**	**928**	*0.2*
OECD Asia	**280**	**400**	**404**	**469**	**494**	**508**	**531**	**560**	*1.2*
Japan	126	172	176	177	181	176	170	169	0.1
South Koria	48	80	81	120	133	142	158	175	2.9
Australia/New Zealand	106	148	147	172	180	190	203	216	1.4
Total- OECD	**2,550**	**2,442**	**2,476**	**2,680**	**2,774**	**2,938**	**3,175**	**3,436**	**1.2**
Non-OECD									
Non-OECD Europe & Eurasia	**1,028**	**530**	**543**	**595**	**668**	**735**	**795**	**856**	**1.7**
Russia	447	240	251	263	296	330	335	382	1.6
Other	581	290	292	333	372	405	440	474	1.8
Non-OECD Asia	**1,507**	**2,041**	**2,167**	**3,352**	**3,988**	**4,585**	**5,191**	**5,855**	**3.7**
China	1,124	1,413	1,531	2,535	3,045	3,530	4,043	5,645	4.2
India	256	431	431	583	686	775	842	887	2.7
Other Non-OECD Asia	127	177	206	234	257	280	306	323	1.7
Middle East	**6**	**16**	**16**	**16**	**16**	**17**	**17**	**19**	**0.8**
Africa	**152**	**187**	**203**	**266**	**293**	**304**	**312**	**321**	**1.7**
Central & South America	**27**	**34**	**35**	**47**	**53**	**63**	**68**	**74**	**2.8**
Brazil	17	22	24	30	35	39	42	45	2.4
Other Central & South America	10	12	11	17	18	24	25	29	3.5
Total Non-OECD	**2,720**	**2,808**	**2,964**	**4,276**	**5,018**	**5,704**	**6,383**	**7,125**	**3.3**
Total World	**5,270**	**5,250**	**5,440**	**6,956**	**7,792**	**8,642**	**9,558**	**10,561**	**2.5**

Source: International Energy Outlook 2006. Energy Information Administration, USA.

coal accounts for 30%, and lignite accounts for 17%. Quality and geological characteristics of coal deposits vary significantly by region and even within individual coal seams. For example, the International Energy Agency in 2003 estimated that the average calorific value of lignite in major coal producing countries varies from a low of 1,084 kcal/kg in Greece to a high of 3,100 kcal/kg in Canada.

$$\text{USA's share} = 267.6 \times \frac{100}{997.7} = 26.8\%$$

$$\text{Russia's share} = 173.0 \times \frac{100}{997.7} = 17.3\%$$

$$\text{China's share} = 126.2 \times \frac{100}{997.7} = 12.65\%$$

$$\text{India's share} = 101.9 \times \frac{100}{997.7} = 10.2\%$$

Global use of coal is projected to increase by 2.4 billion tons between 2003 and 2015 and by another 2.7 billion tons between 2015 and 2030 (Table 1.7). Japan is an exception, where the electricity sector continues to be dominated by natural gas and nuclear power generation. World coal consumption is projected to increase from 5,440 million tons in 2003 to 7,792 million tons in 2015, at an average annual rate of 2.2% (Table 1.7). Of the coal produced worldwide in 2003, 67% was shipped to electricity producers, and 30% to industrial consumers. With higher prices for oil and natural gas making coal more competitive, the IEO 2006 projection for world coal use in 2025 is 16% higher than in IEO 2005. Consequently, coal's share of total energy use is projected to rise from 24% in 2003 to 27% in 2030 and world coal consumption continues to exceed world natural gas consumption throughout the projections (Table 1.1). The largest increase in coal use worldwide is projected for China and India, where coal supplies are plentiful. Together China and India will account for 86% of the rise in non-OECD coal use and 70% of the world increase in coal demand over the projection period.

Consumption of coal for non-electricity purposes is projected to double between 2003 and 2030. The strongest growth in non-electricity consumption is projected for the non-OECD economies, averaging 3.9% per year in IEO 2006 reference case. Robust economic growth in many of the non-OECD countries is expected to boost demand for electricity to run newly purchased home appliances, for air-conditioning, cooking, space and water heating, and refrigeration.

Coal consumption in the OECD countries (Table 1.7) is projected to rise at a relatively even pace in the reference case: from 2.5 billion tons in 2003 to 2.8 billion tons in 2015, and to 3.4 billion tons in 2030. The increase represents an average growth of 1.2% per year over the entire period and a slightly higher rate of 1.4% per year from 2015 to 2030. Much of the increase of 0.9 billion ton in coal consumption projected for the OECD countries from 2003 to 2030 is the result of the expected strong growth in US coal demand. In 2003, the United States consumed 1.1 billion tons of coal, accounting for 92% of coal consumption in North America and 44% of the OECD total. With substantial coal reserves, the total electricity generation in the USA depends heavily on it; natural gas cannot compete with it.

Coal consumption in non-OECD countries is projected to increase by 140% in the IEO 2006 reference case, from 3.0 billion tons in 2003 to 7.1 billion tons in 2030, led by strong economic growth and rising demand for energy in China and India. China and India are very large countries in terms of both population and land mass, and both have substantial quantities of the remaining coal reserves of the non-OECD countries. Together, they account for 23% of world coal reserves and 67% of non-OECD coal-reserves, and for 70% of the projected increase in world coal consumption. Strong economic growth (averaging 6.0% per year in China and 5.4% per year in India from 2003 to 2030) is projected for both countries and much of the increase in their demand for energy, particularly in the industrial and electricity sectors, is expected to be met by coal.

In India, almost 70% of the growth in coal consumption is expected to be in the electric power sector and most of the remainder in the industrial sector. In 2003, India's coal-fired power plants consumed 1.3 quadrillion kcal of coal, representing 69% of the country's total coal demand. Coal use for electricity generation in India is projected to grow by 2.7% per year, to 2.6 quadrillion kcal in 2030, as an additional 94 gigawatts of coal-fired capacity is brought on line. As a result, India's coal-fired generating capacity is expected to more then double in the IEO 2006 projections, from 67 gigawatts in 2003 to 161 gigawatts in 2030. More than 50 gigawatts of new coal-fired capacity is to be added during India's eleventh plan period ending in March 2012.

Currently, there are no coal-to-oil projects under construction in India; however , a domestic coal-to-liquids industry would represent a viable means of increasing domestic oil supply. The Indian Oil Corporation is evaluating the construction of a coal liquefaction facility with an output of 13,000 kilolitres per day, which would be based on the coal-to-liquids technology developed by South African Company Sasol and would use both domestic and imported coal as feedstock.

1.3 ENERGY: CONSUMPTION AND DEMAND

On a global scale, energy demand in the industrial sector grows most rapidly; it is projected to grow at an average rate of 1.8% per year from 2004 to 2030. Slower growth is projected for the buildings sector, while residential energy use will rise by an average of 1.4% per year and commercial use by 2.0% per year (Table 1.8). In contrast, the use of energy for transportation will grow at the same rate as the use of industrial energy, but faster than energy use in the buildings sector. The higher oil prices are largely responsible for the slower growth in energy demand by the transportation sector.

The world marketed energy from all sources is projected to increase over the next two decades on an average of 2.0% per year from 2003–2030 (Table 1.1). World energy consumption is projected to increase by 71% with fossil fuels continuing to supply much of the energy used worldwide. Although, increasing world oil prices dampen the growth in demand for oil, the total world energy use continues to increase as a result of robust economic growth. Worldwide, total energy use is expected to increase from 446.7 quadrillion Btu in 2004 to 701.6 quadrillion Btu in 2030.

The use of all energy sources increases through 2030. Fossil fuels will continue to supply much of this increment. Oil will remain the dominant energy source, given its importance in the transportation and industrial end-use sectors; however, world oil prices indicate that oil's share in the energy market is lessened decreasing in the projection as other fuels replace oil wherever possible. Renewable energy sources will become more economically competitive with fossil fuels,

Table 1.8 Total world delivered energy consumption by end-use sector and fuel, 2004–2030 (quadrillion Btu)

Sector/Fuel		Projections					Average annual per cent change, 2004–2030
	2004	2010	2015	2020	2025	2030	
Residential							
Liquids	10.9	11.5	12.2	12.7	12.9	13.0	0.7
Natural gas	18.4	20.2	21.7	22.8	23.7	24.9	1.2
Coal	3.5	3.7	4.0	4.0	3.9	3.8	0.3
Electricity	14.3	17.8	20.4	22.6	24.6	26.6	2.4
Renewables	0.5	0.7	0.7	0.7	0.7	0.7	1.2
Total	**47.7**	**53.9**	**59.0**	**62.7**	**65.8**	**69.0**	**1.4**
Commercial							
Liquids	5.3	5.5	5.8	5.9	6.0	6.2	0.6
Natural gas	6.7	7.2	7.8	8.3	8.8	9.3	1.2
Coal	0.7	0.8	0.9	1.0	1.0	1.0	1.2
Electricity	11.7	14.6	17.0	19.2	21.5	24.0	2.8
Renewables	0.1	0.2	0.2	0.2	0.2	0.2	1.8
Total	**24.5**	**28.3**	**31.7**	**34.6**	**37.5**	**40.7**	**2.0**
Industrial							
Liquids	55.5	61.0	64.5	67.9	70.9	74.2	1.1
Natural gas	45.2	51.5	56.6	63.1	68.3	73.3	1.9
Coal	35.4	39.5	45.3	50.4	55.9	62.0	2.2
Electricity	25.3	28.6	32.2	36.1	39.9	44.3	2.2
Renewables	2.2	2.6	2.8	2.9	3.1	3.3	1.6
Total	**163.6**	**183.1**	**201.4**	**220.5**	**238.1**	**257.1**	**1.8**
Transportation							
Liquids	85.9	95.4	104.1	112.9	122.7	133.8	1.7
Natural gas	0.8	1.0	1.1	1.2	1.3	1.3	1.8
Coal	0.2	0.2	0.2	0.2	0.2	0.1	−2.4
Electricity.	0.8	0.9	1.0	1.1	1.2	1.2	1.7
Total	**87.7**	**97.5**	**106.3**	**115.4**	**125.3**	**136.5**	**1.7**
All end-use sectors							
Liquids	157.6	173.4	186.5	199.3	212.5	227.2	1.4
Natural gas	71.3	79.9	87.2	95.4	102.0	108.8	1.6
Coal	39.7	44.2	50.4	55.6	60.9	66.8	2.0
Electricity	52.1	61.9	70.6	79.0	87.2	96.1	2.4
Renewables	2.9	3.5	3.7	3.9	4.1	4.3	1.5
Delivered energy	323.5	362.9	398.4	433.2	466.7	503.3	1.7
Electricity-related losses	123.2	148.2	161.0	173.8	187.0	198.3	1.8
Total	**446.7**	**511.1**	**559.4**	**607.0**	**653.7**	**701.6**	**1.8**
Electric power							
Liquids	10.6	10.5	11.0	11.3	11.6	11.6	0.4
Natural gas	32.2	40.7	47.1	51.6	56.5	61.6	2.5
Coal	74.7	92.2	101.2	111.6	122.0	132.4	2.2
Nuclear	27.5	29.8	32.5	35.7	38.1	39.7	1.4
Renewables	30.3	36.9	39.8	42.7	46.0	49.2	1.9
Total	**175.2**	**210.1**	**231.6**	**252.9**	**274.2**	**294.5**	**2.0**

Table 1.8 *Continued*

Sector/Fuel	2004	Projections 2010	2015	2020	2025	2030	Average annual per cent change, 2004–2030
Total energy consumption							
Liquids	168.2	183.9	197.6	210.6	224.1	238.9	1.4
Natural gas	103.4	120.6	134.3	147.0	158.5	170.4	1.9
Coal	114.5	136.4	151.6	167.2	182.9	199.1	2.2
Nuclear	27.5	29.8	32.5	35.7	38.1	39.7	1.4
Renewables	33.2	40.4	43.4	46.5	50.1	53.5	1.9
Total	**446.7**	**511.1**	**559.4**	**607.0**	**653.7**	**701.6**	**1.8**

Source: International Energy Outlook 2007. Energy Information Administration, USA.

and their use expand as rapidly as consumption of natural gas and coal. Higher fossil fuel prices also support renewed interest in expanding the use of nuclear power to generate electricity.

The most rapid growth in energy demand from 2004 to 2030 is projected for the non-OECD nations. They account for three-fourths of the increase in world energy use. The energy demand is expected to grow at an average of 3.4% per year for non-OECD Asia (which includes China and India), 2.8% per year for Central and South America, 2.6% per year for Africa, 2.5% per year for the Middle East, and 1.4% per year for non-OECD Europe and Eurasia (Table 1.9). The increases result from projections of strong regional economic growth. Non-OECD energy use is projected to surpass OECD energy use by 2015, and in 2030 the total energy demand in non-OECD nations will exceed that in the OECD countries by 34%. Much of the growth in energy demand among the non-OECD economies occurs in non-OECD Asia, which includes China and India. Total primary energy consumption in the non-OECD countries grows at an average annual rate of 3.0% between 2003 and 2030. The OECD nations, for the most part, are more mature energy consumers with well-established infrastructures, and their economies are usually moving away from energy intensive industries towards services. Consequently, total OECD energy demand increases by only an average of 0.9% per year over the projection period, as compared with an average increase of 2.7% per year for the non-OECD nations.

In the OECD, where population growth is generally slow or negative in many countries over the projection period, the slowest growth in energy use is projected for the residential sector, at 0.6% per year; and the fastest growth is in the industrial sector, averaging 1.2% annually. For the non-OECD regions as a whole, strong growth in demand for energy is projected for every end-use sector, ranging from 2.3% per year in the transportation sector to 3.2% per year in the commercial and industrial sectors.

Economic growth is among the most important factors to be considered in projecting changes in the world energy consumption. The outlook for medium- to long-term economic growth depends on the underlying demographic and expected productivity trends in each economy. These in turn depend on population growth, labour force participation rates, productivity growth, and national savings and capital accumulation.

Table 1.9 World total energy consumption by region, high world oil price case, 1990–2030 (quadrillion Btu)

Region/Country	History			Projections					Average annual per cent change, 2004–2030
	1990	2003	2004	2010	2015	2020	2025	2030	
OECD									
OECD North America	**100.8**	**118.3**	**120.9**	**129.3**	**134.8**	**142.7**	**149.9**	**157.6**	**1.0**
United States[a]	84.7	98.3	100.7	105.5	109.9	115.8	121.6	127.7	0.9
Canada	11.1	13.5	13.6	15.6	16.0	17.0	17.6	18.3	1.1
Mexico	5.0	6.5	6.6	8.2	8.9	10.0	10.7	11.6	2.2
OECD Europe	**69.9**	**79.5**	**81.1**	**83.6**	**84.1**	**85.3**	**86.2**	**86.9**	**0.3**
OECD Asia	**26.6**	**36.9**	**37.8**	**39.4**	**40.5**	**42.4**	**43.8**	**45.3**	**0.7**
Japan	18.4	22.2	22.6	23.2	23.2	23.7	24.0	24.3	0.3
South Korea	3.8	8.7	9.0	9.4	10.2	11.2	11.9	12.8	1.4
Australia/New Zealand	4.4	6.0	6.2	6.8	7.1	7.6	8.0	8.2	1.1
Total OECD	**197.4**	**234.7**	**239.8**	**252.3**	**259.4**	**270.4**	**279.9**	**289.8**	**0.7**
Non-OECD									
Non-OECD Europe and Eurasia	**67.2**	**47.9**	**49.7**	**55.2**	**60.7**	**67.0**	**70.3**	**72.1**	**1.4**
Russia	39.0	28.8	30.1	33.2	36.3	39.4	41.3	42.2	1.3
Other	28.3	19.2	19.6	22.0	24.4	27.6	29.0	29.9	1.6
Non-OECD Asia	**47.5**	**88.2**	**99.9**	**129.4**	**149.6**	**174.2**	**197.0**	**219.7**	**3.1**
China	27.0	49.7	59.6	81.6	94.0	110.1	125.1	140.8	3.4
India	8.0	14.4	15.4	17.9	20.8	24.3	27.7	30.8	2.7
Other Non-OECD Asia	12.5	24.0	24.9	29.9	34.8	39.9	44.2	48.2	2.6
Middle East	**11.3**	**19.9**	**21.1**	**26.6**	**29.9**	**32.9**	**34.9**	**36.9**	**2.2**
Africa	**9.5**	**13.3**	**13.7**	**16.7**	**18.5**	**20.6**	**22.4**	**23.9**	**2.2**
Central and South America	**14.5**	**21.7**	**22.5**	**27.3**	**30.5**	**33.8**	**36.7**	**39.5**	**2.2**
Brazil	5.8	8.7	9.1	11.0	12.2	13.5	14.7	16.2	2.3
Other Central and South America	8.8	13.0	13.5	16.3	18.3	20.4	22.0	23.2	2.1
Total Non-OECD	**150.0**	**191.0**	**206.9**	**255.2**	**289.1**	**328.5**	**361.3**	**392.1**	**2.5**
Total World	**347.3**	**425.7**	**446.7**	**507.4**	**548.5**	**598.9**	**641.3**	**681.9**	**1.6**

Source: International Energy Outlook 2007. Energy Information Administration, USA.

1.3.1 Residential Sector

The following information is derived from a combination of Tables 1.8 and 1.9, and Figs. 1.3 and 1.4 Households in OECD nations use energy more intensively than those in non-OECD nations, primarily because of their higher income levels (Fig. 1.3). Total residential electricity use in the OECD region accounts for about 80% of the total projected growth is in demand for residential energy in OECD nations.

Household energy use increases more rapidly in the non-OECD countries than in the OECD countries due to robust economic growth and expanding population. China and India account for nearly 40% of the total increase in residential energy use in the non-OECD countries through 2030, as their economies continue to grow strongly through the projection period.

1.3.2 Commercial Sector

The commercial sector, or the services sector, consists of business institutions and organisations that provide services, as opposed to those in manufacturing and agriculture.

Energy consumption by the commercial sector in the OECD nations increases in the reference case by an average of 1.2% per year from 2004 to 2030 (Fig. 1.4). Natural gas continues to displace petroleum products and coal as the preferred commercial heating fuel in the OECD region. Economic growth and commerce are expected to increase rapidly in the non-OECD nations. Under these circumstances, energy use by the commercial sector in non-OECD countries will nearby double between 2004 and 2020, with an average annual rate of increase of 3.7%.

Electricity demand for commercial applications is projected to grow rapidly in the non-OECD nations. The largest increases in commercial electricity demand are projected for the nations with the rapidly growing economies, particularly China and India. Increasing commercial activity is expected to lead to growth in demand for fossil fuels in the non-OECD nations. Natural gas is an economically attractive choice for commercial water heating, space heating, and cooking in non-OECD countries, especially in China and India, which together are expected to account for around 80% of non-OECD commercial coal use from 2003 through 2030.

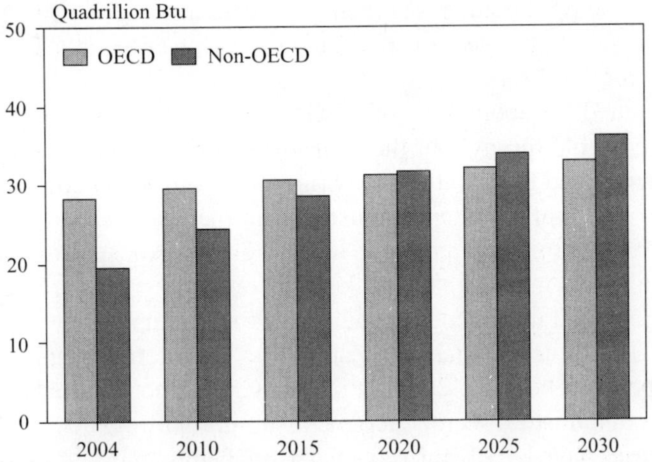

Figure 1.3 OECD and Non-OECD residential sector delivered energy consumption, 2004–2030 (quadrillion Btu)

Source: International Energy Outlook 2007. Energy Information Administration, USA.

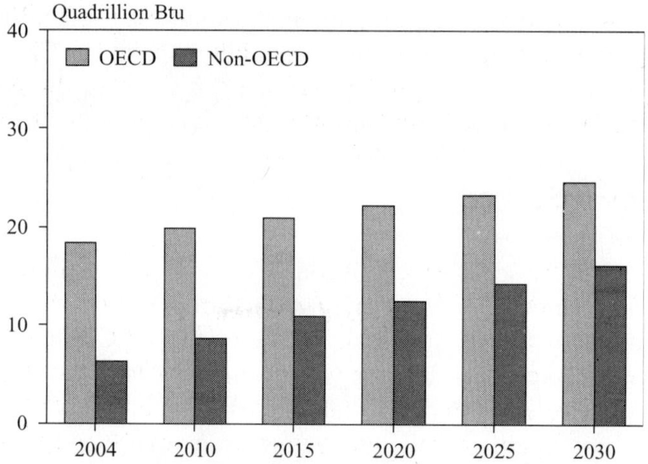

Figure 1.4 OECD and Non-OECD commercial sector delivered energy consumption, 2004–2030 (quadrillion Btu)
Source: International Energy Outlook 2007. Energy Information Administration, USA.

1.3.3 Industrial Sector

Energy is consumed in the industrial sector by several industries, including manufacturing, agriculture, mining and construction, and for a wide range of activities, such as process and assembly use, heating and cooling, and lighting. Energy demand by the industrial sector varies across regions and countries of the world, based on the level and mix of economic activity, technological development and population growth, among other factors. The industrial sector is the largest of the end use sectors, consuming 50% of delivered energy worldwide in 2004. Industrial energy consumption increases in all countries and regions; however, its growth rate in the OECD region—0.6% per year on an average—is slower than the 2.5% average projected for the non-OECD region.

Electricity accounted for about 16% of OECD industrial sector energy use in 2004 and its share remains fairly stable throughout the projection period (Figure 1.5). Oil and natural gas were the most heavily used fuels in the industrial sector of OECD countries in 2004, together accounting for two-thirds of the energy consumed in the sector. The two fuels maintain their overall share in 2030, but consumption of natural gas is expected to grow almost twice as rapidly as that of oil. Coal makes up the bulk of the remaining industrial energy consumption, while renewables remain a minor energy source for the sector.

In the non-OECD countries, energy consumption by the industrial sector is projected to increase by 2.5% per year between 2004 and 2030. China, the Middle East and India have the most rapid increase in industrial sector energy consumption from 2004 to 2030. The economies of the OECD countries have largely moved away from heavy, energy-intensive industries (such as steel and cement) towards a greater emphasis on light manufacturing and service activities,

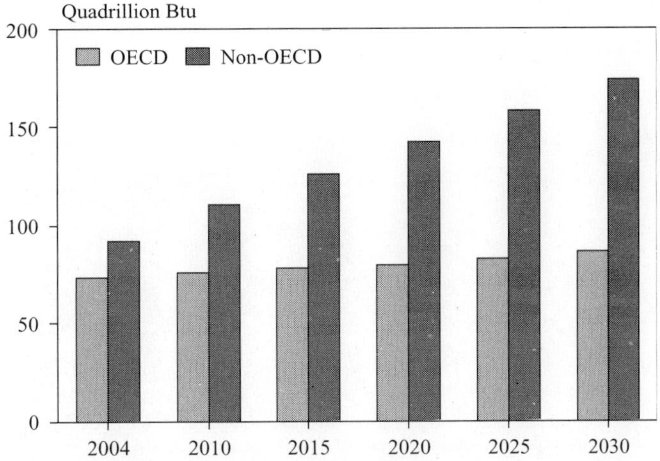

Figure 1.5 OECD and Non-OECD industrial sector delivered energy consumption, 2004–2030 (quadrillion Btu)
Source: International Energy Outlook 2007. Energy Information Administration, USA.

but the economies of many of the non-OECD countries and regions have growing intensive, heavy manufacturing sectors.

Electricity is expected to become an increasingly important component of delivered energy demand of the industrial sector in the non-OECD economies. Oil, coal, and natural gas were the most heavily used fuels in 2004 and are projected to remain so in 2030. Oil use in the non-OECD industrial sector is projected to increase at a slower rate than natural gas or coal use.

1.3.4 Transportation Sector

Energy use in the transportation sector includes fuels consumed for the purpose of moving people and goods by road, rail and air. The road transport component consists of both light-duty vehicles (automobiles, sport utility vehicles, minivans, small trucks and motorcycles, among other small vehicles) and heavy-duty vehicles (large trucks used for moving freight and buses for mass transit).

Petroleum products continue to dominate energy use in the transportation sector; and barring any widespread increase in the penetration of new technologies, the use of alternative fuels is expected to remain relatively modest through 2030. Oil is the dominant fuel for transportation, and currently, there are no other fuels that compete on par with oil. The IEO 2006 reference case projects a 1.4% average annual growth rate in demand for petroleum for transportation from 2003 to 2030. Much of the projected growth in demand for petroleum products is in the transportation sector of the non-OECD economies. Two-thirds of the increase in world consumption of liquid fuels is due to the transportation section. North America and Non-OECD Asia are the largest users.

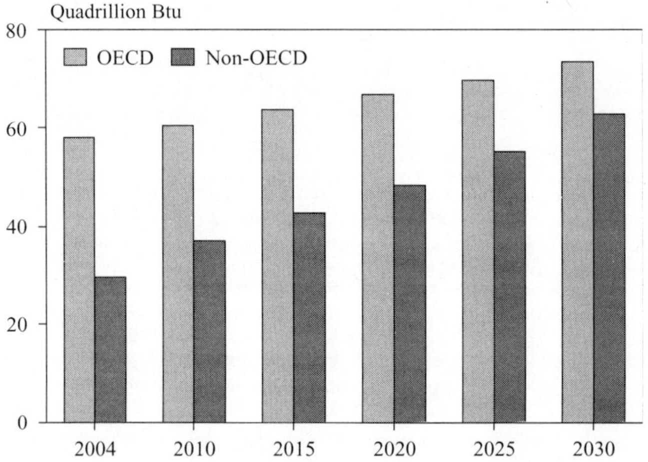

Figure 1.6 OECD and Non-OECD transportation sector delivered energy consumption, 2004–2030 (quadrillion Btu)
Source: International Energy Outlook 2007. Energy Information Administration, USA.

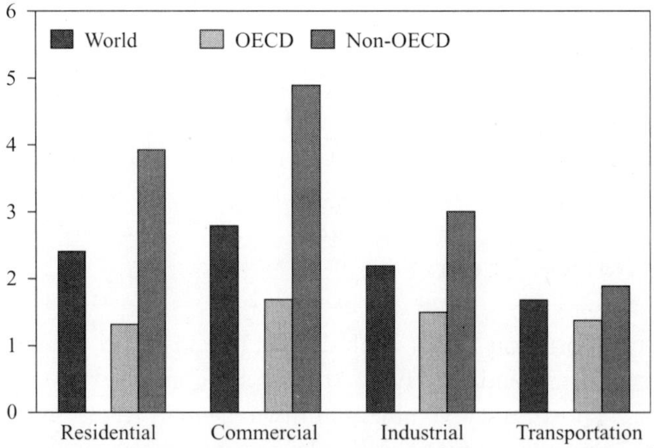

Figure 1.7 Average annual change in end-Use sector electricity demand, 2004–2030 (per cent per year)
Source: International Energy Outlook 2007. Energy Information Administration, USA.

In general, the transportation sector of the OECD economies is fully established, with extensive infrastructure that includes highways, airport facilities and rail systems. Energy demand for transportation in the OECD region is projected to grow at an average annual rate of 0.9%, from 57.9 quadrillion Btu in 2004 to 63.7 in 2015 and to 73.4 in 2030 (Figs.1.6 and 1.7). Petroleum products remain the largest source of energy for transportation in the OECD countries. Although usage of gasoline is expected to be higher than that of diesel, the product mix for road travel is dominated by diesel fuel. Most countries are expected to keep taxes on diesel fuel lower than those for gasoline through 2030, encouraging a switch to diesel in the mid-term. Fast-paced growth in air travel is expected to bring about robust growth in demand for aviation fuels in the region.

Energy demand in the non-OECD transportation sector as a whole grows at an average annual rate of 2.9%, from 29.8 quadrillion Btu in 2004 to 42.7 in 2015 and to 63.1 in 2030. As in the OECD economies, the growth in energy use by the transportation sector is led by greater demand for air travel. Expanding ownership of private automobiles and the increasing role of trucking in freight transportation play a significant role in pushing up the demand for energy. In 2004, the non-OECD economies accounted for about 34% of world energy use for transportation. In 2030, their share is projected to be 46%, as the gap between the transportation energy consumption in the non-OECD and OECD economies is expected to narrow substantially over the projection period.

Non-OECD Asia is poised for strong and significant growth in transportation sector energy use, with China and India being the primary contributors to the growth. Infrastructure development is the major hurdle faced by both China and India. India has an effective and extensive rail system but its highways system needs development. Work has started on a highway project to connect India's major cities and it is expected to be completed in the next five years. Similar developments are occurring in China as well.

In India, energy demand in the transportation sector is projected to grow at an average rate of 3.3% a year. This could grow even faster if all of the new highway projects currently under consideration are completed.

1.4 ELECTRICITY: CONSUMPTION AND GENERATION

1.4.1 Consumption of Electricity

The net world electricity consumption is projected to more than double from 2015 to 2030. Most of the growth occurs in the non-OECD nations, where electricity use increases on an average by 3.9% per year from 2003 to 2030, as compared with 1.5% per year in the OECD nations (Fig. 1.7). Coal and natural gas remain the most important fuels for electricity production throughout the projection period.

In 2004, nearly 60% of the total electricity consumption in the OECD economies was in the buildings sector (residential and commercial). The remainder was consumed in the industrial sector, except for a small amount (1%) used by mass transit trains and buses. Overall, net electricity consumption in OECD countries is projected to increases by 50% from 2004 to 2030, mostly as a result of increasing penetration and telecommunications technologies. Electricity consumption in the non-OECD economies is projected to grow at an average annual rate of 3.9% from 2004 to 2030. The average annual growth rate translates to a near tripling of net electricity consumption in the non-OECD nations over the projection period. In 2003, non-OECD economies consumed 40% of the world's electricity; in 2030, their share is projected to be 56%. The large growth is mainly due to gross domestic products and population growth.

From 2003 to 2030, residential electricity consumption for the non-OECD economies as a whole is projected to grow from 23% to 30% of the total net electricity consumption. In absolute terms, nearly four times as much electricity will be consumed in the residential sector in 2030 than in 2003, supporting a major transformation in living standards as electric lighting, appliances, and new technologies become available to an increasing share of the world's population. Electricity consumption growth in the non-OECD industrial sector is somewhat slower than in the buildings sector, despite the rapid adoption of consumer electronics and computers for business use. The industrial sector's share of total non-OECD electricity demand is projected to decline from 61% in 2003 to 54% in 2030, even as industrial use of electricity more than doubles.

The increase in residential electricity consumption in non-OECD Asia region is 6.5% per year. It is by far the fastest in the world, driven by population growth (see Appendix 1.2) and rising living standards. In 2030, residential electricity consumption in the region is projected to increase to nearly four times its 2004 level. In the commercial and industrial sectors, electricity consumption is projected to grow strongly, at an average annual rate of 4.8% to 1,291 and 5,653 billion kWh, respectively in 2030. An extensive expansion of installed generating capacity will be required to meet the world's electricity demand. In the reference case, worldwide installed electricity generating capacity is projected to grow from 16,424 billion kilowatt-hours in 2004 to 30,364 in 2030, at an average rate of 2.0% per year (Figs. 1.8 and 1.9).

1.4.2 Fuels Used for Generation of Electricity

The mix of primary fuels used to generate electricity has changed a great deal over the past two decades on a worldwide basis. Natural gas and renewable energy sources are expected

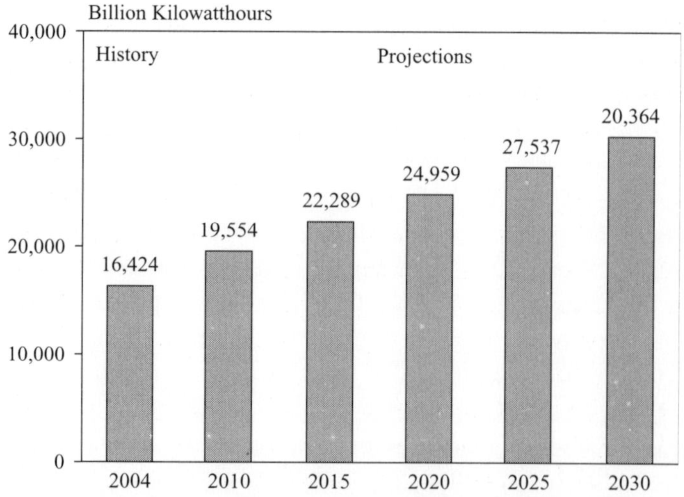

Figure 1.8 World electric power generation, 2004–2030 (billion kilowatt-hours)
Source: International Energy Outlook 2007. Energy Information Administration, USA.

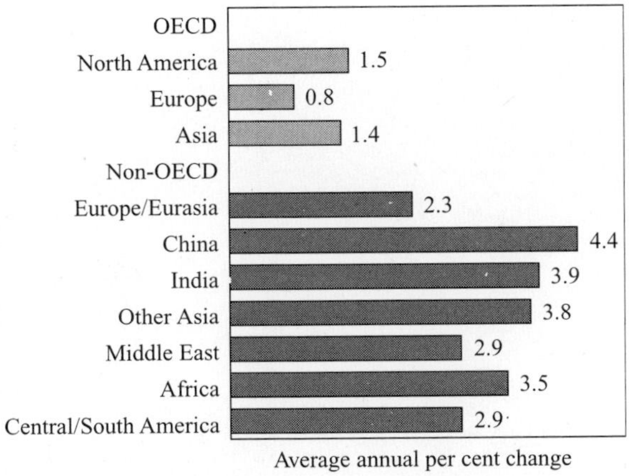

Figure 1.9 Annual growth in electricity generation by region, 2004–2030
Source: International Energy Outlook 2007. Energy Information Administration, USA.

to increase their individual share of total world electricity generation in the projections (Fig. 1.10). The natural gas share of world electricity markets increases from 19% in 2004 to 22% in 2030 and the renewable energy share rises from 18% in 2004 to 20% in 2030, before slightly declining to 19% in 2030. The relative environmental benefits and efficiency of natural gas make the fuel an attractive alternative to oil- and coal-fired generation.

Coal has remained a dominant fuel, although electricity generation from nuclear power increased rapidly from the 1970s through the mid-1980s, and natural gas-fired generation grew rapidly in the 1980s and 1990s. In contrast, in conjunction with the high world oil prices, after the oil embargo of 1973–74 and the Iranian revolution in 1979, the use of oil for electricity generation has been slowing since the mid-1970s. High world oil prices encourage switching from oil-fired generation to natural gas and nuclear power and have reinforced the role of coal in world electric power generation. Similarly, owing to the relatively high fossil fuel prices of recent years there is renewed interest in nuclear power and also in making energy from renewable sources more competitive economically. Rising fossil fuel prices have also allowed renewable energy sources to compete economically in the electric power sector (Figures 1.9 and 1.10). Consumption of hydroelectricity and other grid-connected renewable energy sources is expected to expand by 2.4% per year—approximately the same as the rates of growth projected for natural gas and coal—and the renewable energy share of the world's total energy consumption from 8% in 2003 to 9% in 2030.

Natural gas

Natural gas is the preferred energy source for electric power generation. In the IEO 2007 reference case, natural gas-fired generating capacity is projected to increase by approximately 2.5% per year from 2004 to 2030, as compared with 2.2% per year for coal and 1.9% per year for renewables (Table 1.8). Natural gas-fired combined cycle capacity is an attractive choice for new power plants because of its fuel efficiency, operating flexibility, relatively short

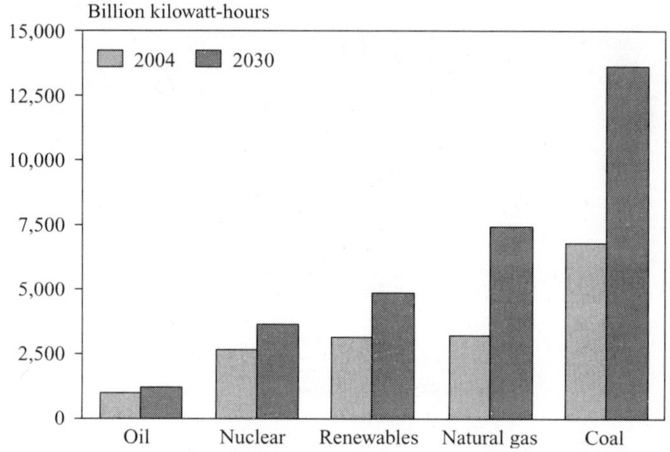

Figure 1.10 World electricity generation by fuel, 2004 and 2030 (billion kilowatt-hours)
Source: International Energy Outlook 2007. Energy Information Administration, USA.

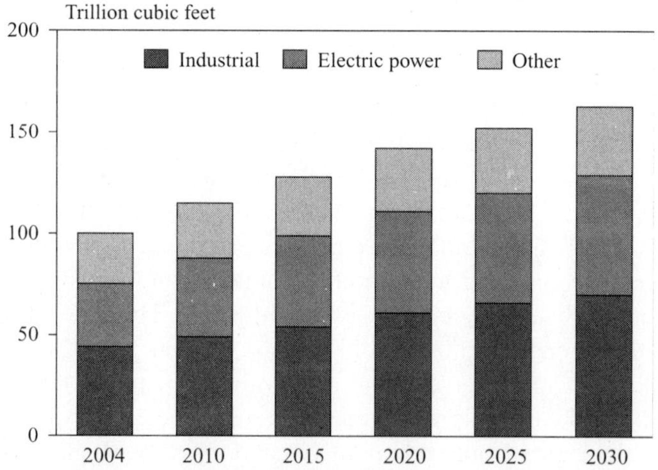

Figure 1.11 World natural gas consumption by end-use sector, 2004–2030 (trillion cubic feet)
Source: International Energy Outlook 2007. Energy Information Administration, USA.

consumption times (months instead of the year that coal or nuclear power typically require), and lower investment costs. The major drawback of natural gas capacity is the potential volatility of fuel costs.

At the world level, natural gas consumption rises from 19% of total fuel use for electricity generation in 2004 to 22% in 2030 (Table 1.8). Non-OECD economies on the whole, rely on natural gas and the consumption grows more than twice as fast as in the OECD countries (Fig. 1.11).

In the non-OECD nations, the use of natural gas in generation of total electricity is projected to rise while the use of oil and renewables is expected to reduce. Natural gas-fired capacity grows

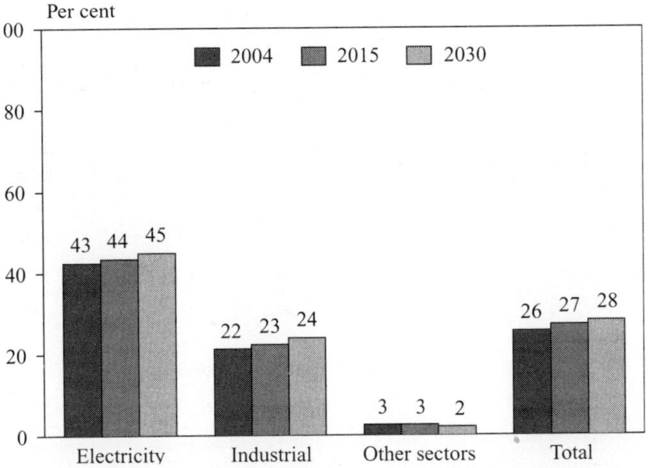

Figure 1.12 World energy consumption of coal by sector, 2004, 2015 and 2030 (per cent)
Source: International Energy Outlook 2007. Energy Information Administration, USA.

most rapidly in non-OECD Asia—especially China and India—and natural gas consumption in the electrical power sector increases by an average of 7.0% per year in China and 7.1% per year in India from 2004 to 2030.

Coal

Coal retains the largest market share in global electricity generation (roughly 40%) in the IEO 2006 reference case, despite losing some of its share to natural gas. Installed coal-fired capacity, as a share of world's total capacity, remains at about 30% (Fig. 1.12). Worldwide, coal-fired capacity is projected to grow by 2.2% per year, slightly faster than the 2.0% average annual increase for all electricity generation capacity. In 2004, the non- OECD economies on the whole relied on coal for roughly 43% of generation, which is slightly more than the OECD economies.

Regional differences in the use of coal for electricity generation are due to the availability of coal resources. Regions with large resources of coal are more likely to use coal for electricity generation, because coal has a lower energy density (energy per weight) and fewer alternative uses than oil or natural gas. These factors help keep coal prices, on an energy basis, lower than oil and natural gas prices. Coal reserves in the United States, China, India and Australia are among the largest in the world, and these countries rely on coal to generate 50% to 80% of their electricity.

Nuclear power

Worldwide, the consumption of electricity generated from nuclear power increases from 3,610 billion kWh in 2003 to 4,030 billion kWh in 2015 (Table 1.10). Higher fossil fuel prices and concerns about security of energy supplies are expected to improve prospects for nuclear power. Non-OECD Asia accounts for 69% (gigawatts) of the increase in non-OECD

Table 1.10 World nuclear energy consumption by region, Reference case, 1990–2030 (billion kilowatt-hours)

Region/Country	History			Projections					Average annual per cent change
	1990	2002	2003	2010	2015	2020	2025	2030	2003–2030
OECD									
OECD North America	650	861	845	923	953	995	989	989	0.6
United States	577	780	764	809	829	871	870	871	0.5
Canada	69	72	71	104	113	113	108	108	1.6
Mexico	3	9	10	10	11	11	11	11	0.2
OECD Europe	743	924	930	895	877	807	748	707	–1.0
OECD Asia	242	393	360	416	464	518	567	610	2.0
Japan	192	280	237	274	289	318	350	371	1.7
South Koria	50	113	123	142	175	200	217	239	2.5
Australia/ New Zealand	0	0	0	0	0	0	0	0	–
Total- OECD	**1,635**	**2,178**	**2,135**	**2,234**	**2,294**	**2,320**	**2,304**	**2,306**	**0.3**
Non-OECD									
Non-OECD Europe & Eurasia	219	254	258	278	323	393	441	443	2.0
Russia	115	134	138	151	190	233	268	321	3.2
Other	104	120	120	127	133	160	155	122	0.1
Non-OECD Asia	38	83	97	187	268	356	434	505	6.3
China	0	25	42	79	130	172	230	304	7.6
India	6	18	16	56	76	99	106	112	7.4
Other	32	40	39	52	62	85	98	89	3.1
Middle East	0	0	0	6	6	6	6	6	–
Africa	8	12	13	14	15	15	15	15	0.7
Central & South America	9	19	20	20	34	32	33	24	0.6
Brazil	2	14	13	13	22	22	23	19	1.2
Other Central & South America	7	5	7	7	12	10	10	5	–1.1
Total Non-OECD	**274**	**368**	**388**	**505**	**646**	**802**	**929**	**993**	**3.5**
Total World	**1,909**	**2,546**	**2,523**	**2,739**	**2,940**	**3,122**	**3,233**	**3,299**	**1.0**

Source: International Energy Outlook 2007. Energy Information Administration, USA.

nuclear capacity generation with an average increase of 6.3% per year from 2003 to 2030. The 51 billion watts of additional installed nuclear generating capacity projected for non-OECD Asia includes 33 billion watts in China and 12 billion watts in India (Table 1.11). Russia accounts for most of the remaining non-OECD additions, adding 22 billion watts nuclear capacity over the period 2003 to 2030. In the IEO 2006 reference case, the world's

Table 1.11 World installed nuclear generating capacity by region, Reference case, 2003–2030 (gigawatts)

Region/Country	History	Projections					Average annual per cent change 2003–2030
	2003	2010	2015	2020	2025	2030	
OECD							
OECD North America	111	117	121	125	125	125	0.4
United States	99	101	104	109	109	109	0.3
Canada	11	14	15	15	14	14	1.2
Mexico	1	1	1	1	1	1	0.2
OECD Europe	134	126	122	111	102	95	−1.3
OECD Asia	59	65	70	76	82	86	1.4
Japan	44	47	49	51	55	56	0.9
South Korea	15	18	22	25	27	30	2.6
Australia/New Zealand	0	0	0	0	0	0	–
Total OECD	**304**	**308**	**313**	**312**	**309**	**306**	**0.0**
Non-OECD							
Non-OECD Europe & Eurasia	38	40	46	56	60	60	1.6
Russia	21	23	27	33	39	43	2.7
Others	18	18	18	22	22	17	0.2
Non-OECD Asia	14	26	36	47	57	66	5.9
China	6	11	18	23	30	39	7.1
India	3	8	11	13	14	15	6.8
Other Non-OECD Asia	5	7	8	11	13	12	3.0
Middle East	0	1	1	1	1	1	–
Africa	2	2	2	2	2	2	0.5
Central & South America	3	3	5	4	4	3	0.4
Brazil	2	2	3	3	3	3	1.0
Other Central & South America	1	1	2	1	1	1	−1.1
Total Non-OECD	**57**	**72**	**90**	**110**	**124**	**132**	**3.1**
Total World	**361**	**380**	**403**	**422**	**433**	**438**	**0.7**

Source: International Energy Outlook 2006. Energy Information Administration, USA.

total installed nuclear capacity rises from 361 gigawatts in 2003 to 438 gigawatts in 2030. The projection for India is 15 gigawatts, while for China it is 39 gigawatts and for Russia it is 43 gigawatts

The world's nuclear-powered generating capacity increases in the IEO 2006 reference case from 361 gigawatts in 2003 to 438 gigawats in 2030, in contrast to projections of declines in nuclear power capacity in past IEOs. The reference case is based on laws and assumes that, for the OECD economies, in the long term, decommissioning of existing nuclear power plants as they reach the end of their operating lives will nearly equal the construction of new ones. This will result in a slight decline in the installed nuclear capacity towards the end of the projection, after peaking in 2020 (Table 1.11). In contrast, rapid growth in nuclear power capacity is projected for the non- OECD economies. The non-OECD economies are expected to add

Table 1.12 World consumption of hydroelectricity and other renewable energy (taken together) by region, Reference case, 1990–2030 (quandrillion Btu)

Region/Country	History			Projections					Average annual per cent change 2004–2030
	1990	2003	2004	2010	2015	2020	2025	2030	
OECD									
OECD North America	**9.5**	**9.8**	**9.9**	**12.2**	**12.6**	**13.1**	**13.8**	**14.4**	**1.5**
United States	6.1	6.0	6.0	7.5	7.8	8.1	8.4	8.6	1.4
Canada	3.1	3.5	3.5	4.0	4.1	4.3	4.7	5.0	1.4
Mexico	0.3	0.4	0.4	0.6	0.7	0.7	0.7	0.7	2.3
OECD Europe	**4.8**	**5.9**	**6.3**	**6.9**	**7.2**	**7.5**	**7.7**	**8.0**	**0.9**
OECD Asia	**1.6**	**1.8**	**1.7**	**1.9**	**2.0**	**2.1**	**2.2**	**2.3**	**1.2**
Japan	1.1	1.2	1.1	1.3	1.3	1.4	1.4	1.5	1.0
South Koria	0.0	0.1	0.1	0.1	0.1	0.1	0.1	0.1	1.3
Australia/ New Zealand	0.4	0.5	0.5	0.6	0.6	0.7	0.7	0.7	1.3
Total- OECD	**15.9**	**17.5**	**17.9**	**21.1**	**21.8**	**22.7**	**23.7**	**24.7**	**1.2**
Non-OECD									
Non-OECD Europe & Eurasia	**2.8**	**2.8**	**2.9**	**3.6**	**4.1**	**4.3**	**4.6**	**4.9**	**2.0**
Russia	1.8	1.6	1.7	2.2	2.5	2.5	2.7	2.8	2.0
Other	1.0	1.2	1.2	1.4	1.6	1.8	1.9	2.0	2.0
Non-OECD Asia	**3.0**	**5.2**	**5.7**	**7.0**	**7.9**	**9.1**	**10.2**	**11.3**	**2.7**
China	1.3	2.9	3.3	4.0	4.6	5.3	5.9	6.6	2.7
India	0.7	0.8	0.9	1.0	1.1	1.4	1.6	1.7	2.5
Other	0.9	1.5	1.5	2.1	2.2	2.5	2.7	3.0	2.8
Middle East	**0.1**	**0.2**	**0.1**	**0.2**	**0.2**	**0.2**	**0.3**	**0.3**	**2.7**
Africa	**0.6**	**0.9**	**0.9**	**1.1**	**1.1**	**1.2**	**1.3**	**1.3**	**1.5**
Central & South America	**3.9**	**5.6**	**5.6**	**7.4**	**8.2**	**9.1**	**9.9**	**11.0**	**2.6**
Brazil	2.2	3.0	3.1	4.2	4.8	5.4	6.1	6.8	3.1
Other Central & South America	1.7	2.5	2.5	3.2	3.4	3.6	3.9	4.1	1.9
Total Non-OECD	**10.3**	**14.5**	**15.3**	**19.3**	**21.6**	**23.9**	**26.3**	**28.8**	**2.5**
Total World	**26.2**	**32.1**	**33.2**	**40.4**	**43.4**	**46.5**	**50.1**	**53.5**	**1.9**

Source: International Energy Outlook 2006. Energy Information Administration, USA.

33 gigawatts of nuclear capacity between 2003 and 2015 and another 42 gigawatts between 2015 and 2030. The largest additions are expected in China, India and Russia.

Hydroelectricity and renewable energy

The use of hydroelectricity and other grid connected renewable energy resources continues to expand over the projection period, increasing by 2.4% per year—approximately the same as

Table 1.13 Renewable energy potential and achievements in India, March 2005

Source/Technologies	Approximate Potential	Achievements		India's position in the world
		As on March 2004	As on March 2005	
Wind power, MW	45000	2488	2980	Fifth
Small hydropower upto 25 MW	15000	1603	1693	Tenth
Biomass power/ congeneration, MW	19500	613	727	Fourth
Biomass gasification, MW	Nil	58	62	First
Solar photovoltaies, MW/km^2	20	151	191	–
Solar water heating, million m^2 cellular area	140	0.8	1.0	–
Wastes to energy, MW	1700	41.5	46.5	–
Biogas plants, millions	12	3.7	3.7	First
Improved biogas chulahs (cooker, stoves), millions	120	33.9	33.9	First

Source: Annual Report 2004/05, Ministry of New and Renewable Energy, Govt. of India.

the growth rates for natural gas and coal demand in the reference case (Table 1.1). Much of the growth in renewable energy sources results from large-scale hydroelectric power projects in non-OECD regions including India (Tables 1.12 and 1.13).

The IEO 2006 projections for hydroelectricity and other renewable energy resources include only on-grid renewables (Table 1,12). Noncommercial biofuels from plant and animal sources are an important source of energy, particularly in non-OECD economies, and the International Energy Agency has estimated that some 2.4 billion people in developing countries depend on traditional biomass for heating and cooking.

APPENDIX 1.1

Regional Definitions

OECD (18% of the 2006 world population)	Organisation for Economic Cooperation and Development: North America, United States, Canada and Mexico, OECD Europe—Austria, Belgium, Czech Republic, Denmark, Finland, France, Germany, Greece, Hungary Iceland, Ireland, Italy, Luxembourg, the Netherlands, Norway, Poland, Portugal, Slovakia, Spain, Sweden, Poland, Portugal, Slovakia, Spain, Sweden, Switzerland, Turkey, and the United Kingdom. OECD Asia—Japan, South Korea, Australia, and New Zealand.
Non-OECD (82% of the 2006 world population)	***Non-OECD Europe and Eurasia*** (5% of the 2006 world population) Albania, Armenia, Azerbaijan, Belarus, Bosnia and Herzegovia, Bulgaria, Croatia, Estonia, Kazakhstan, Kyrgyzstan, Latvia, Lithuania, Macedonia, Malta, Moldova, Romania, Russia, Serbia and Montenegro, Slovenia, Tajikistan Turkmenistan, Ukraine and Uzbekistan. ***Non-OECD Asia*** (53% of the 2006 world population) Afghanistan, Bangladesh, Bhutan, Brunei, Cambodia, China, Fiji, French Polynesia, Guam, Hong Kong, India, Indonesia, Kiribati, Laos, Malaysia, Macau, Maldives, Mongolia, Burma, Nauru, Nepal, New Caledonia, Niue, North Korea, Pakistan, Papua New Guinea, Philippines, Samoa, Singapore, Solomon Islands, Sri Lanka, Taiwan, Thailand, Tonga, Vanuatu and Vietnam.
Middle East (3% of the 2006 world population)	Bahrain, Cyprus, Iran, Iraq, Israel, Jordan, Kuwait, Lebanon, Oman, Qatar, saudi Arabia, Syria, the United Arab Emirates and Yemen.
Africa (14% of the 2006 world population)	Algeria, Angola, Benin, Botswana, Burkina Faso, Burundi, Cameroon, Cape Verde, Central African Republic, Chad, Comoros, Congo (Brazzaville), Congo (Kinshasa), Djibouti, Egypt, Equatorial Guinea, Eritrea, Ethiopia, Gabon, Gambia, Ghana, Guinea, Guinea-Bissau, Ivory Coast, Kenya, Lesotho, Liberia, Libya, Madagascar, Malawi, Mali, Mauritania, Mauritius, Morocco, Mozambique, Namibia, Niger, Nigeria, Reunion, Rwanda, Sao Tome and Principe, Senegal, Seychelles, Sierra Leone, Somalia, South Africa, St. Helena, Sudan, Swaziland, Tanzania, Togo, Tunisia, Uganda, Western Sahara, Zambia and Zimbabwe.
Central and South America (7% of the 2006 world population)	Antarctica, Antigua and Barbuda, Argentina, Aruba, Bahamas Island, Barbados, Belize, Bolivia, Brazil, British Virgin Islands, Cayman Islands, Chile, Colombia, Costa Rica, Cuba, Dominica, Dominican Republic, Ecuador, El Salvador, Falkland Islands, French Guiana, Grenada, Guadeloupe, Guatemala, Guyana, Haiti, Honduras, Jamaica, Martinique, Montserrat, Netherlands, Antilles, Nicaragua, Panama Republic, Paraguay, Peru, Puerto Rico, St. Kitts Nevis, St. Lucia, St. Lucia, St. Vincent/Grenadines, Suriname, Trinidad and Tobago, Turks and Caicos Islands, Uruguay, U.S. Virgin Islands and Venezuela.
Some other groups of countries	***European Union (EU):*** Austria, Belgium, Cyprus Czech Republic, Denmark, Estonia, Finland, France, Germany, Greece, Hungary, Ireland, Italy, Latvia, Lithuania, Luxembourg, Malta, the Netherlands, Poland, Portugal, Slovakia, Slovenia, Spain, Sweden and the United Kingdom.

G8: Canada, France, Germany, Italy, Japan, Russia, the United Kingdom and the United States.

North American Free Trade Agreement (NAFTA) Member Countries: Canada, Mexico and the United States.

Organisation of Petroleum Exporting Countries(OPEC): Algeria, Indonesia, Iran, Iraq, Kuwait, Libya, Nigeria, Qatar, Saudi Arabia, the United Arab Emirates and Venezuela.

Pacific Rim Developing Countries: Hong Kong, Indonesia, Malaysia, Philippines, Singapore, South Korea, Taiwan and Thailand.

Persian Gulf Countries: Bahrain, Iran, Iraq, Kuwait, Qatar, Saudi Arabia and the United Arab Emirates.

APPENDIX 1.2

WORLD POPULATION BY REGION , REFERENCE CASE, 1990–2030 (MILLIONS)

Region/Country	**History**			**Projection**					**Average annual per cent change**
	1990	2002	2003	2010	2015	2020	2025	2030	**2003–2030**
OECD									
OECD North America	366	424	427	457	478	498	518	537	0.9
United States	254	288	291	310	324	337	351	365	0.8
Canada	28	32	32	34	35	36	38	39	0.8
Mexico	84	104	104	113	119	125	129	133	0.9
OECD Europe	497	529	530	544	550	555	559	562	0.2
OECD Asia	187	198	199	202	204	204	203	201	0.1
Japan	124	127	128	128	128	127	125	123	−0.2
South Koria	43	47	47	49	49	49	49	49	0.1
Australia/New	20	24	24	25	27	28	29	30	0.9
Total OECD	1,050	1,151	1,156	1,203	1,232	1,257	1,280	1,300	0.4
Non-OECD									
Europe and Urasia	348	343	343	338	334	330	325	319	−0.3
Russia	148	145	145	140	137	133	129	125	−0.3
Other	200	198	198	198	197	197	196	194	−0.1
Non-OECD Asia	2,748	3,300	3,316	3,592	3,783	3,958	4,108	4,231	0.9
China	1,155	1,296	1,299	1,355	1,393	1,424	1,442	1,446	0.4
India	849	1,064	1,071	1,183	1,261	1,332	1,395	1,449	1.1
Other	743	940	946	1,054	1,129	1,202	1,271	1,336	1.3
Middle East	137	185	187	216	238	260	282	301	1.8
Africa	635	862	869	1,006	1,165	1,229	1,344	1,463	2.0
Central and South America	360	439	442	486	515	542	567	589	1.1
Brazil	150	180	182	198	209	219	228	236	1.0
Other	210	259	260	288	306	323	339	353	1.1
Total Non-OECD	4,228	5,129	5,156	5,538	5,985	6,319	6,626	6,903	1.1
Total World	**5,278**	**6,280**	**6,312**	**6,841**	**7,217**	**7,576**	**7,906**	**8,203**	**1.0**

Bibliography

1. *International Energy Outlook 2006*, June 2006, Energy Information Administration, U.S. Department of Energy, Washington. International Energy Outlook 2007, Energy Information Administration, U.S. Dept. of Energy, Washington.
2. *International Energy Outlook 2007*, June 2006, Energy Information Administration, U.S. Department of Energy, Washington. International Energy Outlook 2007, Energy Information Administration, U.S. Dept. of Energy, Washington.
3. Advisory Board on Energy, Govt. of India, Towards a Perspective on Energy Demand and Supply in India in 2004/5, New Delhi, May, 1985.
4. Huberty MP and Flock WL. *Natural Resources,* New York: McGraw-Hill, 1959.
5. Loftness, RL. *Energy Handbook,* New York: Van Nostrand , 1978.
6. *Oil And Gas Journal*, 30 December,1985.
7. Putnam PG. *Energy in the Future*, Princeton: Van Nostrand, 1985.
8. Reid WT. The Melchet Lecture, 1969. The Energy Explosion, *J. Inst. Fuel*, 43: 45–51 (1970).
9. Sutton GP and Ross DM. *Rocket Propulasion Elements*, Fourth ed. New York: Wiley, 1976.
10. *Energy Statistics Yearbook.* United Nations. New York, 1985.
11. Van Krevelen, DW. *Coal*. Amsterdam: Elsevier, 1961.
12. *World Coal*, Vol 7, No. 7 (Nov./Dec. 1981).

2 Definitions, Units and Measures

2.1 DEFINITIONS

A systematic study of fuels and combustion requires a prior knowledge of certain frequently used terms.

Proximate analysis

The proximate analysis of a fuel indicates the moisture, volatile matter, fixed carbon and ash content of the fuel in terms of percentage by weight.

Moisture means the water expelled from the fuel by specified methods without causing any chemical change to the fuel.

Volatile matter is the total loss in weight minus the moisture when the fuel is heated out of contact with air to a sufficiently high temperature under specified conditions.

Ash is the inorganic residue left when the fuel is completely burnt in air under specified conditions. It is different from the original mineral matter associated with the coal because of changes that take place during incineration.

Fixed carbon is the residue obtained by subtracting the sum of the percentages by weight of moisture, volatile matter and ash from 100. It is essentially carbon containing minor amounts of nitrogen, sulphur, oxygen and hydrogen.

Ultimate analysis

The ultimate analysis of a fuel gives its elementary composition. It is the analysis in terms of the percentage by weight of the elements: carbon, hydrogen, oxygen, nitrogen and sulphur which constitute the pure fuel, free from moisture and inorganic constituents.

Combustion

It is the process by which heat is liberated from fuel by its high temperature reaction with an oxidant which is usually oxygen present in the air. The quantity of heat evolved by the combustion of unit quantity of fuel is its *calorific value* or *heating value*. There are different ways of expressing the calorific value.

Gross calorific value (higher heating value) at constant volume is the quantity of heat liberated by the combustion of a unit quantity of the fuel at constant volume in oxygen saturated with water vapour, the original material and final products of combustion being at a reference temperature of 25°C, and the water obtained from the fuel being in the liquid state.

Gross calorific value at constant pressure is similar to the former expression except that the combustion takes place at constant pressure and not at constant volume. In the laboratory determination, solid and liquid fuels are burnt at constant volume and gaseous fuels are burnt at constant pressure. In ovens and furnaces, however, the combustion takes place at constant pressure. The difference in the two corresponding values is small. For coal, the calorific value at constant pressure exceeds the calorific value at constant volume by about 5.5 kcal/kg.

Net calorific value (lower heating value) at constant volume is the quantity of heat evolved when unit quantity of a fuel is burnt at constant volume in oxygen saturated with water vapour; the original material and final products of combustion being at a reference temperature of 25°C, and the water obtained from the fuel being in the vapour state. The net calorific value is, therefore, less than the gross calorific value by an amount corresponding to the heat of condensation of water vapour, which at 25°C is 583.5 kcal/kg. For hydrogen this is equal to 5,252 kcal/kg or 468.9 kcal/Nm^3 of hydrogen. On approximation, the following formula is used in calculating the net caloric value from gross calorific value of solid and liquid fuels:

$$C_N = C_G - 52.5H \qquad (2.1)$$

where, C_N and C_G are the net and gross calorific value in kcal/kg., respectively, and H is the percentage of hydrogen in the fuel, including hydrogen present in moisture and in water of hydration of minerals for fuels like coal.

For a gaseous fuel, the formula is

$$C_N = C_G - 4.7V \qquad (2.2)$$

Where, C_N and C_G are equal to net and gross calorific value in kcal/Nm^3, respectively, V = volume percentage (as H_2) of total hydrogen of the gaseous fuel, including the hydrogen obtainable from other combustible components as well as water vapour in the gas.

Net calorific value at constant pressure is the same as above, except that the combustion takes place at constant pressure and not at constant volume.

Ignition point or ***ignition temperature*** or ***spontaneous ignition temperature:*** It is the minimum temperature at which a fuel ignites in an oxidising atmosphere without the help of any external source of fire. It is a characteristic property of the fuel and also depends upon other factors like the atmosphere (oxygen or air), method of heating and nature of container.

Flash point and ***Fire point:*** These are characteristic properties of liquid fuels. The *flash* point is the minimum temperature at which the fuel gives enough vapours to produce a flash in the atomosphere when an external point source of fire is brought to it; and fire point is the minimum temperature at which vapours of the fuel produce a continuous flame instead of just a flash in momentary contact with a fire source. Fire point is therefore higher than the flash point. Both these temperatures indicate the fire risk of a liquid fuel. For highly volatile liquid fuels, the flash and fire points are much lower than the ignition temperature.

Rank of coal

It is well recognised that coal is derived from vegetable matter by a natural process giving rise to a transition series of solid fuels. The vegetable matter first decomposes and forms peat. Subsequently, the peat is progressively converted into lignite, bituminous coal and anthracite. The process of conversion of lignite to anthracite is known as *metamorphism* or *coalification*. The position of a coal in this metamorphic series is its rank, or degree of metamorphism, or degree of maturity.

Carbonisation of coal

It is the process of heating coal without contact with air to a sufficiently high temperature, so that the coal undergoes decomposition and yields a residue which is richer in carbon content than the original fuel. A similar but much slower process takes place in nature also. The natural process is distinguished by the term *coalification*. Like coal, many other organic substances are also subjected to the process of carbonisation.

Gasification

It is the process of converting solid or liquid fuels into gaseous fuels. Carbonisation is essentially a partial gasification process. In addition to the main gaseous fuel, some liquid products are also obtained during gasification. The gaseous product may be directly used as a fuel or may be used as a raw material for the production of chemicals, fertilisers and liquid fuels.

Caking and coking

Some coals are capable of producing a lump or cake from a powdery state by carbonisation. These coals are called *caking coals* and the phenomenon is known as the *caking of coal*. When carbonisation of coal is carried out in a commercial coke oven at about 1,000°C, and the resultant lumpy product is sufficiently hard and strong as judged by standard tests, the solid product is called *coke*, and the process is known as the coking of coal. While the process of caking is general, the process of coking is specific. All coking coals are necessarily caking but only some of the caking varieties fall under the category of coking. The main use of hard coke is in the feedstock of blast furnace.

Cracking

It is essentially a thermal decomposition process carried out with or without a catalyst, in which larger molecules are broken down into smaller ones with rearrangement of atoms and groups. The process is disproportionate in nature and forms some complex molecules also. Cracking is universally practiced in petroleum refining and is also employed in petrochemical industries on a wide scale. When cracking is performed in the presence of hydrogen and suitable catalysts, the process is called *hydrocracking*.

Orsat analysis

The composition of a gaseous fuel may be determined by analysis in a standard apparatus known as the orsat apparatus. The orsat analysis normally reports the percentage quantities of carbon dioxide, oxygen, carbon monoxide and nitrogen. Provisions can also be made for using the orsat apparatus for more complex analysis, namely, ascertaining the quantities of hydrogen, methane and unsaturated hydrocarbons. The data of orsat analysis are always on a dry basis. Moreover, sulphur dioxide is reported along with carbon dioxide.

Flue gas

It is the gaseous product of combustion of fuels in ovens and furnaces. When combustion is complete it consists of carbon dioxide, water vapour, nitrogen and excess oxygen. If the fuel contains sulphur, the flue gas will also contain sulphur dioxide, and will be reported with carbon dioxide. The term 'flue gas' is derived from the flues that lead the gases from the oven and furnaces to the chimneys.

2.2 BASIS FOR REPORTING RESULTS OF ANALYSIS

Since fuels are heterogeneous in nature, it is essential to report the analytical data on an accepted basis. The following forms of basis are in vogue, especially for solid fuels which are extremely heterogeneous substances.

- run-of-mine (rom)
- as-received
- air-dried
- dry
- dry-and-ash-free (daf)
- dry-and-mineral-matter-free (dmmf)
- moist-mineral-matter-free or, simply, mineral-free

The coal as obtained from a mine is known as *run-of-mine* coal. When the data are expressed on this coal as a whole, the basis is termed 'rom'. During the handling of the coal in transit, its moisture and mineral matter changes. The coal also undergoes other changes. Therefore, when the customer analyses the coal after it is received, the data expressed on this coal as a whole is termed *as-received*. The data on an as-received basis may often be identical to that on an *air-dried* basis, because the coal undergoes air drying by exposure to the atmosphere during transit. For getting accurate and comparable results the air-dried data are obtained after exposing the coal to an artificial but standard atmosphere at 40°C with 60% relative humidity.

The considerations for any of the other basis are as follows. Coal consists of moisture, mineral matter and pure coal. When the effect of moisture content on the analytical data is removed, it is expressed as *dry coal*. Similarly the effect of mineral matter is excluded by reporting the data on a *mineral-free* basis, taking the pure coal plus moisture as the whole. However, when it is necessary to calculate the data of the pure coal only, the effects of moisture and mineral matter are excluded by using *dry-and-mineral-matter-free* (dmmf) basis. This basis is also known as *unit coal* basis. Since it is difficult to determine the mineral matter content of coal, the *dry-and-ash-free* (daf) basis is often used in place of the dmmf basis. For low ash coals the use of daf basis is justified in assessing the properties of pure coal. For high ash coals (ash content >10%), the difference between the dmmf basis and the daf basis is appreciable and the use of the latter leads to erroneous results. This is particularly so for Indian coals which are high in ash.

2.3 UNITS AND CONVERSION FACTORS

The units and measures of both common and restricted usage in fuel technology are included here. Some of the more frequently used conversion factors are also given.

Gas Constant

$$\begin{aligned} R &= 1.987 \text{ cal/mol K} \\ &= 0.082051 \text{ l atm/mol K} \\ &= 82.05 \text{ ml atm/mol K} = 10.726 \ \frac{\text{ft}^3 \text{ psia}}{\text{mol K}} \end{aligned}$$

Length

1 inch (in) = 2.54 cm
1 foot (ft) = 12 in = 30.48 cm
1 metre (m) = 39.37 in = 3.28 ft
1 mile = 1.6093 km
1 micron = 1 micrometre (μm) = 10^{-6} m = 10^{-3} mm
1 angstrom (Å) = 0.0001 micron = 10^{-6} m

Area

1 square inch = 6.45 cm^2
1 square foot = 929.03 cm^2

1 square metre	= 1550 square inches = 10.764 square feet
1 square mile	= 640 acres = 259 hectare = 2.59 km^2
1 square kilometre	= 0.3856 square mile
1 hectare	= 10,000 m^2

Volume

1 US barrel	= 0.159 m^3 or 0.159 kl
	= 5.615 ft^3 = 34.973 UK gallons
	= 42 US gallons
1 m^3	= 35.315 ft^3 = 219.97 UK gallon
	= 264.17 US gallons
	= 6.290 US barrels
1 l	= 1000.028 cm^3
1 pint	= 0.5682 litre = 1/8 UK gallon

Mass

1 gram (g)	= 15.432 grains (gr) = 0.0353 ounces (oz)
	= 0.002205 pounds (lb)
1 pound	= 453.59 g = 0.4536 kg = 7000 gr
1 grain	= 0.0648 g
1 tonne (metric)	= 1000 kg = 0.984 tons (UK) = 1.102 short ton (US)
1 short ton (US)	= 0.893 tons (UK) = 0.907 metric tonne
1 long ton (UK)	= 20 hundred weights (cwt) = 1.106 tonnes (metric)

Density and concentration

1 g/ml	= 62.426 lb/ft^3
	= 10.022 lb/UK gal
	= 8.345 lb/US gal
1kg/m^3	= 0.06243 lb/ft^3
1lb/ft^3	= 16.019 kg/m^3

Specific volume

1 m^3/kg	= 16.0185 ft^3/lb
1 ft^3/lb	= 0.062428 m^3/kg

Temperature

1°F	= 5/9 deg C
1°C	= 1.8 deg F
(°F-32)5/9	= °C
(°C × 9/5) + 32	= °F

°C + 273	= kelvin (K)
°F + 460	= rankine (R)

Force

1 gram-force (gf)	= 980.665 dyne
1 kg-force (kgf)	= 2.2046 lb-force (lbf)
	= 70.9316 poundal
1 newton (N)	= 10^5 dyne

Pressure

1 atmosphere (atm) at 0°C	= 760 mm Hg = 29.92 in Hg
	= 33.9 ft H_2O = 1.033 kgf/cm^2
	= 14.695 lbf/in^2 = 2116 lbf/ft^2
1 pound-force per square inch (lbf/in^2)	
	= 0.07031 kilogram-force per square centimetre (kgf/cm^2)
1 kgf/cm^2	= 14.223 lbf/in^2
	= 0.036 lbf/in^2 = 1.869 mm Hg
	= 0.07358 in Hg
1 millibar	= 100 N/m^2
1 N/m^2	= 0.01 millibar
	= 10 dyne/cm^2
1 torr	= 1.333 millibar = 1.316×10^{-3} atm
	= 1.360×10^{-3} kgf/cm^2

Heat, work and energy

1 kcal	= 3.968 Btu
	= 2.205 centigrade heat unit (Chu) = 3088 ft lb
	= 4186.8 joules (J)
	= 0.00116 kWh
1 Btu	= 0.252 kcal = 0.5556 Chu
	= 778.17 ft lbf
1 Chu	= 1.8 Btu = 0.454 kcal
1 therm	= 100,000 Btu = 25,200 kcal
	= 29.31 kilowatt-hour (kW h)
1 horsepower (hph)	= 745.7 W = 0.7457 kW h
1 megawatt (MW)	= 1000 kW
1 kWh	= 1.341 horse power-hour (hph)
	= 3.6×10^6 J
	= 859.845 kcal
	= 3412.143 Btu
1 MJ	= 238.846 kcal = 947.74 Btu

1 J	= 10^7 ergs
1 eV	= 1.601×10^{-12} erg = 1.601×10^{-19} J
1 quintillion (Q)	= 1000 quadrillion (q) = 10^{18} Btu
1 q or quad (quadrillion)	= 10^{15} Btu

Power

1 W	= $1.341 \text{ x } 10^{-3}$ hp = 0.10197 m-kgf /s
	= 0.73756 ft-lbf/s
1 hp	= 745.700 W
1 hph	= 0.7457 kWh

Thermal conductivity

Btu in/ h ft^2 °F	= 0.124 kcal/m h °C
	= 0.344×10^{-3} C
1 Btu ft/ h ft^2 °F	= 1.488 kcal/ h m °C
1 kcal/m h °C	= 0.672 Btu/ h ft^2 °F
	= 8.064 Btu in/ h ft^2 °F
1 cal/cm s °C	= 360kcal/ h m °C
	= 241.9 Btu/ h ft °F
	= 2902.90 Btu in/ h ft^2 °F
1 W/cm °C	= 85.985 kcal/m h °C
	= 57.779 Btu/ h ft °F

Rate of heat flow or heat flux

1 Btu/h ft^2	= 2.7125 kcal/h m^2 = 0.31546×10^{-3} W/cm^2
1 kcal/h m^2	= 0.3687 Btu/h ft^2 = 0.1163×10^{-3} W/cm^2
1 Btu/h in^2	= 390.59 kcal/h m^2 = 0.045426 W/cm^2

Heat transfer coefficient

1 Btu/h ft^2 °F	= 4.8791 kcal/h m^2 °C
1 kcal/ h m^2 °C	= 0.20495 Btu/h ft^2 °F
1 W/cm^2 °C(or K)	= 8598.45 kcal/h m^2 °C
	= 1761.1 Btu/h ft^2 °F

Dynamic or absolute viscosity

1 poise	= 1 dyn s/cm^2 = 1 g/cm s
	= 0.010197 kgf s/m^2
	= 2.0885×10^{-3} lbf s/ft^2
1 kgf s/m^2	= 98.067 poise = 0.20482 lbf s/ft^2
1 lbf s/ft^2	= 478.8 poise = 4.8824 kgf s/m^2

Kinematic viscosity and thermal diffusivity

1 stoke	= 1 cm^2/s = 0.36 m^2/h = 3.875 ft^2/h
1 m^2/h	= 2.7778 stokes = 10.764 ft^2/h
1 ft^2/h	= 0.25806 stokes = 0.0929 m^2/h

Calorific value

Solid and liquid fuels

1 Btu/lb	= 0.5556 kcal/kg = 0.0224 therm/ton
1 kcal/kg	= 1.8 Btu/lb = 0.0403 therm/ton
1 J/g	=0.2388 kcal/kg = 0.4299 Btu/lb

Gaseous fuels

1 Btu/ft^3	= 8.899 $kcal/m^3$ at the same temperature and pressure
1 Btu/ft^3 at STP (30 in and 60°F), saturated	
	= 9.490 $kcal/m^3$ at NTP (760mm and 0°C), saturated
1 $kcal/m^3$	= 0.1124 But/ft^3 at the same temperature and pressure
1MJ/m^3	= 238.846 $kcal/m^3$ = 26.837 Btu/ft^3 at the same temperature and pressure
1 $kcal/m^3$ at NTP, saturated	= 0.1054 Btu/ft^3 at STP, saturated

Liquid fuels

1 therm/UK gal	= 5,543.3 kcal/l
10,000 kcal/l	= 1.804 therm/UK gal

Coal equivalents

Sometimes different fuels are expressed in terms of an arbitrary 'Standard Coal' of 7,000 kcal/kg calorific value.

1 kg coal equivalent	= 7,000 kcal = 2.9308 × 10^7 J
1 tonne coal equivalent	= 2.9308 kcal × 10^{16} = 2.9308 × 10^4 GJ
1 kg oil equivalent	= 10,180 kcal = 4.2622 × 10^7 J
1 tonne oil equivalent	= 4.2622 × 10^4 GJ

2.4 INTERNATIONAL SYSTEM (SI) OF UNITS

The international organisation, General Conference of Weight and Measures (CGPM), has developed an international system (SI) of units, to standardise the units of weight and measures all over the world. The SI units consist of seven base units given in Table 2.1 and two supplementary units given in Table 2.2. There are many derived units expressed in terms of base units and supplementary units (Tables 2.3–2.5). There are also guiding principles for the use of prefixes for forming the multiples and sub-multiples of units (Table 2.6).

Table 2.1 The SI base units

Quantity	Name of unit	Unit symbol	Definition
Length	metre	m	The *metre* is the length equal to 1,650,763.73 wave lengths in vacuum of the radiation corresponding to the transition of electrons between the $2p^{10}$ and $5d^5$ levels of the krypton 86 atom.
Mass	kilogram	kg	The *kilogram* is the unit of mass; it is equal to the mass of the *International Prototype Kilogram* (IPK), which is almost exactly equal to the mass of one litre of water.
Time	second	s	The *second* is the duration of 9,192,631,770 periods of the radiation corresponding to the transition between the two hyperfine levels of ground state of caesium 133 atom.
Electricity	ampere	A	The *ampere* is that constant current which, if maintained in two straight parallel conductors of infinite length of negligible circular cross section, and placed one metre apart in vacuum, would produce between these conductors a force equal to 2×10^{-7} newton per metre of length.
Thermodynamic temperature	kelvin	K	* The *kelvin*, unit of thermodynamic temperature, is the fraction 1/273.16 of the thermodynamic temperature of the triple point of water.
Luminous intensity	candela	cd	The *candela* is the luminous intensity in the perpendicular direction, of a surface of 1/600,000 square metre of a black body at the temperature of freezing/platinum under a pressure of 101.325 newtons per square metre. The candela is the luminous intensity, in a given direction, of a source that emits monochromatic radiation of frequency 540×1012 hertz and has a radiant intensity in that direction of 1/683 watt per steradian.
Amount of substance	mole	mol	The *mole* is the amount of a substance system which contains as many elementary entities as there are atoms in 0.012 kg of carbon 12. Note: When the mole is used, the elementary entities must be specified, and may be atoms, molecules, ions, electrons, other particles or specified groups of such particles.

*The unit kelvin and its symbol (K) should be used to express an interval or a difference of temperature.

Table 2.2 The SI supplementary units

Quantity	Name of Unit	Unit Symbol	Definition
Plane angle	radian	rad	The radian is the plane angle between two radii of a circle which cuts off on the circumference an arc equal in length to the radius. One radian is equal to 360°/2π, which is approximately 57° 17' 44.6".
Solid angle	steradian	sr	The steradian is the solid or 3D angle which, having its vertex in the centre of a sphere, cuts off an area of the surface of the sphere equal to that of a square with sides of length equal to the radius of the sphere.There are 4π, or approximately 12.5664 steradians in a complete sphere.

Notes:
(i) There are units for which the CGPM has not stated whether they are base units or derived units.
(ii) It is sometimes convenient to regard a plane angle and solid angle as each being a dimensionally independent physical quantity; their units are then base units. It is sometimes convenient to regard a plane angle as being the quotient of two lengths, and a solid angle as being an area divided by a length squared; their units are then derived units.

Table 2.3 Examples of SI derived units expressed in terms of the base units

	SI Unit	
Quantity	**Name**	**Symbol**
Area	square metre	m^2
Volume	cubic metre	m^3
Speed, Velocity	metre per second	m/s
Acceleration	metre per second squared	m/s^2
Wave number	1 per metre	1/m
Density, mass density	kilogram per cubic meter	kg/m^3
Concentration (of amount of substance)	mole per cubic metre	mol/m^3
Activity (radioactive)	1 per second	1/s
Specific volume	cubic metre per kilogram	m^3/kg
Luminance	candela per square metre	cd/m^2

Table 2.4 SI derived units having special names or symbols

Quantity	Name	Symbol	Expression in terms of other SI Units	Expression in terms of base and supplementary units (see footnote)
Frequency	hertz	Hz	-	1/s
Force	newton	N	-	kg m/s^2
Pressure*	pascal	Pa	N/m^2	kg/m s^2
Energy, work, quantity of heat	joule	J	N m	m^2 kg/s^2
Power, energy flux	watt	W	J/s	m^2 kg/s^2
Quantity of electricity, electric charge	coulomb	C	A s	s A
Electric tension, electric potential	volt	V	W/A	m^2 kg/s^3 A
Electric capacitance	farad	F	C/V	m^2 kg/ s^3 A
Electric resistance	ohm	Ω	V/A	m^2 kg/s^3 A^2
Conductance	siemens	S	A/V	s^3 A^2/m^2 kg
Magnetic flux	weber	Wb	V s	m^2 kg/s^2 A
Magnetic flux density	tesla	T	Wb/m^2	kg/s^2 A
Inductance	henry	H	Wb/A	m^2 kg/s^2 A^2
Luminous flux	lumen	lm		cd sr
Illuminance	lux	lx		cd sr/m^2

Note: In these expressions there are other equally correct methods of indicating multiplication and division for example, force can be shown as kg m s^{-2}

*According to ASTM, the basic unit is megapascal, (MPa) = 10^6 N m^2. One bar = 10^2 Pa = 0.1 MPa

Table 2.5 Further examples of SI derived units [Most of these are usually expressed in terms of Table 2.4. Expressions in terms of the base (and supplementary) units are included.]

Quantity	Names	Symbol (see foot-note)	Expression in terms of base and supplementary units
Dynamic viscosity	pascal second	Pa s	kg/m s
Moment of force	newton metre	Nm	m^2 kg/s^2
Surface tension	newton per metre	N/m	kg/s^2
Heat flux density, irradiance, heat flow rate	watts per square metre	W/m	kg/s^2
Heat capacity, entropy	joule per kelvin	J/K	m^2 kg/s^2 k
Specific heat capacity, specific entropy	joule per kilogram	J/kg	m^2/s^2
Specific energy	joule per kilogram	J/kg	m^2/s^2
Thermal conductivity	watt per metre kelvin	W/m K	m k g/s^3 k
Energy density	joule per cubic metre	J/m^3	kg/m s^2
Electric field strength	volt per metre	V/m	m kg/s^3A
Electric charge density	coulomb per cubic metre	C/m^3	s A/m^3
Electric flux density	coulomb per square metre	C/m^2	s A/m^2
Permitivity	farad per metre	F/m	s^4A^2/m^2 kg
Current density	ampere per square metre	A/m^2	A/m^2
Magnetic field strength	ampere per metre	A/m	A/m
Permeability	henry per metre	H/m	m kg/s^2 A^2
Molar energy	joule per mole	J/mol	m^2 kg/s^2 mol
Molar entropy, molar heat capacity	joule per mole kelvin	J/mol K	m^2 kg/s^2 mol K
Angular velocity	radian per second	rad/s	rad/s
Angular acceleration	radian per second squared	rad/s^2	rad/s^2
Radiant intensity	watt per steradian	W/sr	m^2 kg/s^2 sr
Radiance	watt per square metre steradian	W/m^2 sr	kg/s^2 sr

Note: For each unit in this table, a typical symbol is given. Many other symbols are equally valid, for example J/m^3 =N/m^2 = Pa = Ws/m^3 = VAs/m^3 = VC/m^3. Moreover, sequence is irrelevant, for example, W/m K = W/K m

Table 2.6 Names and symbols for the SI prefixes

Factor	Prefix	Symbol
10^{24}	yotta	Y
10^{21}	zetta	Z
10^{18}	exa	E
10^{15}	peta	P
10^{12}	tera	T
10^{9}	giga	G
10^{6}	mega	M
10^{3}	kilo	k
10^{2}	hecto	h
10^{1}	deca	da
10^{-1}	deci	d
10^{-2}	centi	c
10^{-3}	milli	m
10^{-6}	micro	m
10^{-9}	nano	n
10^{-12}	pico	p
10^{-15}	femto	f
10^{-18}	atto	a

Bibliography

1. *International Energy Outlook 2006*, Energy Information administration, U.S. Department of Energy, Washington, DC 20585, June, 2006.
2. Mullin JW. Recent developments in the change-over to the International System of Units, *J. Inst. Fuel*, 45: 208, 1972.
3. Rose JW and Cooper JR, (eds.). *Technical Data on Fuel*, Seventh edition, London: World Energy Conference, London, 1976.

3 Solid Fuels

3.1 INTRODUCTION

Solid materials that are used as a fuel to produce energy by combustion with oxygen or oxygen-enriched air are known as solid fuels. Both natural and processed fuels are widely used in domestic and industrial appliances. Wood, peat, lignite or brown coal, bituminous coal and anthracite are important natural solid fuels. Charcoal, soft coke, hard coke and carbonised lignite briquettes are solid fuels derived from natural varieties. Other examples of widely used solid fuels are animal dung and agricultural and industrial wastes such as straw, husk, coconut shells, bagasse and sawdust.

3.2 BIOMASS

Atmospheric carbon dioxide is fixed in the form of biomass—the primary product of photosynthesis in the terrestrial and aquatic regions of the earth. The secondary products are human, animal and industrial organic wastes and residues. Both the primary and secondary products constitute the renewable resources of energy used by human beings around the globe. In prehistoric times biomass was perhaps the only source of useful energy; currently it provides 11% of the world's primary energy. The biomass energy potential is ten times the world's annual energy need. More than half of the energy consumption in India is supported by biomass.

3.2.1 Wood and Charcoal

Wood has been used as a fuel from ancient times. It is now used extensively in countries where there are extensive forests, and where other fuels are not cheap or readily available.

Freshly cut wood contains 25%–50% water. It is normally used in an air-dried condition with 10%–15% moisture content. Ash content is about 0.6% or less. The combustible part of wood is mostly made up of cellulose and lignin, which are organic compounds of carbon, hydrogen and oxygen. Resins and waxes are also present. The average composition and calorific value of dry wood are given in Table 3.1. The chemical constitution of dry wood is fairly constant from different sources.

Although ash content is low, the oxygen content of wood is very high. This makes even dry wood a fuel of low calorific value. However, its ease of ignition, free burning quality and easy disposal of ash are the points in its favour.

Table 3.1 Average composition and properties of hardwoods

Group-wise composition, per cent air-dried	
Cellulose	45–55
Lignin	25–35
Hemicelluloses	4–6
Fats, waxes and resins	0.5–2
Water	10–15
Ultimate analysis and other properties, per cent dry basis	
Carbon	50
Hydrogen	6
Oxygen	43.5
Ash	0.5
Density, kg/m^3	650
Calorific value, kcal/kg	4,600

Wood can be used as a raw material for the carbonisation and gasification processes. Charcoal is the solid product of carbonisation (so called destructive distillation), and producer gas is the gaseous product of the gasification of wood. Charcoal also results from the incomplete burning of wood.

The carbonisation of wood is characterised by several temperature regimes.

1. All loosely bound water evaporates between 100ºC and 170ºC.
2. Gases containing carbon monoxide, carbon dioxide and condensable vapours evolve between 170ºC and 270ºC.
3. An exothermic step starts at 270–280°C, which can be detected by the spontaneous generation of heat and the increase in temperature.

The evolution of the oxides of carbon ceases, but more of the condensable vapours form. Without any outside source of heat, the temperature continues to rise slowly until it reaches 400ºC to 450ºC. A final temperature of 600ºC in practice is achieved by supplying external heat.

Charcoal is the solid product left after the carbonisation of wood. On cooling the gases and vapours give a non-condensable gas (wood gas) and the liquids settle into two layers. The upper layer is pyroligneous acid, which is an aqueous solution of many chemicals, and the lower layer is wood tar. Except for its use as a supplementary plant fuel, wood gas may be considered a waste. Acetic acid, methanol and acetone, in various grades of purity, are recovered from the pyroligneous acid. After acetic acid and water are recovered from the pyroligneous acid, wood spirit is obtained. This is further processed for the recovery of methanol. Before the advent of synthetic processes, wood spirit was the only source of methanol. Wood tar is often used as a supplementary plant fuel. It is however a potential source of many chemicals.

Although any wood gives charcoal on carbonisation, some selection is necessary to get a quality product. Dense charcoal is obtained from the species of mature dense wood with narrow annular rings, while rotten and old wood yields a light product. Wood carbonisation may be practised in open pits, kilns or metal retorts. The primitive open pit method is now practically obsolete; it accounts for a small fraction of the total production. No product other than charcoal is recovered. The yield of charcoal is low—about 15% on dry basis and it is of inferior quality.

In most kilns charcoal is the only product, but its yield and quality vary. There are many designs of charcoal kilns in vogue in India and elsewhere. Paraboloidal mud-walled kilns are more common in India. No by-products are recovered. A kiln of this design is built in a paroboloidal shape of radius 2.5–3 m, height 2.5 m and capacity 25–35 m^3 of stocked wood. Billets of wood are stacked on the ground in such a way that a paraboloid results with the apex at the top (Fig. 3.1). Two interconnecting passages are left in the stack—a vertical central passage acting as the chimney and a horizontal passage at the bottom for introducing fire to the centre. The kiln is covered first with a thick layer of grass, leaves and turf and then plastered with earth or a mixture of earth and charcoal dust. The initial firing is done with grass and twigs. Later the wood billets are partly burned to supply heat for the process. The time taken for carbonisation is about 7 to 10 days. The fire is extinguished with water and the kiln is allowed to cool for a week or more before the charcoal is taken out.

Pits and kilns are located in the forests. Retorts are however used in factories. The yield of charcoal in retorts is much higher and the by-products can also be recovered. Some typical yields of by-products of wood carbonisation in industrial retorts are given in Table 3.2.

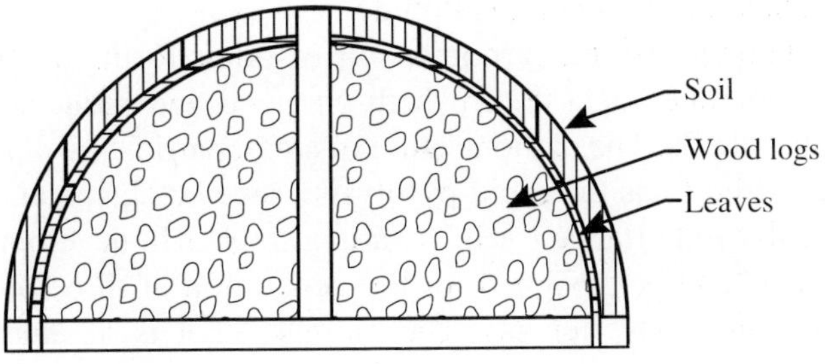

Figure 3.1 Paraboloidal charcoal kiln

Table 3.2 Products of wood carbonisation

Products	Yields, per cent air-dried
Charcoal	25–30
Pyroligneous acid	
• acetic acid	8–12
• wood spirit (crude)	10–20
• water	10–20
Wood tar	8–10
Wood gas	20–23

There are four types of charcoal retorts.

1. Externally fired batch retorts,
2. Externally fired continuous vertical retorts,
3. Internally heated batch vertical retorts,
4. Internally heated continuous vertical retorts.

Internally heated retorts are of recent design, and use forced recirculation of heated inert gases given off during carbonisation, have high thermal efficiency and lend themselves to efficient mechanical handling of the wood and charcoal.

Wood charcoal is a source of fairly pure carbon. It easily ignites and burns at low rates. For clean and smooth burning in heating ovens, wood charcoal is a good but costly fuel. It used to be a very important fuel and reducing agent in metallurgical processes. Coke has now replaced it in most cases owing to cost and other technical reasons. Charcoal has however retained its use in some special cases. It is extensively used as a fuel in blacksmiths' and metal workers' forges, and in numerous other cottage industries. Gasification plants may also use charcoal as the fuel material. During the Second world war, road vehicles using producer gas obtained from wood charcoal were used in many countries, including India. In the post Second world war period, people in industrialised countries rediscovered the excellent properties of charcoal for barbequing. This is now a major economic factor in the charcoal industry. With suitable treatment wood charcoal is widely used as an absorbent in gas purification, solvent recovery and liquid purification. It has a high surface area of 150–450 m^2/g. It remains the preferred material for the manufacture of carbon disulphide.

Charcoal is also obtained by the carbonisation of other forms of biomass, for example, sawdust, nutshells, sugarcane bagasse (crushed, chopped and reduced to small fragments), bark, twigs and coconut shells. The carbonisation of small particles is very fast and this type of charcoaling is also referred to as the rapid pyrolysis process. The small-scale rapid pyrolysis technology has particular importance for the charcoal-maker in developing countries. Almost all rapid pyrolysis plants can be designed for the generation of three products, namely, solid charcoal, pyrolysis oil and converter gas. The biomass char is obtained as a coarse or fine powder and has to be agglomerated or cast into briquettes for most applications. There is one charcoal blast furnace at Mysore Iron and Steel Limited (MISL) at Bhadravati in India with a capacity of 25,000 tonnes of pig iron per annum. The charcoal consumption is one tonne per tonne of pig iron. The supply of charcoal is now mostly met from the kilns of the neighbouring forests. The ferroalloy plant of MISL also uses charcoal as fuel. Some 2 billion people in the world use firewood and charcoal for cooking and heating.

3.2.2 Wood Wastes

Small-sized wood wastes, particularly sawdust and wood shavings, find use as fuel. These can be successfully burnt along with coal and coke. Fuel briquettes can be made by binding the wastes with an inflammable material like pitch. Wood wastes can also be converted into charcoal and producer gas.

3.2.3 Bagasse

The fibrous material left after the crushing and extraction of juice from sugar cane is known as *bagasse*. One of the chief uses of this material is as fuel in sugar factories. In its air-dried condition it contains 45%–50% moisture and 1% ash, and has a calorific value of 1,500 kcal/kg. It has been estimated that one tonne of bagasse at 49% moisture burns with 100% excess air to give the same heat as 160 l of furnace oil. On storing for some months, bagasse undergoes fermentation and dries to about 15% moisture. The fermented variety is a better fuel than the raw air-dried bagasse. It has a higher calorific value, 3,800 kcal/kg, ignites easily and has free burning quality. India produces 60 million tonnes of raw bagasse annually, out of which about 70%–75% is burnt as fuel; a small quantity is sold for pulp and paper making, and about 20% is surplus.

3.2.4 Dry Leaves and Twigs

These are collected from the forests and elsewhere and meet the fuel requirement of the poor in India to a great extent. The moisture, ash and calorific value of some non-fossil solid fuels are given in Table 3.3.

During the Second world war, when petroleum supplies were scarce, Brazil burned coffee and Argentina burned wheat and corn as fuel.

Power alcohol from farm wastes may be economical in those countries where sugar crops are easily raised and the cost of imported petroleum is high.

3.2.5 Other Agricultural Wastes

Besides bagasse, a number of other agricultural wastes also find use as fuel. The total availability of these in India per year is: rice straw 100 million tonnes, wheat straw 50 million tonnes, maize

Table 3.3 Properties of some non-fossil domestic fuels, air-dried

Fuel	Moisture, per cent	Ash, per cent	Calorific value, kcal/kg
Casuarina leaves	8.4	1.3	4400
Coconut shell	9.3	0.2	4270
Cow dung	8.5	21.7	3290
Cow dung cake	4.3	33.2	3140
Date palm leaves	10.5	4.4	4070
Groundnut shells	12.1	1.3	4200
Mango leaves	9.8	18.0	3390
Palmyra leaves	9.4	1.3	4200
Rice husk	11.6	19.5	3440
Sal wood	10.4	0.3	4480
Sawdust from sal wood	9.8	1.5	4460

straw and cob 20 million tonnes, cotton stalk 24 million tonnes, and jute stick 2.5 million tonnes. The straws are used as cattle feed and fuel. The jute stick is mainly used as a very pleasant and free-burning fuel. Rice husks are higher in ash (>10%) than the above material. It is used in ovens and furnaces. Coconut shells and fibres are also employed as fuels. It is estimated that more than 250 million tonnes of agricultural and forest residues are available in India.

3.2.6 Cattle Dung

Unfortunately, cattle dung is widely used as fuel in India. More than 100 million tonnes of dry cattle dung are burnt every year as domestic fuel. It is equivalent to more than 40 million tonnes of coal. Raw cattle dung is made into briquettes with straw, leaves, twigs or rice husks and sun dried before being used as fuel. The more rational use of cattle dung is in the production of biogas and by product manure. So long as alternative cheap fuels are not available in abundance, the misuse of cattle dung will continue, especially in villages and small towns.

3.3 PEAT

Peat is a naturally-occurring solid fuel consisting of partly decomposed plant material that has accumulated in situ under marshy conditions. Peat bogs grow at slow, but measurable rates. In general, peat accumulates in an active swamp at the rate of about 3 m in 2,500 years. It is associated with very large content of water. In fact, the amount of solid matter in peat bogs is 10%, or less. Near the surface of the deposit, peat is light brown in colour and highly fibrous in nature. With an increase in the depth, the colour becomes darker and finally black, when the vegetable origin and structure is not so obvious. A part of the water content of freshly won peat can be drained off while a much larger part is removed by drying in air. The air-drying operation may require 40–50 days.

Peat is not regarded as coal; it represents the first stage in the conversion of vegetable matter into coal. The other major stages are lignite, bituminous coal and anthracite. All these fuels together constitute the humites. Large deposits of peat are found in Russia, Germany, Poland, Finland, the USA, Sweden, Norway, Ireland, Canada, Indonesia and Scotland. Russia has 60% of the world's total reserves of peat. In the Republic of Ireland, peat is the principal fuel. The only significant deposit of true peat in India is in the swamps of the Nilgiri hills in southern India at an elevation of 2,000 m. Peat-like material also occurs at depths of 5–10 m on the two sides of the Hooghly river in and around Kolkata.

The composition and properties of peat vary widely from place to place, depending on the nature of the original plant material and the agencies and extent of decay. Raw peat consists of decayed plant material and peat humus. The content of montan wax may vary from 3% to 12%. The average properties of peat are shown in Table 3.4. The lower layers of peat usually have a higher ash content than the upper layers.

Peat is an important fuel in those countries having large deposits of peat. It is extracted by an old well-established method of hand-cutting and also by the mechanised excavation method.

Table 3.4 Average properties of peat

Proximate analysis

Per cent, air-dried				Per cent, dmmf*	
Moisture	Ash	Volatile matter	Fixed carbon	Volatile matter	Fixed carbon
15–25	3–10	50–55	25–30	65–70	30–35

Ultimate analysis, per cent dmmf

Carbon	Hydrogen	Oxygen	Nitrogen	Sulphur
55–60	6–6.5	30–35	1.5–2	0.6–1

Calorific value, kcal/kg dmmf = 4500–5000

* dry and mineral matter free

Table 3.5 Lignite resources of states as on 1 April 2007 (million tonnes)

State	Geological reserves
Tamil Nadu and Pondicherry (including NLC)	31327.02
Rajasthan	4311.43
Gujarat	2662.75
Jammu and Kashmir	27.55
Kerala	9.65
W. Bengal	1.15
All India	38339.55

Source: Geological Survey of India, News, 2007. http://www.gsi.gov.in.

Its main use is as domestic fuel, as it is easy to ignite and burns freely to give a long pleasant flame. Peat is better used in the form of briquettes. It is also largely used in steam boilers, power stations and gas producers. The low temperature carbonisation of peat is practised for getting peat coke and by-products. Peat coke is a valuable fuel for some metallurgical processes. Peat is also used as a fertiliser or for making fertilisers. However, about 75% of the world production of peat is used in heat generation.

3.4 LIGNITE OR BROWN COAL

Lignite occurs in a number of forms distinguished by their physical characteristics:

1. Woody or fibrous brown coal with the clear structure of plant tissues,
2. Earthy brown lignite, compact but friable,
3. Brown coal showing cleavage and slight woody structure.

A distinction is sometimes made between the terms lignites and brown coals on the basis of the woody structure, which is predominant in the former and not obvious in the latter. However, the two terms are often synonymous in English-speaking countries.

Lignite is characterised by high moisture content ranging from 30% to 50%. On exposure to the atmosphere, the brown colour darkens and the moisture content reduces to an equilibrium

Table 3.6 Properties of lignite

Proximate analysis

Locality	Per cent, air-dried				Per cent, dmmf	
	Moisture	Ash	Volatile matter	Fixed carbon	Volatile matter	Fixed carbon
Neyveli	10–30	5–0	40–45	30–35	52–60	40–48
Palana	25–37	4–8	35–40	30–38	45–58	42–55
Nichahom	10–25	40–54	27–28	15–16	60–61	39–40
Umarsar	18–19	15–16	33–34	30–31	53–54	46–47

Ultimate analysis

Locality	Per cent, dmmf				
	Carbon	Hydrogen	Oxygen	Nitrogen	Sulpher
Neyveli	70–73	4.6–5.5	22–26	0.6–1.0	0.6–1.5
Palana	72–75	4.5–5.5	21–23	0.8–0.9	2.0–2.5
Nichahom	67–69	5.0–5.5	23–25	1.0–2.0	0.3–1.0
Umarsar	68–69	5.0–5.5	23–24	0.8–1.0	1.5–2.0

Calorific value, kcal/kg

	Air-dried	dmmf
Neyveli	4000–5500	6500–6600
Palana	3800–4900	6900–7000
Nichahom	2500–2600	6500–6600
Umarsar	4500–5000	6500–7350

value of 12% to 20%. On drying, lignite shrinks and breaks up readily in an irregular manner. Hence, it cannot be transported far from the mine. It is also likely to ignite spontaneously as it adsorbs oxygen readily and must therefore not be stored in the open without care.

As already stated, there are extensive deposits of lignite in Russia, the USA, East Germany, Australia, Canada and India. Table 3.5 gives the lignite resources in Indian states. The major deposits of true lignite are in Tamil Nadu and Pondicherry (31,327.02 million tonnes: Neyveli alone accounting for about 4,200 million tonnes), Rajasthan (4311.43 million tonnes), Gujarat, Jammu and Kashmir and Kerala. The all India figure is about 38 billion tonnes. The lignite deposits in many areas are relatively near the earth's surface and are quite thick. The deposit at Neyveli occurs about 60 m below the ground level, is 15–20 m thick and is won by quarrying.

As in the case of peat, the composition and properties of lignite also vary widely. The immature varieties of woody brown coals cannot be easily distinguished from the mature peat. Likewise, the more mature brown coals are similar to sub-bituminous coal. The proximate analysis, ultimate analysis and calorific value of typical lignite of India are given in Table 3.6. The carbon content is 70%–75% and the oxygen content is 21%–26%. The volatile matter is often over 50% dmmf and in a large number of cases the ratios of volatile matter to fixed carbon are approximately 1:1. The high oxygen content is responsible for relatively low calorific value (6500–7000 kcal/kg dmmf). On an air-dried basis, the value may be as low as 2,500 kcal/kg

Table 3.7 Year-wise lignite production

Years	Neyveli Lignite Corporation Limited	Gujarat Mineral Development Corporation Limited	Gujarat Industries Power Company Limited	Rajasthan States Mines and Minerals Limited	Gujarat Heavy Chemicals Limited	Total
1990–91						14.072
1991–92						15.987
1992–93						16.621
1993–94	14.238	3.86				18.098
1994–95	15.464	3.846				19.31
1995–96	17.205	4.941				22.146
1996–97	17.456	5.184				22.64
1997–98	18.109	4.943		0.179		23.231
1998–99	18.168	5.002		0.249		23.419
1999–00	17.552	4.351	0.35	0.222		22.475
2000–01	18.172	4.558	1.3	0.217		24.247
2001–02	18.369	4.849	1.318	0.277		24.813
2002–03	18.624	5.504	1.417	0.473		26.018
2003–04	20.556	5.24	1.484	0.678		27.958
2004–05	21.567	6.651	1.571	0.548	0.074	30.411
2005–06	20.435	7.221	1.723	0.687	0.162	30.228
2006–07	21.014	7.989	1.66	0.463	0.159	31.285

Source: *Coal Directory of India*, Part-I-Coal Statistics, Ministry of Coal, Coal Controller's Organisation, Govt. of India, Kolkata.

due to the high moisture and ash content. The ash of lignites is generally low. This is true for India also where the ash of coal is pretty high. However, Nichahom lignites in Kashmir have a very high ash content up to 50%.

Montan wax is present in lignites ranging from 1% to 10%. Neyveli lignite yields up to 5% montan wax. The quality of montan wax of Indian lignites is poor.

Compared to bituminous coal, raw lignite is an inferior fuel owing to its high moisture, low calorific value, small size and bad weathering properties. Lignite is of economic importance in those places where it is available and other fuels do not occur in abundance. Such a situation exists in South India. Neyveli lignite is used in the generation of electricity in thermal power stations and for the production of carbonised briquettes which are used as smokeless fuel. Kashmir lignite is also made into briquettes and used as fuel. The other important uses of lignite are in gas production and metallurgical furnaces. The tar obtained by the low temperature carbonisation of lignite is a valuable raw material for the production of synthetic petroleum and other liquid fuels. Lignite is extensively used in the manufacture of producer gas. It is also gasified into synthesis gas for ammonia production.

The year-wise lignite production is given in Table 3.7. The production of lignite in India was as high as 31.285 million tonnes in 2006–07. Neyveli Lignite Corporation (NLC) accounted for the major share—67.2%.

3.5 SUB-BITUMINOUS COAL OR BLACK LIGNITE

In many countries this fuel is regarded as a variety of mature lignite resembling true coal in colour and appearance. In India it is regarded as a separate class and the term sub-bituminous coal is preferred. It is black in colour with a dull, waxy lustre. It is denser and harder than lignite and has a lower moisture content of 12% to 25%. Most sub-bituminous coals appear banded like bituminous coal. The bands are parallel to the bedding plane, but are poorly jointed and easily split into slabs instead of breaking into rectangular lumps. Like lignite, sub-bituminous coal disintegrates on exposure to atmosphere and is therefore difficult to transport.

Sub-bituminous coal is available in the USA, Russia, Germany, Canada, Australia and India. Some of the tertiary coals in Assam, Kashmir and Rajasthan are of the sub-bituminous type.

The sub-bituminous coal has 70%–78% carbon, 4.5%–5.5% bydrogen and about 20% oxygen (all on a dmmf basis). The air-dried moisture is 10%–20%. The volatile matter is 40% (dmmf) and above. The calorific value is 6800–7600 kcal/kg, dmmf. It ignites easily and may be used in raising steam for various purposes. If low in sulphur, it may also be used for manufacturing gaseous fuels.

3.6 BITUMINOUS COAL

This is the most common and widely used variety of solid fossil fuel. The raw coal ordinarily sold in the market under the Hindi name of *koela* belongs to this class. The term 'bituminous' is a misnomer. Coal does not contain true bitumen. Moreover the so-called bitumen extractable from coal with the help of organic solvents cannot define coal and its properties. The name is perhaps due to the fact that it burns with a smoky yellow flame similar to that of bitumen, and that the pitch obtained from coal tar is of a bituminous nature. The corresponding terms for bituminous coal in German and Russian imply stone-like coal and do not contain any equivalent for bitumen. The term coal ordinarily refers to bituminous coal. In the wider sense coal covers all the solid fossil fuels.

Bituminous coal is black and usually banded. The bands are parallel to the bedding plane. The coal breaks along vertical joints or *cleats* into rectangular, columnar and cubical places. Sometimes the fracture is conchoidal. The lustre varies from bright to dull. Bituminous coal is denser and harder than lignite and sub-bituminous coal, and does not disintegrate into *slacks* on exposure to atmosphere. Owing to the good heating qualities and ease of handling, bituminous coal is the major fuel in most countries.

Coal occurs in strata called *seams* which are bound between the upper and lower layers of rocks, and have longitudinal and lateral extensions. Coal seams are parallel to the bedding planes which, may or may not be parallel to the earth's surface. The whole of the coal-bearing rocks in a region is collectively known as the coal measures. The seams vary widely in thickness, longitudinal and lateral extension and depth of occurrence. For comfortable mining a seam should be at least a metre in thickness. The thickest coal seam (156 m) of India has been

Table 3.8 Coal resources of states as on 1 January 2005 (million tonnes)

State	Proved	Indicated	Inferred	Total
Andhra Pradesh	8263	6079	2548	16926
Arunachal Pradesh	31	40	19	90
Assam	279	27	34	340
Bihar	0	0	160	160
Chhatisgarh	9373	26191	4411	39975
Jharkhand	35417	30439	6348	72204
Madhya Pradesh	7513	8815	2904	19232
Maharashtra	4653	2309	1620	8582
Meghalaya	117	41	301	459
Nagaland	4	1	15	20
Orissa	15161	30976	14847	60984
Uttar Pradesh	766	296		1062
West Bengal	11383	11876	4554	27813
All India	**92960**	**117090**	**37797**	**247847**

Source: *Annual Report 2004–05*, Ministry of Coal 2005, Govt. of India, New Delhi.

Table 3.9 Coal resources by category as on 1 January 2007 (million tonnes)

Type of coal	Proved	Indicated	Inferred	Total
Coking				
Prime coking	4614	699	–	5313
Medium coking	11774	11601	1880	25255
Blendable/semi-coking	482	1003	222	1707
Sub-total	16870	13303	2102	32275
Non-coking	81050	105687	36158	222897
Total	**97920**	**118992**	**38260**	**255172**

Source: *Annual Report 2006–07*, Ministry of Coal 2005, Govt. of India, New Delhi.

discovered in Deocha-Panchmi-Dholkati (Rajmahal coalfield, West Bengal), while the deepest coal mine (720 m shaft) is at Chinakuri in the Raniganj (West Bengal) coalfield.

Extensive deposits of bituminous coal occur in the USA, Russia, Germany, China, UK, Poland, India and Canada. Inexplicably, about 95% of the world's known coal lies in the northern hemisphere. Asia, including the old USSR, and North America possess slightly more than four-fifths of the world's coal in almost equal amounts. India's coal resources (proved, indicated and inferred) in its various states are given in Table 3.8. Out of the grand total of about 248 billion tonnes, Jharkhand has the largest share of 29%, closely followed by Orissa (24.6%). West Bengal accounts for about 11% only.

Category-wise, non-coking coal contributes the maximum of 87% (Table 3.9). Prime coking coal's share is only 2%, medium coking coal's share is 10% and blendable semi-coking coal's share is less than 1%. Naturally India has to depend on imports of suitable coal for its coke plants in the iron and steel industries.

The world's proved reserves of bituminous coal and anthracite were about 998 billion short tons in 2004. The main beneficiaries are the USA (27%), Russia (17%), China (13%), South

Table 3.10 World recoverable coal reserves as of 1 January 2003 (billion short tons)

Region/Country	Bituminous and Anthracite	Sub-bituminous	Lignite	Total
World Total	528.8	298.1	170.9	997.7
United States	123.7	110.3	33.5	267.6
Russia	54.1	107.4	11.5	173.1
China	68.6	37.1	20.5	126.2
India	99.3	0.0	2.6	101.9
Other Non-OECD Europe and Eurasia	50.1	18.7	31.3	100.1
Australia and New Zealand	42.6	2.7	41.9	87.2
Africa	55.3	0.2		55.5
OECD Europe	19.5	5.0	18.8	43.3
Other Non-OECD Asia	1.4	2.0	8.1	11.5
Brazil	0.0	11.1	0.0	11.1
Other Central and South America	8.5	2.2	0.1	10.8
Canada	3.8	1.0	2.5	7.3
Other	1.8	0.4	0.1	2.1

Source: International Energy Outlook 2006, Energy Information Administration, USA.

Africa (5%), Australia (9%), West Germany (5.7%) and Poland (5.1%). India's share is 10%. The world's production of coal (including small quantities of anthracite) was 5370 Mt in 2006. Some of the countries leading in coal production are: USA. China, India, Australia and South Africa.

The production of coal in India for Industrial purposes dates back to 1775 when the first coal mines were opened in Sitarampur in the Raniganj coalfield. In those days the mined coal used to be transported to Kolkata partly by river and partly by road. The production of coal in India was 282 Mt in 2005–2006 as against 26 Mt in 1947 and 70 Mt in 1973.

There are two modern methods of mining: underground mining and open cast (or strip) mining from the surface. Modern coal mines are highly mechanised with sophisticated equipment.

Underground mining

Underground mining has two variants:

1. *Board-and-pillar* (or pillar and stall), which is used especially in medium to thick seams at relatively shallow depth
2. *Longwall*, a method adaptable to most mining conditions but which is the sole technique for mining at great depths.

In the board-and-pillar method, tunnels (galleries) are driven by blasting along and across the extension of the coal seam. The galleries divide the seam into many blocks knows as coal pillars. Coal is won by driving galleries and by depillaring to the extent possible by sand stawing. The coal is moved underground by conveyor belt and brought to the surface by haulage along an incline, or by vertical lifts in a pit.

Extraction of coal from the longwall face requires the driving of pairs of roads which link with pairs of access roads into the area under mining. These access roads in turn link with a pair of mine shafts (pits). Shafts and roads are in pairs for the forced circulation of air ventilation. The main machinery on a longwall face consists of a cutter–loader, a conveyer and a set of powered supports. The supports are for the roof so that there is no loss of coal locked up in pillars.

Open cast mining

The technology employed in strip mining is very advanced. It is based mainly on a huge dragline, with a bucket scooping up tonnes of coal at a time. Large shovels, bulldozers and bucket-wheel excavators are also used. The top soil is first removed and placed on one side exposing the coal seams, which are then progressively extracted. Intervening din bands are separately removed. Strip mining involves the restoration of the land after extraction of the coal.

More than half the coal in India is raised by open cast mining. The productivity per man-shift in open cast mining is more than five times that in underground mining. The age old board-and-pillar technique is being increasingly mechanised and will continue to remain the mainstay of underground coal production in India. The longwall powered support technology is restricted to selected mines, where geomining conditions are conducive to such a method. About 1 Mt of coal was raised by this method in 1985–86.

Coal production in various states for the years 1997–98 and 2005 are given in Table 3.11. It may be seen that the all India production was 154.4, 289.3 and 382.3 Mt, respectively. Considering the production for 2004–05, Jharkhand accounted for 20.4% and Chhatisgarh comes next (18.1%). Orissa was also close (17.3%). If we consider all the years, Madhya Pradesh produced 29.3% out of a total of 289.3 Mt. West Bengal's contribution varied between 12.6% in 1985–86 and 6.2% in 1997–98 and 6.17% in 2004–05.

Table 3.11 Coal production of various states: 1985–86 to 2004–05 (million tonnes)

State	1985–86	1997–98	2004–05	Total per cent
Andhra Pradesh	15.7	28.9	35.3	9.2
Assam	0.9	0.7	0.6	0.2
Bihar	54.3	81.3	–	–
Chhatisgarh	–	–	69.1	18.1
Jharkhand	–	–	78.0	20.4
Madhya Pradesh	42.6	84.8	52.7	13.8
Maharashtra	11.6	26.2	34.5	9.0
Meghalaya	–	3.2	5.6	1.5
Orissa	6.0	42.2	66.1	17.2
Uttar Pradesh	3.9	18.0	16.8	4.4
West Bengal	19.4	18.0	23.6	6.2
All India	**154.4**	**289.3**	**382.3**	**100.0**

Source: *Coal Directory of India* (various years), Coal Controller's Organisation, Ministry of Coal, Govt. of India, Kolkata.

Table 3.12 Sector-wise coal consumption in India (million tonnes)

	2006–2007
Power	216.472
Steel and coke ovens	12.109
Railways	0.00
Cement	10.942
Fertilisers	2.307
(Soft cokes) LTC	0.000
Brick, kilns, textiles, chemicals, paper and other industries	58.024
Colliery use	0.745
Total offtake	**300.599**

Source: *Annual Report 2006–2007*, Ministry of Coal, Govt. of India.

Table 3.12 gives the sector-wise coal consumption in India. The chief consumers of coal are the thermal power and coke plants. The most recent data on year-wise coal production is given in Table 3.13.

It may be noted that the figure 456.37 Mt is for 2007–08. Projecting from the old data in Table 3.6 of the second edition of the book, we find the state of Bihar accounted for 35.5% of coal production. Next came Madhya Pradesh (29.2%). The Raniganj coalfield in West Bengal accounted for only 12.9%.

Table 3.14 shows the properties of Indian bituminous coals. These coals vary widely in composition and properties. The air-dried moisture varies from 14% to 0.5%. On the dmmf basis, the volatite matter varies from about 45% down to about 20%, carbon from under 80% to about 92%, hydrogen from 6% to 4%, oxygen from 15% to 0.5% and the calorific value from 8,900 to 7,500 kcal/kg. The limiting figures given are for the bituminous coals of India and may differ slightly for the coals of other countries. Unlike coals of Europe and America, Indian coals have a rather high percentage of ash ranging between 12% and 40%, the usual range being 15% to 30%. In the tertiary coals of Assam, the ash is often low (3% to 8%) and only in particular areas is it rather high (24% to 26%).

The caking and coking coals form sub-groups of bituminous coal. A demerit of the Indian coals is that 82% of the total resources is non-caking. Again, out of the 18% of the caking coals, only 3.5% is of the prime coking variety and 14.5% comprises medium coking and weakly caking coals. The coals of the super seam (X to XVII) of the Jharia coalfield and the Upper and Lower Karharbari seams of the Giridih coalfield are almost the sole deposits of of truly coking coals in India. There are some coals, for example, the Dishergarh seam in Raniganj, Mahuda in Jharia and Makum in Assam, which are highly caking, but do not produce strong coke.

The tertiary coals of India deserve special mention. Geologically the coals deposited in the tertiary rocks are young and should normally mature to lignite only. However, due to various reasons, the tertiary coals in the north eastern region have matured to the bituminous stage and those in Jammu to the anthracitic stage.

Table 3.13 Year-wise coal production (million tonnes)

Year	Coal India	All India
1973–74	69.96	78.18
1974–75	78.99	88.41
1975–76	88.98	99.68
1976–77	89.48	101.04
1977–78	88.96	100.98
1978–79	90.05	101.97
1979–80	91.44	103.98
1980–81	100.86	114.01
1981–82	108.94	124.93
1982–83	114.68	130.598
1983–84	121.41	140.242
1984–85	130.81	147.438
1985–86	134.11	154.296
1986–87	144.74	165.686
1987–88	159.02	181.285
1988–89	171.50	196.23
1989–90	178.60	203.361
1990–91	189.68	214.06
1991–92	204.16	232.82
1992–93	211.22	242.00
1993–94	216.10	248.69
1994–95	223.06	257.77
1995–96	237.27	273.42
1996–97	250.62	289.32
1997–98	260.55	300.40
1998–99	256.48	296.51
1999–00	260.58	304.10
2000–01	268.14	313.70
2001–02	279.65	327.79
2002–03	290.69	341.27
2003–04	306.36	361.25
2004–05	323.58	382.62
2005–06	343.39	407.04
2006–07	360.91	430.83
2007–08	379.49	456.37

Source: *Coal Directory of India*, Part I – Coal Statistics, Ministry of Coal, Govt. of India, Coal Controller's Organisation, Kolkata.

Bituminous coal is the major commercial fuel in most countries, including India. Its chief use is combustion in domestic ovens, industrial furnaces and boilers, railway locomotives and thermal power stations. Two other important uses are carbonisation and gasification, whereby coal is converted into solid fuels, for example, coke and semi-coke, gaseous fuels like producer gas, water gas and coal gas, and liquid fuels such as coal tar fuels. Coal is also the source of a wide range of chemicals, fertilisers and synthetic liquid fuels.

Table 3.14 Properties of Indian bituminous coals

Coalfield	Air-dried, per cent		dmmf, per cent					Calorific value, kcal/kg, dmmf	Caking properties
	Moisture	Ash	Volatile matter	Carbon	Hydrogon	Nitrogen	Oxygen		
1. Raniganj									
a) Ranigunj measure									
i) Dishergarh, Sanctoria,	2.5–3.5	15–20	39–44	83–85	5.3–5.8	2.0–2.3	8.0–9.0	8100–8500	Caking
ii) Samla, Jambad-Bowlah, etc.	3.0–11.0	13–28	39–42	79–83	5.2–5.3	2.0–2.4	10.0–14.0	7600–8200	Noncaking to caking
b) Barakar measures									
Chanch, Begunia, Laikdih, etc.	0.5–1.7	15–30	25–36	86–90	4.5–5.4	1.8–2.2	4.0–7.0	8400–8800	Caking
2. Jharia									
a) Raniganj measures									
Mahuda,Lohpitic, etc.	1.5–2.2	20–25	36–40	85–87	5.4–5.8	1.7–2.2	3.9–7.5	8400–8600	Caking
b) Barakar measures									
i) I to IX seams	0.6–1.5	18–42	17–28	90–93	4.5–4.9	2.0–2.4	0.5–3.5	8600–8900	Caking to coking
ii) X to XVIII seams	0.6–2.0	15–25	22–35	87–91	4.6–5.4	1.9–2.4	2.0–3.0	8400–8900	Coking
3. Giridih									
a) Upper and lower Karharbari	0.6–1.3	12–22	27–33	89–92	4.7–5.2	1.6–1.9	0.6–2.4	8700–8900	Coking
b) Bhadua, Khandiha, etc.	0.7–1.3	20–24	27–33	89–92	4.6–5.1	1.5–1.9	0.7–1.7	8000–8900	Caking to coking
4. East Bokaro									
a) Jarangdih to Uchitdih	0.8–2.4	13–21	28–36	85–90	4.5–5.4	1.8–2.0	4.0–8.0	8300–8700	Caking
b) Kargali to Karo Bottom	0.7–1.9	17–28	24–37	86–90	4.5–5.4	1.3–2.0	3.6–6.0	8400–8800	Caking
5. West Bokaro									
a) Kuju, Bandra-Chuha	4.2–4.7	14–19	34–37	84–86	4.9–5.1	1.9–2.2	6.0–8.0	8400	Caking
b) IX to X seams, etc.	0.5–2.5	21–36	21–36	86–91	4.6–5.3	1.7–2.0	2.0–6.0	8800	Caking
6. Singranli	6–10	15–39	36–47	75–81	4.4–5.3	1.5–1.8	11–18	7900–7800	Non-caking
7. Korba, Chirimiri	6–9	12–38	28–41	81–85	4.5–5.3	1.8–2.0	8–12	8200	Non-caking
8. Pench-Kanhan Valley									
a) Damva, Rakhikole, etc.	1.2–3.0	18–24	25–37	86–89	5.1–5.5	1.7–1.9	2.5–6.0	8500–8700	Caking
b) Parasia, Eklaria, etc.	2.0–7.0	13–24	34–41	82–85	5.0–5.5	1.9–2.0	5–10	7800–8200	Non-caking
9. Wardha Valley: Ghugus, Chanda, Ballarpur, etc.	6–10	12–30	35–42	80–83	4.5–5.0	1.6–2.0	12–14	7900	Non-caking
10. Talcher	6–8	15–40	35–45	79–82	4.9–5.3	1.6–1.8	10–12	7800–7900	Non-caking
11. Singareni	6–8	15–25	35–40	78–82	4.2–5.1	–	–	7300–8000	Non-caking
12. North-Eastern	2–16	5–20	42–50	75–82	5.4–6.3	–	–	7200–8500	Non- to strongly-caking

3.7 SEMI-ANTHRACITE

Semi-anthracite is intermediate between bituminous coals and anthracites. It is harder than the most mature bituminous coal, and ignites more easily than anthracite to give a short flame changing from yellow to blue. Some of the properties of semi-anthracite are:

Moisture, per cent air-dried = 1 to 2

Volatile matter, per cent dmmf = 10 to 15

Calorific value, kcal/kg dmmf = 8,500 to 8,800

Semi-anthracites do not occur in India.

3.8 ANTHRACITE

Anthracite is the most mature and hardest form of solid fossil fuel. It has sub-metallic lustre, sometimes even a graphitic appearance. It does not soil the hand. Although there is banded structure this is not always obvious. Anthracites are characterised by low volatile matter, 3% to 10% dmmf, and high carbon content, over 92% dmmf. The air-dried moisture is 2% to 4%. The hydrogen content is 2.8% to 3.9% and the calorific value is 8,400 to 8,700 kcal/kg, both on a dmmf basis. Anthracites and semi-anthracites are non-caking.

The most notable deposits of anthracite are in South Wales, United Kingdom and Pennsylvania, USA. True anthracites do not occur in India. The tertiary coals in Jammu and the Lower Gondwana (Permian) coals in Darjeeling (West Bengal) have properties approaching those of anthracite and are known as anthracitic coals. There are about 100 to 150 Mt of such coals in each of the above two deposits. Some of the properties of these coals are given in Table 3.15.

The chief industrial uses of anthracite are in boilers, domestic stoves and metallurgical furnaces. It also finds use in small quantities as a component of coke oven charges. On heating it gives thermo-anthracite which is a useful raw material in the production of carbon electrodes and other carbon articles.

3.9 CANNEL COAL AND BOGHEAD COAL

Cannel coal and boghead coal (torbanite) are examples of sapropelic coals. By their nature they cannot be fitted into the series: peat-lignite-bituminous coal-anthracite which constitutes

Table 3.15 Properties of anthracitic coals in India

Location	**Per cent air-dried**		**Per cent dmmf**				**Calorific value, kcal/kg, dmmf**
	Moisture	Ash	Volatile matter	Carbon	Hydrogen	Oxygen	
Jammu	0.5–2.0	10–35	13–17	91–93	3.9–4.2	0.4–3	8400–8700
Darjeeling	1.0–5.0	19–29	14–28	–	–	–	8500–8700

Table 3.16 Properties of cannel coal and boghead coal

Type of Coal	Per cent air-dried		Per cent dmmf				Calorific value, kcal/kg, dmmf
	Moisture	Ash	Volatile matter	Carbon	Hydrogen	Oxygen	
Boghead	4–8	5–38	65–90	72–79	6–10	815	9000–9500
Cannel	2–5	2–40	45–56	75–84	6–10	5–12	9000–9500

humic coals. Both humic coals and sapropelic coals are of vegetable origin—the former being derived from higher plants and the latter from smaller plant organisms like algae.

Cannel coal and boghead coal have a dull lustre and yield a conchoidal fracture. They are quite massive, tough and unbanded. A splinter of these coals can be ignited with a match. Cannel coal owes its name to the word 'candle' because it burns with a long and steady flame. Cannel is dull black in colour while boghead is dull brown.

These coals have higher volatile matter and higher hydrogen than the corresponding normal (humic) coals. Some typical data are given in Table 3.16. The high hydrogen content is responsible for the high calorific value and also high yields (45% to 65% dmmf) of tarry products on low temperature carbonisation.

Cannel coal and boghead coal usually occur in isolated bands in rocks of normal coal seams. When they occur in a separate seam, this is of limited extent and thickness. These coals do not occur in India. They are unimportant in other countries also.

In earlier years cannel and boghead were used in the UK extensively for the manufacture of town gas—a gas of high illuminating power. The advent of incandescent mantles spoiled this market and the use of these coals has been discontinued.

3.10 NATURAL COKE/SLV FUEL

The intrusion of igneous rocks into coal seams is a common feature of Indian coalfields. Natural coke or *jhama* or special low volatile (SLV) fuel, as it is commonly known in India is the result of carbonisation of coal in situ by igneous intrusion. Jhama occurs in many coalfields of India, for example, at Raniganj, Jharia and Bokaro. Its estimated reserves in the Jharia coalfields are 400 Mt. It is a material of extremely variable composition and physical characteristics. The variation depends on the nature of the intrusion, its proximity to the coal seam and the extent of its effect. Some properties of a few samples of jhama from Jharia coalfield (seams XV and XVIII) are as follows: Moisture 1.4%–2.0%, air-dried; ash 21%–35%, air-dried; volatile matter 5.6%–15.7%, dmmf; carbon 88.7%–94.9%, dmmf and hydrogen 1.4%–4.3% dmmf.

Jhama may find commercial use as a substitute for coke in some cases.

3.11 ORIGIN OF COAL

Chemical and geological studies have conclusively shown that coal is formed from vegetable material such as trees, vascular plants and spores. The larger remains like tree trunks, bark, twigs, leaves and some plant residues can be seen with the naked eye. Small structures such as wood cells, spores and algae are identified with the help of a microscope. Some lower animal organisms might have also taken part in the formation of sapropelic coals and coal shales. The various ranks of coals represent different degrees in the conversion of the original plant material. The method of accumulation of the vegetable matter has also influenced the nature and properties of coals.

There are two theories on the origin of coal, namely the autochthonous (growth in situ) theory and the allochthonous (drift origin) theory. According to the former, the plants grew and decayed in the same area where the coal is found today. According to the other view, the vegetable matter was driven from the original place of occurrence by water into the neighbouring lagoons, lakes or estuaries. Either theory can account for the wide area occupied by coal seams and can also explain their relatively uniform thickness.

The in situ theory is favoured by the following points:

1. In the existing peat bogs, for example, the Great Dismal Swamp in North Carolina, USA, the decayed vegetable matter is accumulating in situ.
2. The underclays of the coal bed and sometimes the rocks associated with coal contain upright fossil roots which are sometimes so numerous as to suggest the occurrence of a fossil tree forest.
3. The coal seams are fairly constant in composition over a wide area.

The drift theory has the following points in its favour.

1. Large quantities of plant material are carried downstream by rivers and sometimes deposited near the estuary, for example, in the deltas of the Mississipi and the Ganga.
2. There is a similarity between coal and sedimentary rock. Many seams are stratified and made up of layers of coal with partings of shale, clay or sandstone, varying from a mere film to an appreciable thickness. Carbonaceous shales often pass literally into coal bands.
3. Tree stems without attached roots in a seam are in favour of the transport character of the trunks and of the vegetable debris generally.

Collected evidences indicate that the majority of coal seams in the world are of the in situ origin while there is little doubt that some coals have been formed by the drift method. The characteristic features of the Gondwana coal seams support the common belief that these coals were formed from plant materials of terrestrial vegetation which were transported into lakes, river valleys and estuaries or even into the sea. Many coalfields consist of coal seams and sedimentary rocks in a sequence that is repeated several times which corresponds to a cyclic process of accumulation of plant debris and cover by sedimentary rocks. Such formations of stratified coal seams and sedimentary rocks are known as *coal measures*.

There were two important stages in the formation of coal from vegetable matter

1. Peat stage or biochemical stage,
2. Metamorphism or dynamo-chemical stage.

Table 3.17 Normal age of coal deposits

Geological System		Approximate mean age, × 10^6 years	Normal rank of coal formed
Era	Period		
Paleozoic	Carboniferous	300	Anthracite
	Permian	245	Semi-anthracite and bituminous
Mesozoic	Triassic	200	Bituminous
	Jurassic	160	Bituminous
	Cretaceous	120	Bituminous and sub-bituminous
Tertiary	Eocene	60	Sub-bituminous and lignite
	Oligocene	40	Lignite
	Miocene	20	Lignite
Quaternary	Pleistocene	<1	Peat

In the first stage the plant material underwent decay under moist conditions by bacterial attack. The decay continued until the absence of an adequate supply of oxygen and the development of exterminative toxins ended the microbe activities. This could have been caused either by rapid and complete burial of the peat deposits under a cover of inorganic sediment or by complete flooding by stagnant water, followed later by a covering of sedimentary earth.

The agencies causing changes in the dynamochemical stage are

- pressure of overburden,
- tectonic pressure caused by severe earth movement, for example, during the folding or buckling of the earth's strata,
- regional temperature increasing by 0.5°C to 3.5°C for every 100 m increase in depth,
- contact with igneous intrusions (molten rocks and lava).

These agencies were partly or wholly operative for a very long time, spread over geological ages. The result has been coals of different maturity or rank. By and large the rank of coal rises with the increase in the age of the geological system of the coal deposits (Table 3.17).

Thus bituminous coals are found in paleozoic and mesozoic deposits and lignite in tertiary deposits. However high tectonic pressure and proximity of lava might bring about accelerated metamorphism as in the deposits of anthracitic coals in Jammu and Darjeeling. The effect of depth of occurrence on the maturity of coals is obvious. Both temperature and pressure rise with the increase in depth. Again the rate of a chemical change doubles for a rise of about 10°C. Hence the coals in the lower seams of coal measures are generally more mature than those in higher seams. This variation of rank with depth is known as *Hilt's law*.

The precise nature of the changes that took place in the gradual conversion of plant materials into coal is not known. It may however be inferred that the bacterial attack in the first stage of coal formation brought about chemical reactions under predominantly oxidising atmosphere, in which almost all the ingredients of plants took part and produced the humic acids and other components of peat. Later peat was buried under mineral rocks and pressed into more compact material. Chemical reactions such as dehydration, decarboxylation and dehydrogenation, and others, took place with the result that H_2O, CO_2, CO, CH_4 and H_2S were removed and the humic acids of peat were converted into mostly neutral substances. The result was lignite. In contrast to the biochemical stage, the lignite formation passed through a reducing atmosphere. The aging of lignite into bituminous coal and then into anthracite was due to further reduction and condensation reactions.

3.12 COMPOSITION OF COAL

Coal consists of an organic mass with some quantities of inorganic substances like water and mineral matter. Coal also contains some adsorbed gases, namely methane, carbon dioxide and carbon monoxide. The organic mass is a mixture of complex organic compounds of carbon, hydrogen, oxygen, nitrogen and sulphur. All efforts to completely isolate and identify the organic compounds of coal have failed. Coal composition is therefore studied by indirect methods.

Coal appears heterogeneous even to the naked eye. With most coals differences in texture are apparent, some bands having a bright and others a dull appearance. These differences are most marked in the case of bituminous coals but can be traced in sub-bituminous coals, in some lignites and, to a lesser extent, in anthracitic coals. The heterogeneity of coal is much more marked under a powerful microscope. An apparently homogeneous part of coal is found to consist of microscopically identifiable distinct components.

Coal petrograph is the study of coal components by visual methods, with or without the help of a micorscope. The components identified with the naked eye are referred to as macroscopic components or simply macro-components (*rock-type*). Similarly, microscopic studies identify microscopic components or micro-components (*maceral*).

Macro-components

There are four types of macro-components in coal: *vitrain*, *clarain*, *durain*, and *fusain*. This terminology is due to Stopes. According to the terminology of Thiessen, these components correspond to *anthraxylon*, *translucent attritus*, *opaque attritus* and *fusain*, respectively.

- *Vitrain* is the bright, glossy and homogeneous part of coal, having a massive texture and showing characteristic vitreous conchoidal fracture.
- *Clarain* is the semi-bright, heterogeneous material with a banded structure. It has a definite and smooth surface when fractured at right angles to the bedding plane and these faces have a pronounced gloss or shine.
- *Durain* is the dull and hard band of coal. It is heterogeneous and has a firm granualar texture. A broken face of durain is never truly smooth but always has a matt surface.
- *Fusain* occurs as patches or wedges of soft, fibrous material resembling charcoal. It is highly friable and can be readily powdered with the fingers. This material is responsible for the dirty and dusty character of ordinary coal. As a rule, fusain forms only a minor fraction of a coal seam, say about 2% to 5%.

Micro-components

A very large number of macerals of coal have been reported by different investigators. The more important of these are:

- *Vitrinite* is a major component of vitrain and one of the two principal components of clarain.
- *Fusinite* is a major component of fusain.
- *Micrinite* is one of the two principal components of durain.
- *Liptinite* or *exinite* is the other principal component of clarain and durain. It is essentially a mixture of the fossilised remains of spores, cuticles and resins.

Indian coals are generally dull in appearance, although they are often interbanded with bright and dull coals as in the seams of the Raniganj and Jharia coalfields. Fusain is present in small amounts, not exceeding 5%. In some coalfields, for example, in East Bokaro, sometimes thin but persistent fusainous bands, locally known as *matti*, occur and can be traced over wide areas. With regard to micro-components, the coals are surprisingly high (over 50%) in vitrinite content in spite of their predominantly dull appearance and apparentlv durainous nature.

The petrographic components of coal have different chemical and physical properties and they can be partly separated by mechanical and chemical methods. This situation has been gainfully exploited in technology, for example, in the petrographic preparation of coal for coke production.

The so-called *rational analysis* of coal determines a number of components on the basis of the differences in their resistance to nitric acid and the solubility in some organic solvents. The components studied by the rational analysis appear to correspond to the micro-components of coal to some extent. The approximate equivalence is shown below:

Ulmins		Vitrinite
Carbonised fibres		Fusinite
Opaque matter		Micrinite
Spore exins and cuticles	Exinite	Liptinite
Waxes, resins and hydrocarbons	Resinite	(Exinite)

The above components are fully developed in bituminous coals. They gradually lose their identity and converge into a homogeneous mass in anthracites. These components have been produced from the original plant material through the peat and lignite stages.

Peat consists of waxes, resins, simple carbohydrates, pectin, hemicellulose, celulose, lignin and humic acids.

Lignite contains waxes, resins, humic acids and a *residual mass* composed of the components corresponding to those of bituminous coal. Humic acids constitute the major part of lignite.

The exact chemical constitution of the organic content of coal is not known. It is however generally believed that these compounds are macromolecular in nature and some structural units repeat in the macromolecules. Each unit has a nucleus of condensed aromatic rings with side-chains of an aliphatic, aromatic and naphthenic (hydroaromatic) nature. The variation in the composition and proportion of the nucleus and side-chains, and in the number of structural units constituting the macromolecules gives rise to the different components as they occur in the coals of different ranks.

For bituminous coals, aromaticity is 70% to 85% of the structure, hydroaromaticity is 10% to 25% and aliphaticity is below 10%.

3.13 ANALYSIS AND PROPERTIES OF COAL

The chemical and physical properties of coal and the principles of analysis are discussed here. The detailed methods of analysis and testing vary from country to country. Those specified in India are available in the relevant Bureau of Indian Standards (BIS) publications.

The *proximate analysis* of coal determines moisture, ash, volatile matter and fixed carbon of coal. The *ultimate analysis* or *elementary analysis* determines the carbon, hydrogen, sulphur, nitrogen and oxygen of pure coal (organic mass). Based on the reports of these two analyses, a host of information can be obtained on the properties and probable utilisation of coal.

The *calorific value* is the single most important property of all fuels. The *caking properties* are of importance for bituminous coals only. Other chemical and physical properties of coal include *solubility*, *specific gravity*, *surface area* and *porosity*, *angle of repose*, *grindability*, *specific heat*, *reflectance* and *refractive index*.

3.13.1 Moisture

Owing to its nature, origin and occurrence, coal is always associated with moisture. It is customary to differentiate between external and inherent moisture. When wet coal is exposed to the atmosphere, the external moisture evaporates but the apparently dry coal still contains some moisture which can be removed only on heating above 100°C. External moisture may also be called accidental or free moisture, while *inherent moisture* may be referred to as equilibrium, or air-dried, or hygroscopic moisture. The quantity of external moisture depends mainly on the mode of occurrence and handling of coal, but the air-dried moisture is related to the inherent hygroscopic nature of the coal.

Excessive quantities of free moisture are undesirable in coal from the point of view of the customer. Moisture does not contribute to the heating value of coal and takes away appreciable quantities of heat on conversion to the vapour form. However, a small amount of free moisture is helpful in some cases of handling, combustion and coke production.

In order to get reproducible and correct values of air-dried moisture, it is necessary first to equilibriate the coal sample in a standard atmosphere of specified temperature and humidity before the actual determination is done.

Air-dried moisture

It is usually determined by observing the loss in weight of the coal sample on heating to about 105°C. In the case of peat and lignite which start decomposing at temperatures below 105°C, this method is not applicable and the Dean and Stark* method is followed.

Air-dried moisture content of coal decreases with increasing rank from a value of 25% for lignite to a minimum value of 0.5% for a low volatile (15% VM) bituminous coal. Beyond the minimum, however, the value increases to about 3% for anthracites (Fig. 3.2). For coals with a moisture content of above 3%, it is a useful measure of the rank and hence of the other qualities indicated by the rank (Table 3.18).

If coal is exposed to an atmosphere of increasing humidity at a fixed temperature, it absorbs more and more moisture. The quantity of moisture held by coal at an approximately saturated atmosphere (96% to 99% relative humidity) is called *near-saturation moisture* or *capacity moisture* or *bed moisture*. This is a better measure than the air-dried moisture for the determination of rank of high volatile bituminous, sub-bituminous and lignite coals and has found a place, directly or indirectly, as a parameter in some coal classification systems.

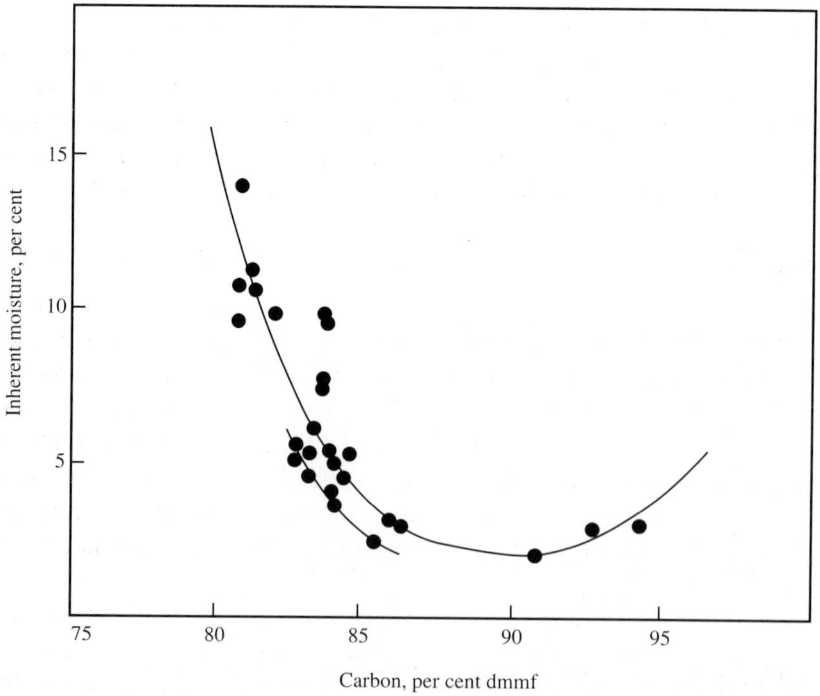

Figure 3.2 Inherent moisture and carbon per cent

* Coal is heated with an excess of toluene when moisture distills over as an azeotrope with toluene at 84°C. The vapours are condensed and separated into two layers. The amount of water is then read off in the graduated receiver.

Table 3.18 Analysis and fusion range of coal ash

Coal field	Analysis of coal ash										Ash fusion range in mildly reducing atmosphere, °C
	SiO	Al_2O_3	Fe_2O_3	TiO_2	CaO	MgO	SO_3	P_2O_5	MnO	K_2O and Na_2O	
Raniganj											
Dishergarh seam, Seetalpur colliery	54.6	21.8	7.4	1.6	5.7	2.8	1.9	1.4	0.1	2.7	1160–1260
Laikdih seam, Laikdih deep colliery	46.0	34.9	6.3	2.4	3.7	2.9	2.4	0.6	trace	0.8	1380–1450
Jharia											
XIV seam, Bararee colliery	45.5	21.8	14.6	2.0	6.2	2.5	2.8	2.7	trace	1.9	1260–1410
XI seam, Sendra Bansjora colliery	57.9	24.0	11.7	1.5	1.1	1.4	1.0	0.4	0.1	0.9	1160–1300
Giridih											
Lower Karharbaree seam, Karharbaree colliery	69.8	18.7	5.6	2.2	1.0	1.0	1.0	0.1	trace	0.6	1260–1420
Kanhan valley											
Damua colliery	55.0	25.8	8.9	1.6	3.7	0.6	0.7	1.6	trace	2.1	1260–1420

To avoid the effect of variable mineral content, moisture is also expressed in parts per 100 parts of organic mass or dmmf coal.

3.13.2 Ash and Mineral Matter

Coal contains inorganic mineral substances which are converted into ash by chemical reactions during the combustion of coal. Ash and mineral matter of coal are therefore not identical. Depending upon the nature of the source, the mineral matter in coal is called inherent and extraneous. The inorganic material of the original vegetable substances is responsible for inherent mineral matter. The extraneous mineral matter is due to

1. The substances which got associated with the decaying vegetable material during its conversion into coal,
2. Rocks and dirt getting mixed up during mining and handling of coal.

The former type of extraneous mineral matter is in a fine state and intimately associated with the organic mass of coal. Hence, it is difficult to remove this from coal by mechanical methods. The second type of extraneous mineral matter is more amenable to coal cleaning methods. The inherent mineral matter cannot be removed by any mechanical means. Fortunately, the inherent mineral matter content is low (about 2%–3% or less) in all coals. Indian coals suffer from the great disadvantage that the mineral matter content is both high and an intimately associated type due to its drift origin.

The bulk of the mineral matter of coal is due to clay or shale consisting of alumino-silicates of different composition. Other major constituents may be calcite (calcium carbonate) and pyrites, or related matter. When coal burns, shale and other hydrated minerals lose water of hydration, while sulphides, sulphates and carbonates usually decompose, or oxidise, leaving their basic radicals free to combine with excess of silica, if any. The net result is a loss of weight so that the ash of coal is always less than the mineral matter content.

The composition of coal ashes varies widely. The data for some Indian coals are given in Table 3.17. Generally about 90% or more of coal ash consists of silica, alumina, iron oxide and lime. The remaining 10% or less consists mainly of magnesium, sodium and potassium oxides and titania as basic constituents, and sulphates and phosphates as acidic constituents. Gallium, germanium, nickel, beryllium, boron and chlorine are among the trace elements of coal ash.

When heated, coal ash does not melt sharply at any temperature, but begins to soften at a much lower temperature than that at which it becomes molten and flows. This necessitates the determination of the fusion range of temperatures in place of an ash fusion point. The ash fusion characteristic of coal depends on the composition of the ash. Higher alumina and silica content make the ash refractory, while lime, magnesia, iron oxides and alkali oxides have a fluxing effect and lower the ash fusion temperatures. The ash of Indian coals is generally refractory in nature and the fusion range is mostly above 1300°C.

The quantity of mineral matter or ash has no relation with the rank or maturity of coal. Among the macro-constituents of coal, vitrain is of low ash (1% to 2%), clarain is higher (3% to 5%), and durain is still higher (5% to 8%). Fusain may have high ash. The mineral matter does not contribute to the calorific value of coal. On the contrary it creates many difficulties in the efficient utilisation of coal. In a furnace grate the ash may seriously restrict the passage of air and lower the rate of combustion. High ash leads to large heat losses and carbon losses and creates such problems as boiler deposits and clinkering. It has been estimated that 1% rise in coal ash is equivalent to 0.3% to 4% fall in boiler efficiency. High ash is therefore undesirable and harmful. The content of mineral matter should be reduced by the methods of coal cleaning. This is particularly important for coals used for making metallurgical grade coke. High ash coke seriously reduces the efficiency of blast furnace operations. Coal cleaning is vitally important for high-ash Indian coking coals.

Apart from the total percentage of ash, its composition and fusion range also influence the efficiency of coal utilisation. In general, clinkering troubles are augmented by the lower fusion temperatures. On the other hand, coals of refractory ash, for example, Indian coals, are a problem to use in slagging furnaces and slagging gas generators where ash is removed in the molten state. A non-clinkering ash contains more than 45% of silica and 35% alumina and less than 15% ferric oxide and 10% calcium oxide.

The coal ash may be used for extracting valuable rare elements. It is reported that certain coal-based gases produce flue dusts that provide most of Britain's germanium.

Enormous quantities of fly ash are produced from coal-based thermal power stations all over the world. Fly ash is useful in the production of cement and concrete, bricks, filling material and lightweight aggregates. A plant at Ennore in Tamil Nadu produces 70 million bricks annually from cement, lime and fly ash.

The ash of coal is directly determined in the laboratory by completely burning the coal. It is necessary to know the accurate content of the mineral matter to determine the content of a pure organic mass of coal. Some methods are available for this purpose. The Radmacher and Mohrhauer method essentially consists of dissolving out the mineral substances by treatment with hydrofluoric acid and hydrochloric acid, and finding the loss of weight.

Empirical formulae are available for calculating the mineral matter content from the ash values. The Parr formula is

$$\mathrm{MM} = 1.08\mathrm{A} + 0.55\mathrm{S} \tag{3.1}$$

and the KMC (King, Maries and Crossley) forumula is

$$\begin{aligned}\mathrm{MM} = {} & 1.09\mathrm{A} + 0.5\mathrm{S}_{\mathrm{pyr}} + 0.84\mathrm{CO}_2 - 1.1\mathrm{SO}_{3,\mathrm{ash}} \\ & + \mathrm{SO}_{3,\mathrm{coal}} + 0.5\mathrm{C1}\end{aligned} \tag{3.2}$$

where, MM = mineral matter, A = ash, S = total sulphur, S_{pyr} = pyritic sulphur, CO_2= carbon dioxide from carbonates, $S0_{3,ash}$ = sulphate in ash, $SO_{3,coal}$ = sulphate in pure coal, and Cl = chlorine, all in per cent.

The KMC formula is regarded as the standard of reference, but it involves too many analytical steps. The Parr formula is quite simple. It recognises only two losses of consequence during conversion of mineral matter into ash, namely loss of water of hydration (O.O8A) and loss in the conversion of pyrites into oxides. It further assumes that all the sulphur in coal is in the inorganic form.

Indian coals in general are low in sulphur and carbonate. A simplified version of the Parr formula is used for these coals:

$$MM = 1.1A$$

The formula assumes that the difference in the weights of mineral matter and ash is due to the water of hydration which is taken to be 10% of the ash value (0.1A).

3.13.3 Volatile Matter and Fixed Carbon

Volatile matter and fixed carbon are not the constituents of coal. They represent the volatile and non-volatile products of thermal decomposition of coal under specified conditions. Volatile matter does not include the moisture in coal but it contains water that is formed from the hydrogen and oxygen of coal during the decomposition. When volatile matter is reported on air-dried and daf (dry and ash-free) basis, it also includes the part of the mineral matter, for example, water of hydration, which escapes into the gaseous state. On dmmf basis however, it represents only the volatile products from the organic mass. Fixed carbon does not include the ash. It is the non-volatile residue of the organic mass. Furthermore the term 'fixed carbon' is a misnomer: It contains the non-volatile parts of other elements also. Nonetheless, long usage has shown that the quantities of volatile matter and fixed carbon are of great scientific and technological value.

The volatile matter and fixed carbon, both expressed as a percentage on dmmf basis add up to 100 parts of pure coal (organic mass). The higher the volatile matter, the lower the fixed carbon. There is a correlation between volatile matter and maturity of coal. With the increase in the maturity or rank of coal, its volatile matter content decreases. However, there seems to be a critical value of volatile matter, 32%–33% dmmf, above which this relationship is not valid. Different macerals of the same coal yield different amounts of volatile matter. The highest yields are from the exinites and the lowest the fusinite. Vitrinite occupies an intermediate position.

The physical nature of the residue left after the volatile matter test is an approximate measure of the caking properties of coal.

The moisture, ash, volatile matter and fixed carbon data of typical Indian coals are given in Table 3.14.

3.13.4 Carbon

The carbon of coal is not to be confused with its fixed carbon. In anthracites the volatile matter is very small and the values of fixed carbon and total carbon are almost equal. In other coals fixed carbon is less than total carbon. The carbon content is determined by Liebig's method by completely burning the coal in a current of pure oxygen and finding the amount of resultant carbon dioxide. Correction is made for the inorganic carbonates. The carbon content depends on the nature of coal. It increases with increase in the rank (Tables 3.14 and 3.19). Among the macerals of the same coal, the exinite and vitrinite have almost similar carbon content while fusinite has higher content.

Table 3.19 Classification of Indian coals, their properties and utilisation (IS: 770-1977)

Class	Type	Symbol	Nature	Basic parameters			Other properties				
				CV kcal/kg (dmmf)	VM per cent dmmf	GK coke type	Moisture (60% RH) part/100 parts unit coal	Carbon % dmmf	Hydrogen % dmmf	Code number	Utilisation
Lignite	Consolidated	L	Non-caking	6150–7300	>50	A	20	67–73	4.5–5.5	100, 200	Combustion, gasification, briquetting, bydrogenation
Sub-bituminous	High volatile	SB	Non-caking	6950–7500	35–50	A	10–20	76–79.5	4.5–5.1	210, 220, 230	Gasification, combustion
Bituminous	High volatile (NE coals)	B_1	Non-caking to strogly caking	7500–8500	>45	A–G_3^+	2–9	75–82.5	5.3–6.3	300, 400, 500, 501–3, 601–3, 701–4, 310, 410, 510, 511–15, 611–15, 711–15	Hydrogenation, blending, combustion
	Medium volatile	B_2	Non-caking	7500–8250	27–43	A–B	7–11	79.5–82.0	4.7–5.2	320, 420, 520, 330, 430, 530, 340, 440, 540, 450	Combustion, hydrogenation, gasification
	High volatile	B_3	Weakly caking	8250–8400	33–43	C–D	5–7	82.5–83.5	5.0–5.4	521, 531, 621, 631, 721, 731	LTC, hydrogenation, blending
	High volatile	B_4	Medium to strongly caking	8250–8500	33–43	E–G_3^+	2–5	83.5–87.5	5.0–5.8	622–5, 722–5, 632, 732–5, 832–5	Blending, hydrogenation
	Medium volatile	B_5	Weakly to medium caking	8500–8700	22–33	C–F	<2	86.5–88.0	4.7–5.0	741–2, 841–2, 851–2	Blending
	Medium volatile	B_6	Strongly caking (prime coking coals)	8500–8900	22–23	G–G_3^+	<2	88–90.5	4.8–5.2	843–5, 943–5, 853–5, 953–5	Metallurgical coke making
	Low volatile	B_7	Weakly to medium caking	8500–8900	18–22	C – G	<2	90.5–91.5	4.5–4.9	861–2, 961–3	Blending, combustion
	Low volatile	B_8	Non-caking to weakly caking	8250–8700	15–18	A –D	<2	91.5–92.0	4.5–5.	670, 770, 870–1	Combustion, domestic use
Anthracite	Semi–anthracite	SA	Non-caking	8250–8700	10–15	A	<2	92.0–93.0	3.7–4.2	670, 770, 870–1, 680, 780, 880.	Carbon artefacts, combustion, domestic use
	Anthractie	A	Non-caking	8500–8700	<10	A	–	>93	<3.7	890	– do –

3.13.5 Hydrogen

Hydrogen is determined together with carbon by completely burning the coal. From a knowledge of the weight of water formed, the hydrogen content is calculated. Corrections are made for the moisture of coal and the water of hydration of minerals. The content of hydrogen in coals from peat to bituminous stage varies between 4.5% and 6.5% and is not related to the rank (Tables 3.14 and 3.19). Beyond the bituminous stage, the hydrogen content sharply decreases to a value of 1%–2% in anthracite. Among the micro-constituents, the exinites are characterised by high hydrogen and the fusinites by low hydrogen content. Cannel coal and boghead coal also have high hydrogen content. The hydrogen of coal is responsible for the production of many useful materials, for example, gaseous hydrogen, methane and coal chemicals in the carbonisation and gasification industries.

3.13.6 Nitrogen

The Kjeldahl method is used to determine the amount of nitrogen in coal. The sample is digested with oleum containing a catalyst when nitrogen is converted into ammonium sulphate. The ammonia is then estimated and the nitrogen content calculated. The nitrogen content does not bear any relation to the rank of coal (Table 3.14). In most coals it is between 1% and 2%. Lower values down to 0.5% and upper values upto 2.5% are also found. The nitrogen in coal is useful in the carbonisation industries which recover a part of it as valuable by-products like ammonia, ammonium sulphate and pyridine bases.

3.13.7 Sulphur

Sulphur is commonly present in coal in three forms: pyrites, organic sulphur and sulphate. The occurrence of sulphides and elementary sulphur in coal have been reported in some cases. Only organic sulphur should be included in the ultimate analysis. Since most Indian coals contain 0.7% or less of total sulphur, it is usually shown as sulphur in the ultimate analysis, without any corrections.

Total sulphur is determined by Eschka method or Bomb method. In the former, the entire content of sulphur is converted into soluble sulphates by heating the coal with an oxidising mixture of magnesium oxide and sodium carbonate and then estimating the sulphate. In the Bomb method the total sulphur is converted into soluble sulphates by heating the coal with an oxidizing mixture of magnesium oxide and sodium carbonate and then estimating the sulphate. In the Bomb method the total sulphur is also converted into the sulphate form during the determination of calorific value in Bomb calorimeter. The pyritic and sulphate sulphur in coal are determined by the usual methods of analytical chemistry and the organic sulphur is then calculated by subtracting the inorganic sulphur from the total sulphur.

Although Indian coals are usually low in sulphur (0.7% or less), many coals of the North-Eastern coalfields have high sulphur content of around 4%. In general, the coals of most countries of the world are found to contain between 1% and 3% sulphur. Higher and lower values are also reported. It has been found that the high sulphur content is generally due to pyrites. However, North-Eastern coals form an exception where much of the sulphur is in the organic form.

The sulphur content of coal has no relation to its rank or composition. Although elementary sulphide, pyritic and organic forms of sulphur contribute to the calorific value of coal, it is an undesirable constituent. It emits corrosive sulphur dioxide during combustion and obnoxious hydrogen sulphide and mercaptans during carbonisation and gasification. A part of the sulphur can be recovered as a by-product in the coking and gas-making industries. Since coke always retains a considerable part of sulphur in coal, metallurgical industries demand coals of low sulphur content in coke making.

3.13.8 Oxygen

The oxygen in coal is usually determined by the method of difference. The carbon, hydrogen, nitrogen and sulphur are expressed in per cent on a dmmf basis and their sum is subtracted from 100. The disadvantage of this procedure is that the errors of all the other determinations are reflected on the oxygen value. Direct methods are also available for the determination of oxygen, but these are laborious and time-consuming. The oxygen content of coal has a close relation to the rank. The former decreases with the increase in the latter (Table 3.14). Among the macerals of the same coal, vitrinite has the highest oxygen content, fusinite the lowest, and exinites the intermediate value.

3.13.9 Phosphorus

Phosphorus occurs in small quantities in coal. Its determination is important when the coal is to be used in the production of metallurgical coke. Indian coking coals have normally less than 0.15% phosphorus which is not a problem for common use in iron and steel production. The coking coals of Giridh are even lower in phosphorus, 0.01% to 0.04%, and suitable for producing coke for making ferro alloys. There are certain areas in the coalfields of Madhya Pradesh and North-Eastern India where again the coal contains 0.01% or less of phosphorus.

3.13.10 Calorific Value

This is the basic property of fuels, indicating the quantity of heat evolved by their complete combustion. Distinction is made between gross and net calorific values, and between gross calorific values at constant volume and constant pressure (see Chapter 2). The more generally used property for coal is the gross calorific value at constant volume. For combustion calculations, however, the net calorific value is often used with advantage. The calorific value of coal is determined in a Bomb calorimeter. The complete combustion is ensured by using powdered coal in a small briquetted form, and pure oxygen at a moderately high pressure of

25 atm. Precision equipment under specified conditions and adoption of various correction factors are necessary for the high accuracy that the test demands.

There are many formulae for the calculation of the calorific value of coal without actually determining the same in the laboratory. One such formula is due to Goutal (1902) and is based on proximate analysis

$$C_G = 82F + aV$$

where C_G = gross calorific value, kcal/kg air dried, F = fixed carbon, per cent air-dried, V = volatile matter, per cent air-dried, and a = a constant, depending upon the value of volatile matter expressed as per cent daf (V′) as shown below.

V′	a	V′	a	V′	a
5	145	20	109	35	94
10	130	25	103	38	85
15	117	30	98	40	80

This formula considers both volatile matter and fixed carbon as contributing to the calorific value of the coal. The fixed carbon of different coals is assumed to be of a fixed composition and hence of fixed calorific value. The composition and calorific value of the volatile matter differ from coal to coal, and are assumed to depend on the nature of coal as indicated by the volatile matter on a daf basis. These assumptions limit the utility of the Goutal formula.

The following formulae have been developed by the Central Fuel Research Institute (CFRI), Dhanbad, Bihar for the calculation of calorific value of Indian coals from their proximate analysis:

For low moisture coals, $M \leqslant 2\%$

$$C_G = 91.7F + 75.6(V - 0.1A) - 60M$$

For high moisture coals, $M \geqslant 2\%$

$$C_G = 85.6[100 - (1.1A + M)] - 60M$$

where M, A, V and F denote moisture, ash, volatile matter and fixed carbon, all in per cent air-dried, respectively.

There are also several formulae that are based upon the ultimate analysis of coal. The original Dulong formula is:

$$C_G = 80.8C + 344(H - O/8)$$

where C, H and O represent carbon, hydrogen and oxygen, respectively, as per cent of the coal.

The above formula assumes that the heat of formation of coal is zero and the only heat giving elements are carbon and surplus hydrogen, the latter meaning the hydrogen in excess of that required to combine with the oxygen of the coal to form water.

On including sulphur as a heat giving element, the Dulong formula is amended as

$$C_G = 80.8C + 345(H - O/8) + 22.2S$$

where S = sulphur, per cent of coal.

All these formulae are of limited use because the heat of formation of coal is neither zero nor constant. The modified Dulong formula is, however, fairly useful for all practical purposes. Seyler has developed a set of formulae relating the calorific value to the carbon, hydrogen and volatile matter. These are found to be quite accurate for many coals, excluding those of India.

The calorific value of coal bears a good relationship with the rank (Tables 3.14 and 3.19). The higher the rank, the higher the calorific value. Anthracites, however, form an exception to this regularity and have lower calorific values than many bituminous coals. This is due to the sharp fall of hydrogen content in anthracites, the hydrogen having more than four times the calorific value of carbon. The high hydrogen content is also responsible for the relatively high calorific values of cannel coal, boghead coal and the exinites of humic coals.

3.13.11 Caking Properties

Caking coals form a part of the bituminous group: Peat, lignite, sub-bituminous coal, semi-anthracite and anthracite are all non-caking. So also are some of the bituminous coals. A measure of the caking properties is necessary for coal classification and for selection of coal for different industries. Caking properties influence not only the production of coke but also the performance of the coal in combustion and gasification.

Caking, swelling, agglutinating and plastic properties are all interrelated. Numerous laboratory tests have been proposed for measuring the propensity of these properties. While some have been developed into standard tests, others have become either obsolete or are of theoretical interest only. The standard tests in vogue in different countries are: Free swelling index, Gray–King (low temperature) Coke type, Caking (agglutinating) index, Roga index and Thickness of plastic layer. The appearance of the residue (coke-button) of the volatile matter test may give an idea of the caking capacity of coal. This simple test is stil in vogue in some countries. The coke-button may indicate five types: pulverulent, sintered, weakly caked, caked and strongly caked.

Free swelling index

This is a simple and very useful test. It is one of the three tests standardised in India. Coal is heated on a flame under specified conditions. The swollen residue is compared with standard profiles and assigned numbers, increasing from 0 to 9 by ½ unit. The higher the number, the better the caking and swelling properties. The limitation is imposed by the maximum possible number 9 and the very highly swelling coals cannot be differentiated from one another.

The caking index is the maximum whole number ratio of sand to coal in a 25 g mixture of the two, which on heating under standard conditions produces a residue capable of supporting a weight of 500 g without producing more than 5% loose grains. It is a measure of the binding or agglutinating property of the coal undergoing carbonisation. The higher the agglutinating propensity, the higher the amount of sand that can be bound by coal and hence the higher the caking index. It is a modified Grey–Campredon test and responds well to high-ash Indian coals. The Bureau of Indian Standards (BIS) has adopted it as one of the three tests for caking properties. The demerit of the test lies in its trial-and-error approach.

Low temperature Gray–King carbonisation assay

Powdered coal is heated under standard conditions, 600°C. The residue is visually compared with standard profiles and a letter A to G or G_1 to G_9 is assigned. This letter represents the Gray–King (LT) Coke Type. 'A' means non-caking. Letters from 'B' to 'G_9' indicate the increasing order of caking and swelling capacity. This is one of the three standard tests in India and is now judged the best suited for Indian coals. If the coal has higher than 17% ash, then the sample has to be washed before testing.

Roga index

This is a variation of the agglutinating index. A mixture of coal and specified anthracite in a fixed proportion is carbonised under standard conditions. The strength of the residue is expressed as a numerical figure called the Roga index.

Sapozhnikov test

This plastometric test measures the maximum thickness of the plastic layer that is formed when coal is carbonised under well-defined conditions.

Whatever test is used, the parameter for the caking properties is significant in defining the nature of coal. The very young coals—peat, lignite and sub-bituminous—and the very mature ones—semi-anthracite and anthracite—are completely devoid of cakability. In the case of bituminous coals, the caking capacity first increases with the rank, attains a maximum and then decreases with further rise in the maturity (Fig. 3.3). The coals of the highest coking capacity are

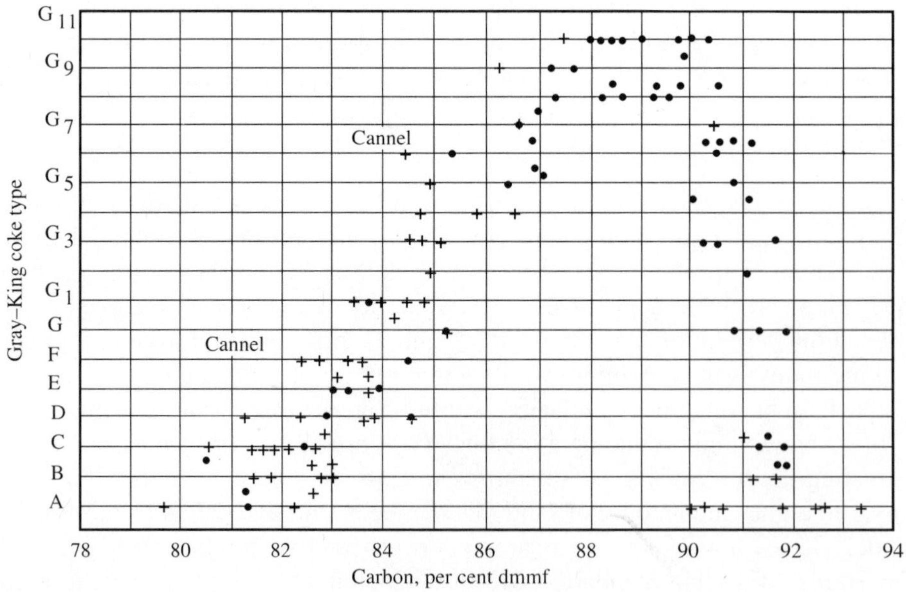

Figure 3.3 Gray–King coke type and percentage of carbon in coal

often called *fat coals*. These coals are rare in India. Among the macerals of the same coal, the fusinite is always found to be non-caking and the exinites are always good caking. The caking properties of vitrinite vary with the rank in the same way as those of the whole coal.

Why some coals are caking and others non-caking is an interesting problem for investigation. The caking properties are perhaps due to a chemical process in which the macromolecular structure of the coal mass is thermally broken down and some products of relatively low molecular weight remain in a softened state for a sufficiently long time in the reaction zone. As a result the entire mass takes the form of a plastic matter which is converted into a solid lump, *resolidified*, on further heating. In this way the powdered caking coals are capable of producing lumpy semicoke and coke. The non-caking coals cannot produce the plastic state and on carbonisation yield only a sintered mass in place of coke.

3.13.12 Specific Gravity and Bulk Density

The proportion and nature of both the organic mass and mineral matter influence the specific gravity of coal. Mineral matter is heavier than pure coal. Therefore, for the same type of coal, the higher ash coals have higher specific gravity. The true specific gravity of average bituminous coals normally varies between 1.27 and 1.45. The following formula is valid for many coals of India.

$$g = k + A/100$$

where, g = specific gravity, k = a constant having an average value of 1.25, and A = percentage of ash. This formula assumes an overall specific gravity of 2.25 for the mineral matter of coal.

Among the coals of different types, the specific gravity seems to increase with maturity: peat 1.15 to 1.25, lignite 1.25 to 1.3, bituminous coal 1.27 to 1.45 and anthracite 1.4 to 1.7.

The bulk density of a material is the overall density of it kept in large quantities. It depends upon the true density of the material, the pore volume of the particles (or lumps) and the voidage between the particles (or lumps). It is therefore a measure of the stowage capacity or the amount which can be stored in a given volume.

The bulk density of packed coal depends on many factors, the most important of which are the size analysis of the coal, the size of the container and the total moisture content of the coal. Other variables are the shape of the particles or lumps and the method of packing. Depending upon the various factors, the bulk density of loosely packed dry coal may vary as

Lumpy coal(sized or free from fines)	: 600–670 kg/m^3
Lumpy coal (with fines)	: 690–830 kg/m^3
Coal dust	: 480–580 kg/m^3
Pulverised coal	: 400–500 kg/m^3

The true density is 1300 kg/m^3

The bulk density of small coal decreases with increase of free moisture content and then reaches a minimum value which may be 16% lower than for air-dried coal and which may occur at about 6% moisture content. After the minimum, the bulk density again increases with

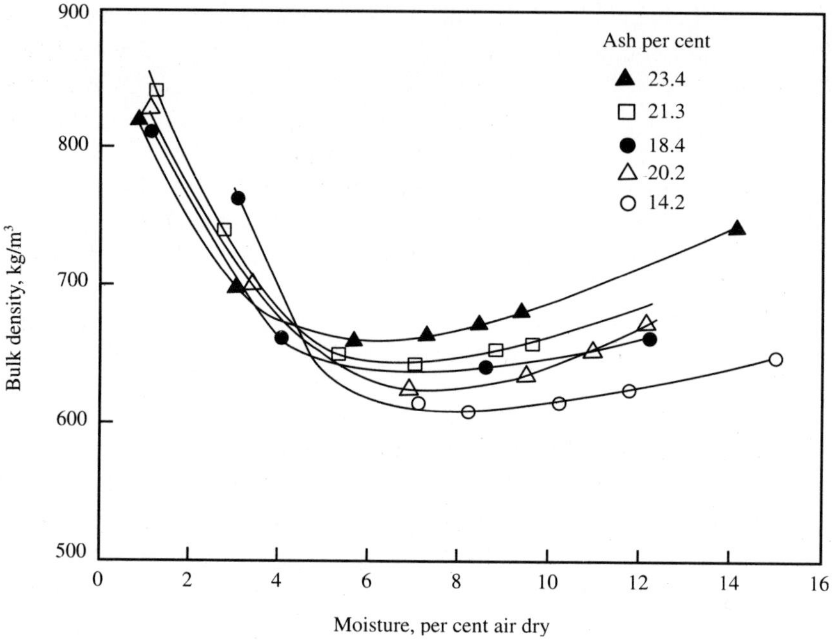

Figure 3.4 Moisture and bulk density of coal

rise in free-moisture content but never attains the original value of the dry coal (Fig. 3.4). The variation of bulk density with the free moisture content is very important in making the charge for coke ovens.

3.13.13 Angle of Repose

The angle that a heap of coal forms with the horizontal is of importance in its storage and in its flow in conveyers and feed hoppers. It is not a reproducible figure and may vary by 2 to 3 degrees depending upon the roughness of the surface, moisture content and other factors. In general, the higher the size, higher the angle of repose. This means that a larger bulk of bigger size coal may be stored in a heap of given height. Approximate values of angle of repose for different sizes of dry coal in bulk are:

Size of coal, mm	18–30	12–8	6–12	0–6
Angle with horizontal, degrees	41	40	38	32

3.13.14 Porosity, Surface Area and Heat of Wetting

Coal is a porous substance with a fairly high internal surface area. When it is brought in contact with a suitable organic liquid such as methanol, the latter enters the very fine pores and the surface of coal is wetted. The process is exothermic and heat of wetting is released. There are

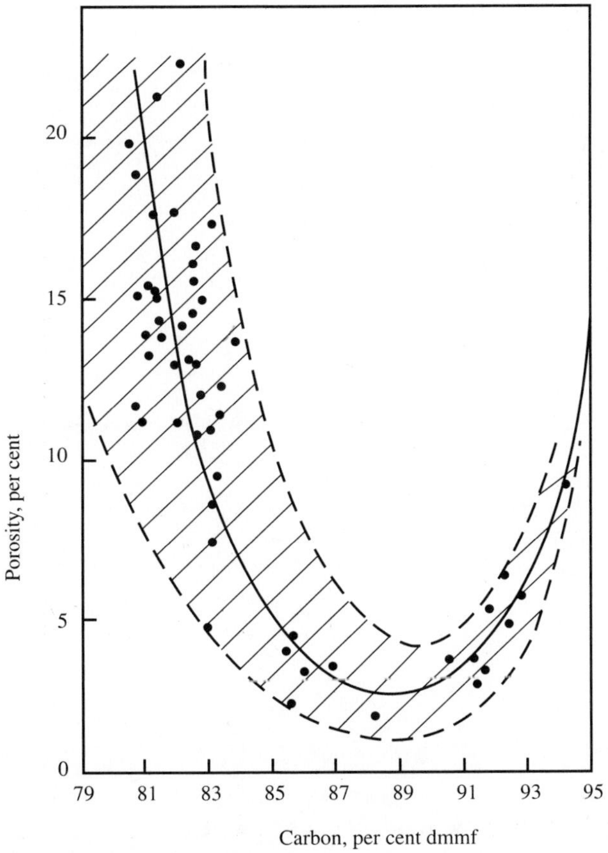

Figure 3.5 Porosity and percentage of coal in carbon

various methods for the determination of the porosity and surface area of coal. Heat of wetting is also a measure of surface area.

Porosity (per cent pore volume), surface area (m^2/g coal) and heat of wetting (cal/g coal) are found to vary with the rank of coal in the same manner as the moisture varies with the rank. There is initially a fall in the values till a minimum of 89% to 90% is obtained for coal carbon. The values again rise in the anthracitic region (Figs 3.2, 3.5 and 3.6). Some average data are given below.

	I	II	III
Carbon, per cent dmmf	80	89	93
Heat of wetting, cal/g dmmf	18–20	2–3	6–8
Surface area, m^2/g dmmf	180–200	20–30	60–80
Porosity, per cent dmmf	18–20	2–3	6–8

Approximately, 1 cal/g heat of wetting = 10 m^2/g surface area = 1% porosity.

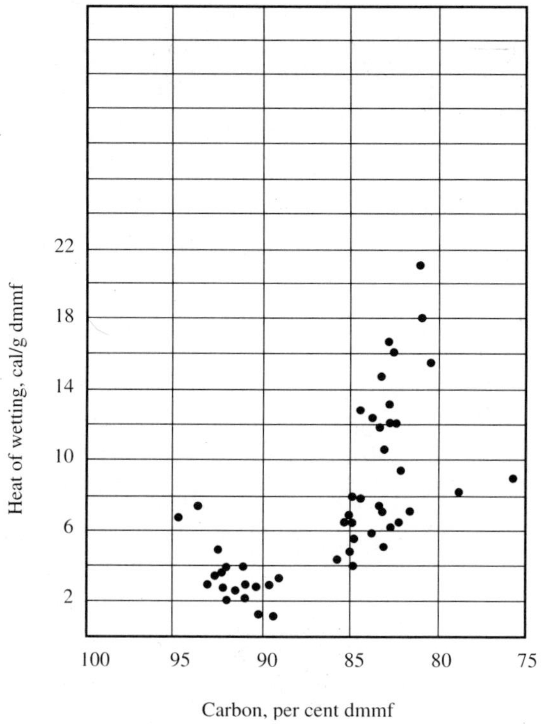

Figure 3.6 Heat of wetting and percentage of carbon in coal

3.13.15 Refractive Index and Reflectance

Both these optical properties are found to increase with increase in coal rank. For vitrain-rich samples, the following values have been reported.

Carbon, per cent dmmf	75	80	84	91	95
Refractive index	–	1.6	–	1.9	–
Reflectance, per cent incident light	0.5	0.67	0.92	2.23	4.17

Among the macerals of the same coal, fusinite and micrinite have very high reflectance and exinite has abnormally low reflectance, the vitrinite having intermediate values. Reflectance is a good parameter for the determination of coal rank.

3.13.16 Specific Heat

The specific heat of coal increases with increase in the volatile matter and decrease in the carbon to hydrogen ratio. Typical data on a daf basis at normal temperature are:

Bituminous coal	0.24–0.26 cal/g°C
Anthracite	0.22–0.23 cal°C

The coal ash has a value of 0.16—0.17 cal/g°C. The relationship between specific heat and moisture content, and between specific heat and ash percentage is linear. The specific heat of a coal of given moisture and ash value can thus be calculated with the help of simple arithmetic.

3.13.17 Grindability

There are two methods of determining the ease of grinding coal to fine sizes, namely, (i) ball mill method and (ii) Hardgrove method. The former measures the amount of work done in grinding a pre-sized material to a given fineness. The Hardgrove method measures the increase of surface produced by the application of a standard amount of work and expresses the result as Hardgrove grindability index, G, which ranges between 20 and 100 for most coals.

$$G = 13 + 6.93W$$

where W = weight of coal in grams passing through a 200-mesh sieve after 50 g of coal of size 16 to 30 mesh are ground in a standard mill for 60 revolutions. The use of a 200-mesh sieve in determining Hardgrove index is significant because in the pulverised coal, firing the size below 200-mesh is often specified. A high value of G indicates a soft and easily grindable coal. It has been established that the Hardgrove grindability index of coal initially increases with the rank, reaches a maximum of about 105 for bright coals of 89% to 90% carbon, and then falls sharply to about 35 for anthracites (Fig. 3.7). The average Hardgrove index for Indian coals is 55 to 65, but coking coals of Jharia and Giridih often show higher values up to 75.

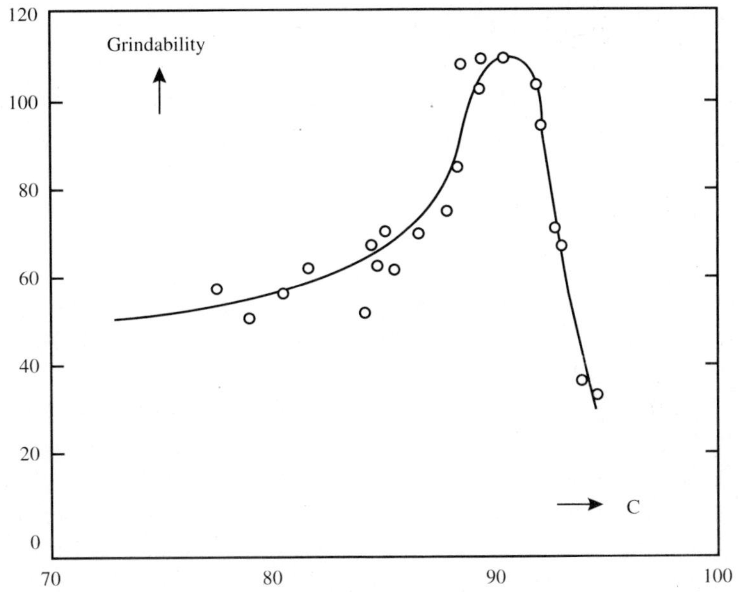

Figure 3.7 Hardgrove grindability and percentage of carbon in coal

3.13.18 Solubility

Peat is slightly soluble in water owing to the presence of simple carbohydrates and pectin. Lignite and other coals are completely insoluble in water, cold or hot. Peat and lignite are however soluble in alkali. These fuels contain humic acid which dissolves forming humates in alkali solutions. They get precipitated on acidification of the solutions. Higher rank coals normally do not contain humic acids but produce them on mild oxidation. Oxidised bituminous coal is therefore partly soluble in alkali solutions. Dilute hydrochloric acid dissolves the hemi-celluloses present in peat. The concentrated acid dissolves most of the mineral matter of coals. The so-called demineralised or ultrapure coal may be prepared by treating coal with hydrochloric acid. Nitric acid has an oxidising effect on the organic mass of coal and produces various aliphatic and aromatic carboxylic acids. Concentrated sulphuric acid decomposes the cellulose present in peat.

Simple organic solvents like benzene, mixture of benzene and alcohol or chloroform dissolve the waxes and resins which constitute 5% to 20% of peat and lignite and, generally, 1% or less of higher rank coals. High boiling solvents like anthracene, phenanthrene, anthracene oil and tetralin, and basic solvents like pyridine, quinoline, monoethanolamine and ethylene diamine can extract considerable portions of coal. The organic mass of a medium rank bituminous coal can be dissolved in tetralin or anthracene oil to the extent of 85% to 95% on treatment at about, 350°C to 400°C in an autoclave. The coal extracts obtained in such large yields are of industrial importance, for example, in the production of liquid fuels or carbon electrodes. The solubility of anthracite in the above mentioned solvents is negligible.

3.14 ACTION OF HEAT ON COAL

Studies on the behaviour of coal on heating provide information on the composition and nature of coal, and supply data for commercial carbonisation process. When a sample of powdered coal is heated out of contact with air, it loses occluded gases consisting of methane, ethane, nitrogen and carbon dioxide at temperatures below 100°C. Moisture is evolved between 100°C and 150°C. The organic mass of coal then starts decomposing with the evolution of gaseous and vaporous products. The decomposition is called active when oil appears in the vaporous products. Peat and young lignites start decomposing at 100°C or less. The initial temperature of decomposition of bituminous coals is 200°C to 300°C while active decomposition starts at 300°C to 375°C for these coals.

The evolution of volatile products of thermal decomposition of bituminous coals is marked by two peaks or maxima, one corresponding to the primary devolatilisation in the temperature range 350°C to 550°C and the other corresponding to the secondary devolatilisation around 700°C (Fig. 3.8). Pyrogenic water, primary tar and gases evolved during the primary, while gases (mainly hydrogen) are evolved during the secondary devolatilisation (Fig. 3.9). Peat and lignite show pronounced primary devolatilisation and insignificant seconday devolatilisation. Anthracite is characterised by the absence of primary devolatilisation and occurrence of

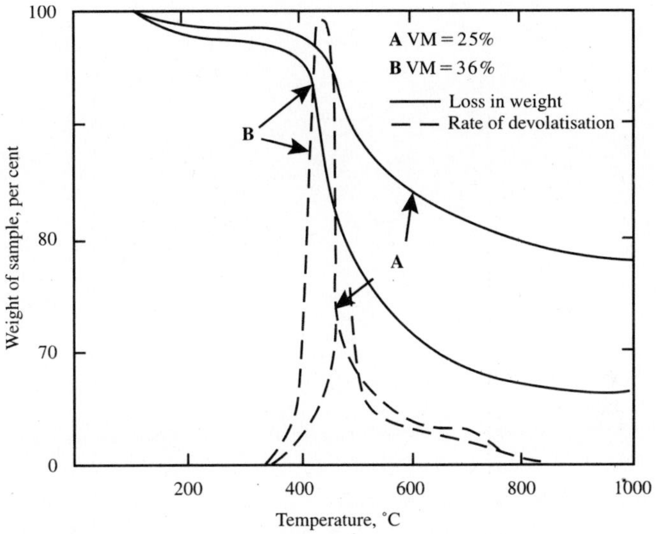

Figure 3.8 Devolatalisation of coal

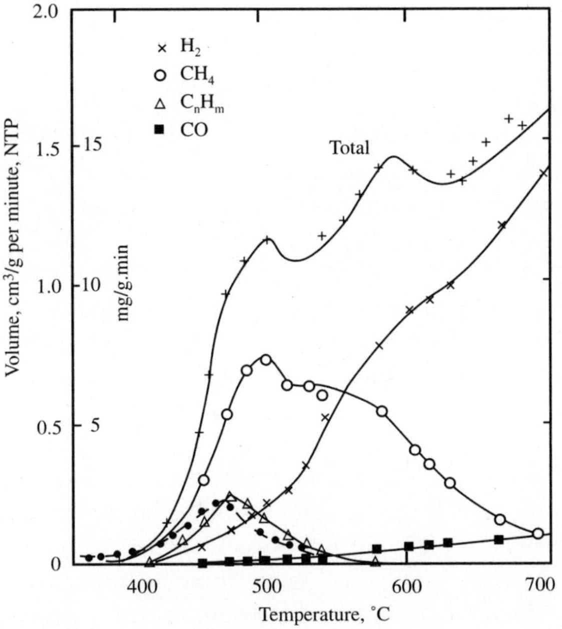

Figure 3.9 Evolution of gases during carbonisation of coal: Coking coal, 1.8°C per minute, heating rate. The dashed line represents rate of weight loss against temperature.

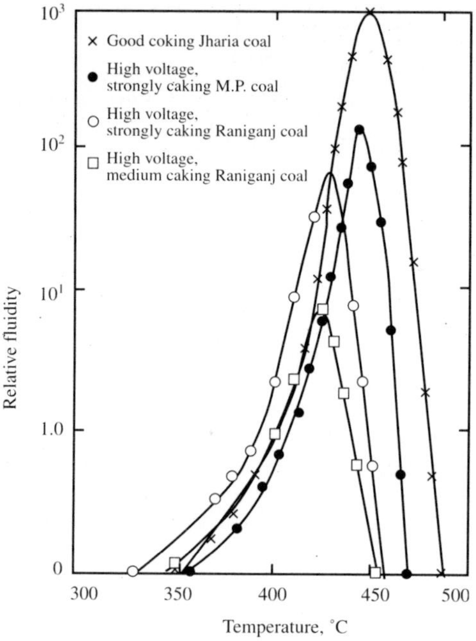

Figure 3.10 Fluidity and heating temperature of coal

secondary devolatilisation only. All the three processes—drying, primary devolatilisation and secondary devolatisation—are endothermic in nature. However, some investigators have also observed exothermic processes during the thermal decomposition of coal.

While coal undergoes decomposition on heating, the residue becomes richer in carbon content and exhibits some characteristic physical and chemical changes. In the case of caking coals, the residue passes through a plastic state in the range 300ºC to 350ºC to 500ºC to 550ºC. The fluidity of the plastic mass initially increases, attains a maximum and then decreases to zero (Fig. 3.10). The last point corresponds to the resolidification of the plastic mass into semi-coke. This is also the upper limit of the primary devolatilisation process. On further heating, secondary devolatilisation takes place and semi-coke is converted into true coke. This conversion is marked by a sudden rise in the electrical conductivity of the solid materials. Therefore, one method to ensure proper coking is to check the electrical conductivity of the carbonaceous residue. If coke is heated further, significant changes take place around 2000ºC when graphite-like materials result. This process is known as *graphitisation* and is used in the production of graphite electrodes. Non-caking coals are not amenable to graphitisation.

The porosity of coal decreases on heating and attains a minimum in the plastic state. At resolidification, porosity again rises considerably. The porosity of coke is 40% or above. This property ensures smooth burning of coke in furnaces. Because of the simultaneous formation of the plastic state and volatile products of thermal decomposition, the carbonaceous residue exhibits an initial contraction and fall in porosity followed by swelling, dilatation and rise in porosity. This phenomenon forms the basis of many tests for the caking property of coal.

The experiments conducted for determining the yields of various products of thermal decomposition of coal are known as *carbonisation assay*. There are two standard procedures: Gray–King and Fischer. The basic principle of these assays is to heat coal under specified conditions and collect the gas after recovering the condensable products, namely tar and liquor. The latter term is used to denote the resultant aqueous solution of ammonia and other dissolved substances. The coke is left in the retort.

3.15 OXIDATION OF COAL

The organic mass of coal is quite susceptible to oxidation. Even atmospheric oxygen brings about oxidation of coal with evolution of heat. The effect is a fall in the caking capacity, calorific value and tar yield of the coal. Under adverse circumstances the coal may catch fire in the process. The rate of oxidation varies inversely with the rank. Oxidation of coal is enhanced by known oxidising agents such as potassium permanganate solution. Initially alkali-soluble materials like humic acids are formed. With the progress of oxidation, one gets benzene carboxylic acids, aliphatic acids and finally carbon dioxide and water.

The technique of controlled oxidation of coal may be industrially used in the following cases: (i) production of benzene carboxylic acids, (ii) fuel cells, and (iii) preparation of coke oven charge from highly swelling coals. Experiments on coal oxidation have been of great help in elucidating the nature and composition of coal.

3.16 HYDROGENATION OF COAL

Mild hydrogenation reduces the oxygen content of coal and raises its caking capacity. This can be achieved by catalytic hydrogenation with molecular hydrogen or hydrogen donating reagents like tetralin and decalin. Under the action of heat and high pressure, hydrogen converts coal into liquid products. This forms the basis of the Bergius process of liquefaction of coal. Once considered uneconomical, this method now appears economically promising for the production of liquid fuels from coal (see Chapter 5). Hydrogenation experiments have also thrown some light on coal constitution. The susceptibility of coal to hydrogenation decreases with rank. Anthracites resist hydrogenation.

3.17 CLASSIFICATION OF COAL

The primary function of any classification is to place together things which are of like nature and to separate those which are of a different nature. Coal defies any universal classification system owing to the following reasons:

1. Coal is a heterogeneous mixture of organic mass, moisture and mineral matter in uncertain proportions.

2. The organic mass is itself heterogeneous and consists of a number of macerals, as a result of which, the coals of the same ultimate analysis do not necessarily have the same properties, while coals of different composition may, in some respects, behave alike.
3. The macerals differ from coal to coal with respect to both quality and quantity; the quality of the macerals is determined by the initial plant material and the coalification process, both of which again vary from place to place and time to time.
4. The mineral matter of coal varies widely in amount and composition; both these factors greatly influence the properties of the whole coal.

Inspite of the above, coal systematics is a vast subject comprising a large number of proposed classification systems. While some systems are of scientific and academic interest only, others have been found to be of use in the commerce and industry of different countries.

A distinction may be made between the terms classification and grading. The term grade in India generally refers to the commercial degree of purity of coal, as assessed from the ash and moisture. Coal classification systems refer to the nature of the organic mass.

The primary basis of coal classification is the nature of the plant debris. Accordingly, there are three types of coals:

1. Humic coals or humites derived from the higher plant kingdom, for example, peat, lignite, bituminous coal and anthracite;
2. Sapropelic coals or sapropelites derived from small plant organisms, for example, cannel coal and boghead coal;
3. Liptobiolites derived from some selected parts of higher plants.

A well-known example of the last type of coal is the *papier kohl* of the Moscow basin which originated from a concentration of cuticles in the Palaeozoic age and bears a superficial resemblance to accumulated sheets of brown paper. Since only humites are of industrial importance, their classification is discussed below.

The most important variable in the nature of coal is its rank. A number of characteristic properties are a function of rank and some of them are used for determining it. Volatile matter, calorific value and carbon content—singly or in combination—are used to divide coals into major classes or groups on the basis of rank. Reflectance is also a good rank parameter. For all practical purposes this classification into groups is satisfactory for the lignite, sub-bituminous coal and anthracite. However, it is necessary to sub-divide the coals of the bituminous stage into sub-classes or sub-groups on the basis of the caking capacity, as judged from one or more standard tests.

When two parameters representing the rank of coal are plotted on a graph, the points never fall on a single line. However, nearly all the points fall within a narrow band of a distinctive shape, see Seyler's chart (Fig. 3.11). The band is wider for the young coals and narrows down towards the anthracites. This has been termed the *coal band*. The band width is a measure of the variation of hydrogen in the coal due to the variation in its petrographic composition.

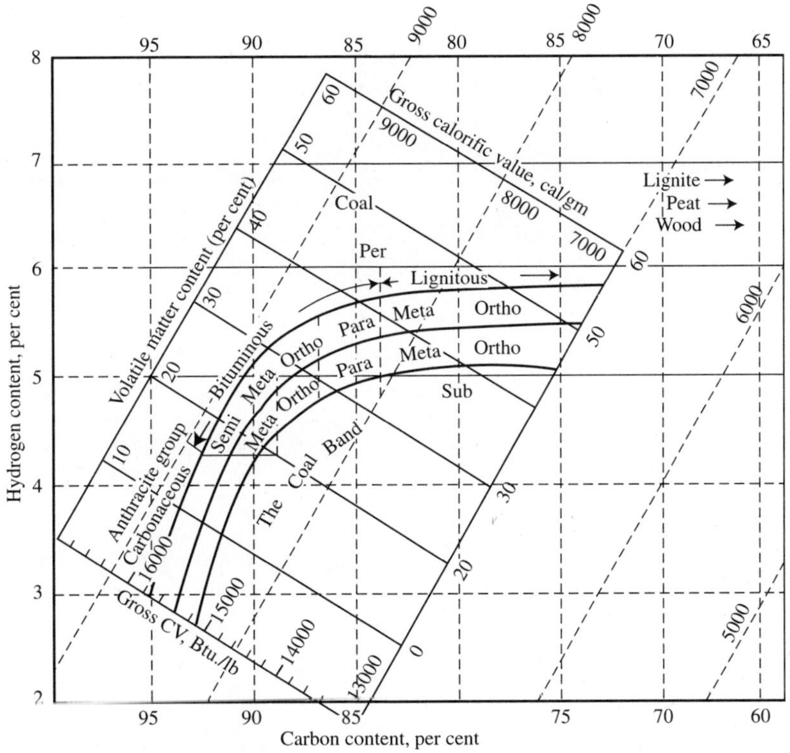

Figure 3.11 Seyler's coal classification chart (simplified)

3.17.1 Regnault–Grüner–Bousquet

The first real attempt at coal classification was made by Regnault in France in 1837 on the basis of carbon, hydrogen and oxygen. In 1874 Grüner modified this by adding volatile matter and caking capacity. Further amendments were made by Grüner's son and Bousquet in 1911 (Table 3.20).

This classification recognises six classes of coal but does not include lignites or any coal containing less than 75% carbon on a daf basis. The descriptive terminology used by Regnault is still in vogue in many countries in some form or other. However, the limiting values of the classification parameters are not applicable to coals of other countries. Various countries have standardised their coal classification systems on the basis of Regnault–Grüner classification.

3.17.2 Frazer

The coals of the USA were divided into four groups by Frazer in 1877 on the basis of fuel ratio (FR), that is, the ratio of fixed carbon to volatile matter. The groups were termed: anthracite

Table 3.20 Regnault–Grüner–Bousquet classification of coal

No. of class	Character of coal	Carbon, per cent, daf	Hydrogen, per cent, daf	Fixed carbon, per cent, daf	Volatile matter, per cent, daf	Nature of coke (VM test residue)	Calorific value, cal/g, daf
I	Dry, long flame, Non-caking	75–80	4.5–5.5	55–60	40–45	Powdery or slightly coherent	8000–8500
II	Fat, long flame	80–85	5.0–5.8	60–68	32–40	Caked, but friable	8500–8800
III	Fat, properly named	84–89	5.0–5.5	68–74	26–32	Caked, moderately compact	8800–9300
IV	Fat, short flame	88–97	4.5–5.5	74–82	18–26	Caked, very compact, lustrous	9300–9600
V	Lean coals—anthracite	90–93	4.0–4.5	82–90	10–18	Powdery or slightly coherent	9200–9600
VI	Anthracties	93–95	4.0–2.0	90–92	8–10	Noncoherent, powdery	9000–9200

(FR = 100 to 12), semi-anthracite (FR = 12 to 8), semi-bituminous (FR = 8 to 5) and bituminous (FR = 5 to 0). This system failed mainly due to much overlapping of the groups.

3.17.3 Grout and Ralston

In 1907 Grout plotted the carbon, hydrogen and oxygen contents of a large number of American coals on a triaxial diagram and found that the humic coals lie on a narrow band, with cannel coals, rich in hydrogen, placed well above the band. In 1915, Ralston extended the findings of Grout. Coals of equal volatile matter (*isovols*) and equal calorific value (*isocals*) were represented by straight lines inclined at an angel to the C, H and O axes. The effects of carbonisation and oxidation could also be shown by similar lines on the triaxial diagram. This classification is of scientific interest but difficult to use in practice (Fig. 3.12).

3.17.4 Seyler

In 1900 Seyler proposed a coal classification system on the Parr basis, that is, the basis of ultimate analysis, the assumption being C+H+O = 100 (Table 3.21). Coals were divided into four main groups or species according to carbon content (anthracitic, carbonaceous, bituminous and lignitous) and the species were sub-divided into five varieties or genera according to hydrogen content (per-bituminous, bituminous, semi-bituminous, carbonaceous and anthracitic). The bituminous species was divided into three subspecies, namely ortho-bituminous, meta-bituminous and para-bituminous. The prefix ortho indicates true, typical or normal. Coals containing more carbon than those of the ortho-type were given the prefix, meta, and those with a lower percentage of carbon were indicated by the prefix, para. In a similar way the lignitous species was divided into ortho-lignitous and meta-lignitous. Again, coals containing more hydrogen than the normal were called per-hydrous and indicated by

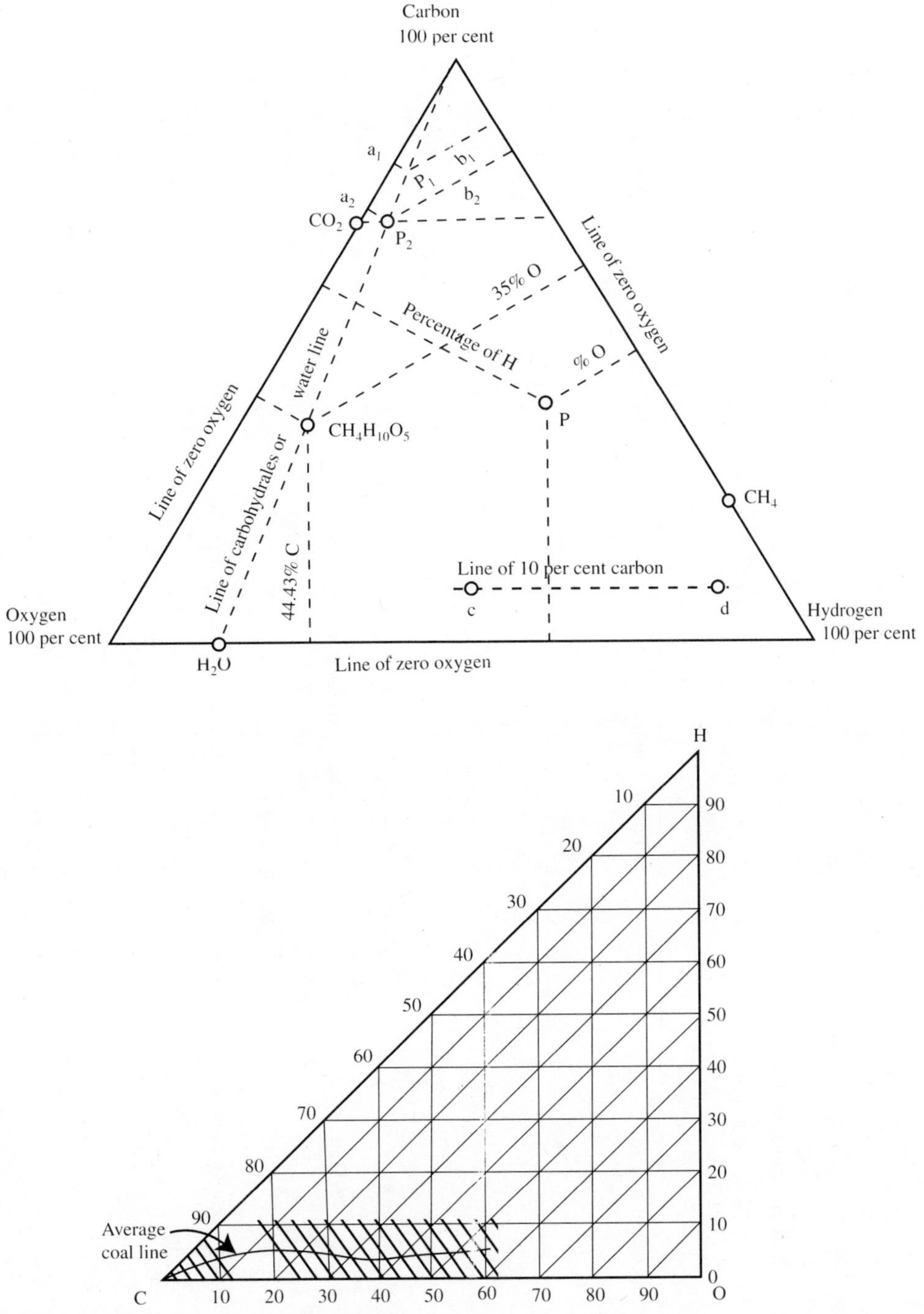

Figure 3.12 Grout–Ralston triaxial diagram

Table 3.21 Seyler's classification of coal (1900)

Carbon	Anthracitic	Carbonaceous	Bituminous			Lignitous	
	93.3%	93.3%–91.2%	meta 91.2%–89.0%	ortho 89.0%–87.0%	para 87.0%–84.0%	meta 84%–80%	ortho 80%–75%
Per-bituminous genus, hydrogen over 5.8%	–	–	Per-bituminous (per-meta-bituminous)	Per-bituminous (per-ortho-bituminous)	Per-bituminous (per-para-bituminous)	Per-lignitous	
Bituminous genus, hydrogen 5.0%–5.8%	–	Pseudo-bituminous species	Meta-bituminous	Ortho-bituminous	Para-bituminous	Lignitous Meta-	Ortho-
Semi-bituminous genus, hydrogen 4.5%–5.8%	–	Semi-bituminous species (ortho-semi-bituminous)	Sub-bituminous (sub-meta-bituminous)	Sub-bituminous)	Sub-bituminous (sub-para-bituminous)	Sub-lignitous Meta-	Ortho-
Carbonaceous genus, hydrogen 4.0%–4.5%	Semi-anthracitic species	Carbonaceous species (ortho carbonaceous)	Pseudo-carbonaceous (sub-metabituminous)	Pseudo-carbonaceous (sub-ortho-bituminous)	Pseudo-carbonaceous (sub-para-bituminous)	–	–
Anthracitic genus hydrogen under 4%	Ortho-anthracite (true anthracite))	Pseudo-anthracite (sub-carbonaceous)	Pseudo-anthracite (sub-meta-bituminous)	Pseduo-anthracite (sub-ortho-bituminous)	Pseudo-anthracite (sub-para-bituminous)	–	–

the prefix, per, and those with less hydrogen than the normal were called sub-hydrous and indicated by the prefix, sub.

As with the Regnault–Grüner classification, the lowest carbon content included in the Seyler classification was 75.0%, thus excluding the lignites. The Seyler classification has however the important advantage that the ultimate analysis is correlated with the petrographic composition of the coals. Thus, members of the bituminous or ortho-hydrous coals contain much exinite and the sub-hydrous coals contain much fusinite. The low hydrogen in the semi-bituminous and anthracitic coals is the result of coalification.

Seyler also successfully correlated the ultimate analysis of coals with the technological properties, namely, volatile matter and calorific value. These correlations were used in the construction of Seyler's coal classification chart (Fig. 3.11). This renowned chart has carbon and hydrogen as rectangular coordinates, with further axes indicating calorific value and volatile matter at right angles to each other, but inclined at about 60° to the carbon axis. Oxygen, moisture and swelling number are indicated by suitable scales.

Seyler has drawn a narrow band on the chart between the carbon limits of 97% to 70% and hydrogen 2.0% to 5.8%. This band is supposed to include all normal bright coals within this range of carbon content. Per-hydrous coals fall above the band and sub-hydrous coals fall below the same. Isovols and isocals are superimposed on the band. The isovols are approximately parallel to each other and at right angles to the isocals.

The isocals are also found to be lines of equal air requirements for the combustion of the coal. These have been correlated with flue gas volumes, calorific intensity and flame temperature under the conditions of adiabatic combustion.

Despite its scientific excellence, Seyler's outstanding coal classification system is not used in industrial practice, the procedure for ultimate analysis being too lengthy, and the system itself being too complicated.

3.17.5 Classification of Indian Coals

A Coal Grading Board was constituted by the Government of India in 1924 to arrange for the grading and certification of exported coal. The grades were originally meant for the Damodar Valley fields for export purposes. Table 3.22 gives the various grades of Indian coals in operation.

The classification of Indian coals was first published in 1955 on the advice of the Planning Commission and subsequently revised several times. The current classification recognises four classes: lignite, sub-bituminous, bituminous and anthracite and twelve types consolidate (lignite), high volatile (sub-bituminous), high volatile (bituminous), medium volatile, high volatile (2 types), medium volatile (2 types), low volatile (2 types), semi-anthracite and anthracite (Table 3.18). Coals of North-Eastern India, grouped together as B_1, constitute a special formation, exhibiting widely different coal characteristics. All coals are suitable for combustion and many for hydrogenation. The more common uses are indicated in the classification.

Calorific value, volatile matter and Gray–King coke type are the three basis parameters of this classification. Moisture, carbon and hydrogen are the other properties included here. The three-digit basic code numbers of different coals are shown in a separate column.

Table 3.22 Grading of Indian bituminous coals (1981)

Serial No.	Class	Grade	Specification
1.	Non-coking coal produced in all states other than Assam, Andhra Pradesh, Meghalaya, Arunachal Pradesh and Nagaland	A	HU* exceeding 6200 kcal/kg
		B	HU exceeding 5600 kcal/kg but not exceeding 6200 kcal/kg
		C	HU exceeding 4940 kcal/kg but not exceeding 5600 kcal/kg
		D	HU exceeding 4200 kcal/kg but not exceeding 4940 kcal/kg
		E	HU exceeding 3360 kcal/kg but not exceeding 4200 kcal/kg
		F	HU exceeding 2400 kcal/kg but not exceeding 3360 kcal/kg
		G	HU exceeding 1300 kcal/kg but not exceeding 2400 kcal/kg
2.	Non-coking coal produced in Assam, Andhra Pradesh, Meghalaya, Arunachal Pradesh and Nagaland		Not graded
3.	Coking coal	Steel grade I	Ash content not exceeding 15%
		Steel grade II	Ash content exceeding 15% but not exceeding 18%
		Washery grade I	Ash content exceeding 18% but not exceeding 21%
		Washery grade II	Ash content exceeding 21% but not exceeding 24%
		Washery grade III	Ash content exceeding 24% but not exceeding 28%
		Washery grade IV	Ash content exceeding 28% but not exceeding 35%
4.	Semi-coking and weakly coking coals	Semi-coking I	Ash plus moisture content not exceeding 19%
		Semi-coking II	Ash plus moisture content exceeding 19% but not exceeding 24%
5.	Hard coke	By-product premium	Ash content not exceeding 25%
		By-product ordinary	Ash content exceeding 25% but not exceeding 30%
		Beehive premium	Ash content not exceeding 27%
		Beehive superior	Ash content exceeding 27% but not exceeding 31%
		Beehive ordinary	Ash content exceeding 31% but not exceeding 36%

Source: Burecan of Indian Standards, New Delhi IS:770(1977) classification and codification of Indian coals and lignite.

*HU = Useful heat value (calorific value).

= 8900 – 138 (ash + moisture)

Note:

1. In the case of coal having moisture less than 2% and VM less than 19%, the HU is the value arrived at as above reduced by 150 kcal/kg for each 1% reduction in VM content below 19%.
2. Moisture at 40°C and 60% relative humidity (RH) is used in evaluation of HU.

The code system is clearly shown in Table 3.19. The group numbers increase from bottom to top (first digit, calorific value) or from left to right (second digit, volatile matter). The number of supplementary parameters also increase from bottom to top in a given group. The third digit is on the basis of coke type. The fourth digit is provided by the supplementary parameter of maximum thickness of plastic layer (MTPL) for caking coals or moisture content for non-caking coals.

Suppose a caking coal with a three-digit basic code number, say 855, has MTPL subgroup number of 6. The integrated code number of such a coal will be 8556 which when decoded expresses the following properties:

Calorific value, kcal/kg dmmf	8,505 to 8,700
Volatile matter, per cent dmmf	22.1 to 28.09
G–K/(LT) coke type	G_3 or above
MTPL, mm	26 and above

The above properties of the coal strongly indicate that it is likely to be a typical prime coking coal. On the other hand, if the basic code number of the coal remains the same, that is 855, but the MTPL, sub-group is 3 instead of 6, its four-digit integrated code number will be 8533, which would indicate a coal likely to be somewhat inferior to 8556 in respect of its cokability.

Again, suppose a non-caking coal with a basic code number, say 330, has a moisture sub-group number 4, the integrated code number will be 3304 and when decoded will expresse the following properties:

Calorific valus, kcal/kg dmmf	7,507 to 7,800
Volatile matter, per cent dmmf	33.1 to 37.0
G–K/(LT) coke type	A
Moisture, parts per 100 parts dmmf	9.1 to 11.0

The above properties indicate that the coal is a fairly high moisture, high volatile, low rank, non-caking coal.

All high volatile non-caking coals giving coke types A and B obviously have 0 as the third number and are thus readily recongnisable in terms of the three-digit basic code number. But all such non-caking coals are not equivalent in their intrinsic/technological properties, and herein lies the importance of the supplementary parameter—moisture—which subdivides such coals in terms of different moisture ranges.

Increasing group number means increasing desired property level depending on use, that is lower moisture, higher CV, lower VM, coke type towards G and higher MTPL.

The total coal resources of India are estimated to be 158.5 billion tonnes. As high as 82% of these are estimated to be non-caking and only 18% caking. Prime coking coals constitute only 3.5% of the total reserves. These low reserves of prime coking and other caking coals require that India should adopt all possible technological methods of conserving coal resources and also import low-ash prime coking coals from abroad.

EXAMPLES

Example 1

A XV seam Jharia coal from the Bararee colliery gave the following proximate analysis: moisture 1.6%, ash 15.7%, volatile matter 27.8% and fixed carbon 54.9%. Calculate its ash on a dry basis and volatile matter on daf and dmmf bases.

Solution

The proximate analysis is given on the air-dried basis:

M	=	1.6
A	=	15.7
VM	=	27.8
FC	=	54.9
Total		100.00

$$\text{Therefore, percentage of ash on dry basis} = \frac{A}{100 - M} \times 100$$

$$= \frac{15.7}{100 - 1.6} \times 100$$

$$= 16$$

Coal does not contain 'ash'. Ash comes from the mineral matter (MM) content. On an average 10% of mineral matter is lost when coal is burnt and mineral matter is converted into ash (A).

$$10\% = \frac{10}{100} = 0.1$$

Similarly MM = 1–1 × A

$$\text{Volatile matter, per cent daf} = \frac{VM}{100 - M - A} \times 100$$

$$= \frac{27.8}{100 - 1.6 - 15.7} \times 100 = 33.6$$

$$\text{Volatile matter, per cent dmmf} = \frac{VM - 0.1A}{100 - M - 1.1A} \times 100$$

$$= \frac{27.8 - (0.1 \times 15.7)}{100 - 1.6 - (1.1 \times 15.7)} \times 100 = 32.3$$

Alternatively, the volatile matter can be calculated through the fixed carbon, since these two together make-up the whole coal on daf and dmmf bases.

$$\text{Fixed carbon, per cent daf} = \frac{FC}{100 - M - A} \times 100$$

$$= \frac{54.9}{100 - 1.6 - 15.7} \times 100 = 66.4$$

Therefore, volatile matter, per cent daf = 100 – 66.4 = 33.6

$$\text{Again, fixed carbon, per cent dmmf} = \frac{FC}{100 - M - 1.1A} \times 100$$

$$= \frac{54.9}{100 - 1.6 - 1.1 \times 15.7} \times 100 = 67.7$$

Therefore, volatile matter, per cent dmmf = 100 – 67.7 = 32.3

Note: The mineral matter of coal contributes to the volatile matter by an amount approximately equal to 10% of the ash. Therefore, necessary corrections to the volatile matter value have to be made while calculating on dmmf basis. Fixed carbon does not need this correction.

Example 2

The following data are available for a Chanch coal of the Raniganj coalfield.

Proximate analysis, per cent air-dried				**Ultimate analysis, per cent dmmf**					**Calorific value, kcal/kg airdried**
Moisture	Ash	Volatile matter	Fixed carbon	Carbon	Hydrogen	Sulphur	Nitrogen	Oxygen	
1.3	16.0	30.1	52.6	86.52	5.27	0.64	2.37	5.20	**68.90**

Calculate its calorific value using at least two different empirical formulae and compare the computed value with the experimental.

Solution

Goutal and CFRI formulae will be used for calculating calorific value from the proximate analysis, and modified Dulong formula for calculating from the ultimate analysis.

Goutal formula

C = 82F + aV

Where a is a constant depending upon the volatile matter expressed as per cent daf (V′).

$$\text{Volatile matter, per cent dmmf} = \frac{30.1}{100 - 1.3 - 16.0} \times 100 = 36.4$$

Using the known values of a for given values of V from the table on page

$$\text{The value of constant, a} = 94 - \left[\frac{94 - 85}{38 - 35} \times (36.4 - 35)\right]$$

$$= 94 - \left[\frac{9}{3} \times 1.4\right]$$

Calculation of constant, a

V′	a	V′	a	V′	a
5	145	20	109	35	94
10	130	25	103	38	85
15	117	30	98	40	80

where V′ is VM, per cent daf

Therefore, calorific value, kcal/kg air-dried = 82 × 52.6 + 89.8 × 30.1
= 7016

This is higher than the experimental value by (7016 – 6890) or 126 kcal/kg

$$\text{Therefore, error} = \frac{126}{6890} \times 100 = +1.83\%$$

CFRI* Formula

Since $M < 2\%$

Therefore, $C_G = 91.7F + 75.6\,(V - 0.1A) - 60M$
$= 91.7 \times 52.6\,(30.1 - 0.1 \times 16.0) - 60 \times 1.3$
$= 6900$ kcal/kg air-dried

This is higher than the experimental value by 10 kcal/kg only.

$$\text{Therefore, error} = +\frac{10}{6890} \times 100 = +0.15\%$$

Modified Dulong formula

$$C_G = 80.8C + 345\left(H - \frac{O}{8}\right) + 22.2S$$

*CFRI stands for Central Fuel Research Institute, a CSIR organisation, where the author worked for two years (1956–58). It is located at Jealgone, near the township of Dhanbad, Bihar.

Therefore, calorific value, kcal/kg dmmf

$$= 80.8 \times 86.52 + 345\left(5.27 - \frac{5.20}{8}\right) + 22.2 \times 0.64 = 8598$$

Therefore, calorific value, kcal/kg air-dried

$$= 8598\,(100 - \mathrm{M} - 1.1\mathrm{A}) \times \frac{1}{100}$$

$$= \frac{8598\,(100 - 1.3 - 1.1 \times 16.0)}{100}$$

$$= 6973$$

This figure is higher than the experimental value by 83 kcal/kg.

Therefore, error $= \dfrac{6973 - 6890}{6890} \times 100 = +1.2\%$

Thus it is found that all the calculated data are higher than the experimental value. The best agreement is shown by the CFRI formula.

Example 3

Compare the gross (C_G) and net (C_N) calorific values of the Chanch coal of Example 3.2 on air-dried and dry bases.

Solution

Air-dried basis

Moisture = 1.3%
Ash = 16.0%

$$\text{Hydrogen (organic)} = \frac{5.27 \times (100 - 1.3 - 1.1 \times 16.0)}{100}$$

$$C_G - C_N = 53\left(\mathrm{H} - \frac{\mathrm{M}}{9} + \frac{0.0A}{9}\right)$$

$$= 53\left(4.27 + \frac{1.3 + 1.6}{9}\right)$$

$$= 244 \text{ kcal/kg}$$

Therefore, the gross CV is higher than the net by 244 kcal/kg on air-dried basis.

Dry basis

$$\text{Ash} = 16.0 \times \frac{100}{100 - 1.3} = 16.2\%$$

$$\text{Hydrogen (organic)} = 4.27 \times \frac{100 - 1.3 - 1.1 \times 16.0}{100 - 1.3}$$

$$= 4.34\%$$

$$\text{Therefore, } C_G - C_N = 53\left(4.34 + \frac{0.1 \times 16.2}{9}\right)$$

$$= 244 \text{ kcal/kg}$$

Therefore, the gross CV is higher than the net by 244 kcal/kg on a dry basis.

Note that when the net CV is to be calculated on a particular basis, first calculate the gross CV on that basis and then convert it into net CV, taking into account 'total hydrogen' on the basis.

Example 4

From the following data for a Damua coal of the Kanhan valley coalfield, find out what coal it is according to IS classification system.

Solution

Proximate analysis, per cent air-dried

Moisture	Ash	Volatile matter	Fixed carbon	Calorific value, kcal/kg air-dried	Gray–King (LT) coke type	MTPL mm
2.2	16.5	31.7	49.6	6720	F	19

Out of the supplied date, calorific value, volatile matter and coke type will be required to obtain the basic code of the coal.

$$\text{Calorific value, kcal/kg dmmf} = \frac{6720}{100 - 2.2 - 1.1 \times 16.5} \times 100$$

$$= 8443 \text{ kcal/kg}$$

$$\text{Volatile matter, per cent dmmf} = \frac{31.7 - 0.1 \times 16.5}{100 - 2.2 - 1.1 \times 16.5} \times 100$$

$$= 37.7\%$$

(LT) coke type = F

Therefore, basic code of the coal = 722

From Table 3.19 it may be seen that it is a high volatile, medium caking bituminous coal B. Its carbon content is likely to be about 85% and hydrogen content about 5.0% dmmf. Its recommended use is in the preparation of coke oven blend.

$$\text{Moisture, parts per 100 parts dmmf} = \frac{2.2}{100 - 2.2 - 1.1 \times 16.5} \times 100$$
$$= 2.8$$

This is in the range 2 to 5 (Table 3.19).

From the MTPL value of 19 mm it is inferred that it belongs to the subgroup 3. Its integrated code number is 7223 and its blendable property is confirmed.

Bibliography

1. Bureau of Indian Standards, New Delhi
   a. IS: 437 (1979) Size analysis of coal and coke for marketing
   b. IS: 439 (1976) Industrial coke
   c. IS: 770 (1977) Classificatioin and codification of Indian coals and lignites
   d. IS: 1350 Methods of test for coal and coke
      Pt. I (1969) Proximate analysis
      Pt.II (1970) Calorific value
      Pt.III (1969) Sulphur
      Pt.IV (Sec.1-1974) Ultimate analysis – Determination of carbon and hydrogen (Sec. 2-1975) Determination of nitrogen
      Pt.V (1979) Special impurities
   e. IS: 1353 (1959) Methods of test for coal carbonisation caking index, swelling properties and Gray-King assay (LT) coke types
   f. IS : 1354 (1992) Method of test for coke – special test
   g. IS : 1355 (1959) Methods of test for ash of coal and coke
   h. IS : 4286 (1976) Specification for domestic coke
   i. IS : 4311 (1967) Method for the determination of mineral matter in coal
   j. IS : 4433 (1979) Method for determination of Hardgrove grindability index of coal
   k. IS : 5062 Method of test for brown coals and lignites
      Pt.I (1969) Moisture content
      Pt.II (1969) Ash
      Pt.III(1969) Yields of tar, water, gas and coke by low temperature carbonisation
      Pt.IV(1969) Yield of benzene-soluble extract
      Pt.V(1973) Acetone-soluble material in benzene extract
   l. IS : 6345 (1971) Methods of sampling of coal for float-and-sink analysis
   m. IS : 7221 (1974) Methods of determining the moisture holding capacity of hard coals
   n. IS : 9127 Methods of petrographic analysis of coal
      Pt.I (1979) Definition of terms
      Pt. II (1979) Preparation of coal samples

2. Charcoal, *The Wealth of India: A Dictionary of Indian Raw Materials and Industrial Products*, Council of Scientific and Industrial Research, New Delhi, 1951.
3. Coal, *The Wealth of India: A Dictionary of Indian Raw Materials and Industrial Products*, Vol. 2, Council of Scientific and Industrial Research, New Delhi, 1950.
4. Dryden IGC. Chemical constitution and reactions of coal. In Lowry HH, (ed.), *Chemistry of Coal Utilization*, Supp. Vol., 232. New York: Wiley, 1963.
5. Elliot MA (ed.), *Chemistry of Coal Utilisation*, Second Supp. Vol., New York: Wiley, 1981.
   (i) Collcott TG and Smith GB. Mechanical Properties of Coal, 285.
   (ii) Gluskoter HJ, Schimp NF and Ruch RR. Coal Analyses, Trace Elements, and Mineral Matter, 369–424.
   (iii) Habermehl D, Orywal F and Beyer H. Plastic Properties of Coal, 317.
   (iv) Neavel RC. Origin, Petrogrophy and Classification of Coal. 91.
   (v) Sharkey AG, Jr, and Mecartney JT. Physical Properties of Coal and Its Products, 159.
   (vi) Wender I, Heredy LA, Neuworth MB and Dryden IGC. Chemical Reactions and the Constitution of Coal, 425.
6. Emrich W. *Handbook of Charcoal Making,* Dordrecht (Holland): Reidel, 1985.
7. Geological Survey of India, News, 2007. http:/gsi.gov.in Ministry of Coal, Govt. of India, Annual Report 2006–07.
8. Lignite, *The Wealth of India: A Dictionary of Indian Raw Materials and Industrial Products*, Vol. 6, Council of Scientific and Industrial Research, New Delhi, 1962.
9. International Energy Outlook 2007, http:/www.eia.doe.gov. World Coal Institute. http:/www.world-coal.org.
10. Mukherjee AK, Chatterjeee CN and Ghosh S. Coal resources of India – its formation, distribution and utilization, *Fuel Science and Technology*, 1: 19–34, 1982.
11. Roses JW and Cooper JQ, (eds.). *Technical Data on Fuel*, Seventh edition, British National Committee, World Energy Conference, London, 1976.
12. Sarkar GG. *Introduction to Coal Preparation and Practice*, Oxford and IBH Publishing Company, New Delhi-Mumbai-Kolkaka, 1986.
13. Sarkar S, A New Index for the Caking Capacity of Coal, *J. Sc. Inds. Res*, 20D: 401–3, 1961.
14. Van Krevelen DW. *Coal*, Amsterdam: Elsevier, 1961.

4 Processing of Solid Fuels

4.1 INTRODUCTION

Solid fossil fuels need treatment after they have been won and before they can be utilised as a source of energy. Different fuels require different kinds of treatment. Solid fossil fuels are found associated with minerals, water, gases and impurities. The associated materials have to be removed to improve the efficiency of the solid fuel. Solid fuels are usually won in large chunks and therefore have to be sized or made into usable portions. Peat contains large amounts of water and is of no use without a major portion of it being removed. By providing proper drainage in peat bogs, a large quantity of water can be removed. However, the major portion of the moisture in peat is removed by air-drying it by spreading on drying fields. Air-dried peat is suitable for firing directly in domestic ovens and industrial furnaces and boilers. It may also be used in the form of briquettes. The production of peat briquettes involves size reduction in disintegrators, further drying in hot gases, and finally briquetting. Peat is also used in the gasification and carbonisation processes.

Lignite is given treatment such as size reduction, drying (in hot gases or steam), and briquetting and carbonisation, depending on the final form in which it is to be fired in domestic ovens and industrial installations. An important use of lignite is for power generation, in which case it is dried to a low moisture content and then ground to a fine size for pulverised firing. Air-dried lignite is suitable for some gasification processes. Cleaning of lignite in heavy media separators is practised on a limited scale for decreasing the ash when required.

Depending upon its end use, bituminous coal and anthracite may require only size reduction and sizing; alternatively they may be subjected to more complex processing like cleaning, drying, briquetting, carbonisation, gasification and liquefaction. This chapter discusses the various stages in the utilisation of solid fossil fuels with emphasis on bituminous coals.

4.2 COAL PREPARATION

The raw coal produced from the mine (run-of-mine or ROM) is unsuitable for use as such. It is necessary to exercise control over one or more of the properties like size, and ash, sulphur, phosphorus, moisture and chlorine content, depending upon the nature of its use. In the simplest case of coal preparation, it may be sufficient to just separate the fines from the lumps; whereas, in a modern coke plant, multiple operations such as size reduction, size separation, cleaning, drying and mixing are carried out on a very large scale. Coal preparation is costly. Although the treatment of fuel is carried out to raise the efficiency of its utilisation, the techno-economic considerations finally dictate to what extent it should be practised.

ROM coal usually has a size of 500–0 mm. The sized coals may be available in various ranges, such as steam coal: 250–25 mm, rubble: 50–25 mm, smithy: 25–13 mm and slacks: 50–0 mm, 25–0 and 13–0 mm. The screening of coal is carried out on various types of screens that are revolving, vibrating and shaking. The screens are necessarily of very large capacity and all measures are taken to operate them at high speed throughout screening. Inside the mine or at the surface, visible dirt is removed by hand-picking. The same method may be used to separate the coal types into bright and dull. Brights are preferred for domestic use and coke production, while dull coal may be used in furnaces and boilers. Excepting Raniganj coals, the separation of coal types by hand-picking is hardly possible in India. By and large, Indian coals are dull in appearance.

4.2.1 Coal Cleaning

High ash coals need cleaning to raise the efficiency of their utilistion. Depending upon local conditions, the upper limits of coal ash are fixed for acceptance by users in various industries. Indian coals are high in ash content. Therefore, in India, even metallurgical coke producers tolerate as much as 20% ash in coking coals, even though an upper limit of 10% ash is the norm for corresponding manufacturers abroad. Furthermore, thermal power stations in India have to tolerate 30% to 35% ash in boiler coals. No doubt this tolerance is at the cost of efficiency. But there is no other alternative than to depend on indigenous fuel resources. Indian coals have many problems in washing. Washing is practised only when it is essential.

With increased coal production, the better coals are being exhausted at a faster rate; dirtier coals are being raised in greater quantity. The practice of modern mechanised mining also increases the content of rock in the coal.

Clean coal has higher efficiency of utilisation and many other concomitant advantages: economy of transport and storage, less ash disposal problem and lower content of sulphur, phosphorus, chlorine and other harmful elements. The caking properties of some coals also improve on washing owing to a decrease in ash percentage and enrichment in vitrain.

The ease of separation of mineral matter from coal depends upon its nature. These inorganic substances may be grouped into two classes.

1. Substances derived from the coal forming plants,
2. Constituents added to coal from external sources.

The two classes are sometimes known as *inherent* and *extraneous* mineral matter, respectively. A part of the extraneous mineral matter is intimately associated with the organic mass of coal. This is due to their association with coal during the early formation stage. The inherent mineral matter cannot be removed from coal by any mechanical method. Chemical treatment like acid leaching has to be used for this purpose. Fortunately, the content of inherent mineral matter in coal is small—less than 3%.

The extraneous mineral matter in coal can be removed by mechanical means. However, the intimately associated variety can be separated only with difficulty. A coal is easy to wash when

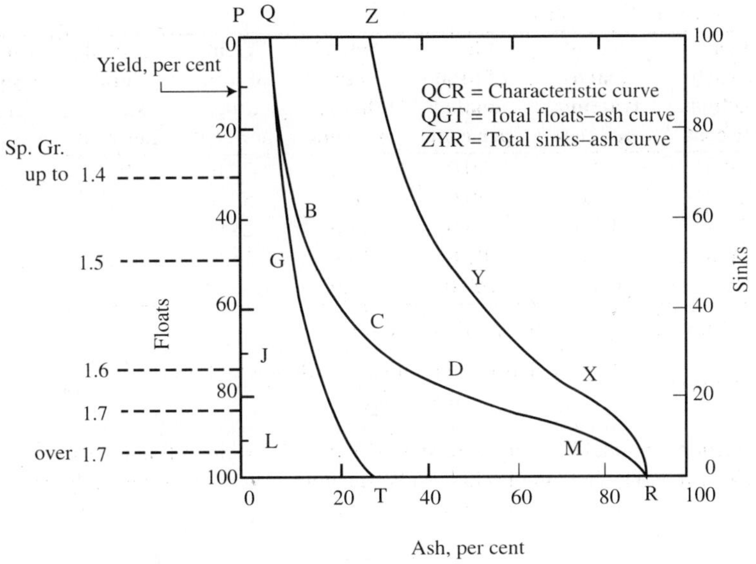

Figure 4.1 Washability curves

the bulk of its mineral matter is due to dirt bands and rock particles which get mixed with the coal during mining, or due to dirt settling on coal during transportation, handling and storage.

Washability curves

The washability characteristics of coal are best evaluated by the float-and-sink test. The test takes advantage of the fact that the associated minerals have much higher specific gravity than pure coal. The specific gravities of pure bituminous coal, shales and pyrites are in the range of 1.28–1.37, 2.0–2.64 and 2.4–4.95, respectively. By using liquids of different specific gravities, the coal sample is divided into a number of fractions in the increasing order of specific gravity, and hence of ash value. The percentage of floats and sinks in each fraction and the ash per cent of the floats and sinks is also determined. Usually three curves are constructed from the results of the float-and-sink test (Fig 4.1 and Table 4.1):

1. Total floats–ash curve,
2. Total sinks–ash curve,
3. Washability characteristic curve or instantaneous ash curve.

From the first two curves (QGT and QYR) the yield and ash of the clean product (float), and the heavy waste (sink) at a certain specific gravity of washing can be found out directly. Similarly, to know the specific gravity at which the cleaning should be done—with a view to producting the maximum yield of clean coals within an upper limit of ash value—the total floats–ash curve is consulted. The yields of sink and its ash are read from the total sinks–ash curve at the same specific gravity. The third curve gives the washability characteristics of coal. The curve is constructed by plotting column 8 versus column 3. It indicates the variation of

Table 4.1 Float-and-sink test results

Specific gravity of separation	Yield of each fraction, per cent	Ash of each fraction, per cent	Yield of total floats, per cent	Ash of total floats, per cent	Yield of total sinks, per cent	Ash of total sinks, per cent	Cumulative yields upto middle of fraction(CMF), per cent
1.40	30.0	9.0	30.0	9.0	70.0	36.1	15.0
1.50	20.0	12.0	50.0	10.2	50.0	45.7	40.0
1.60	26.0	23.0	76.0	14.6	24.0	70.2	63.0
1.70	6.0	47.0	82.0	16.9	18.0	78.0	79.0
>1.70	18.0	78.0	100.0	27.9	–	–	91.0

Note: Yield of total sinks, per cent = 100 – yield of total floats

$$\text{Ash of total sinks, per cent} = \frac{100 \times \text{coal ash} - \text{total floats} \times \text{ash of total floats}}{\text{total sinks}}$$

CMF per cent for fraction F = total floats upto fraction (F − 1) + ½ (yield of fraction F)
Source: Whitaker JW and Sarkar GG. Studies in Coal Washability. Central Fuel Research Institute, Dhanbad, 1950–53.

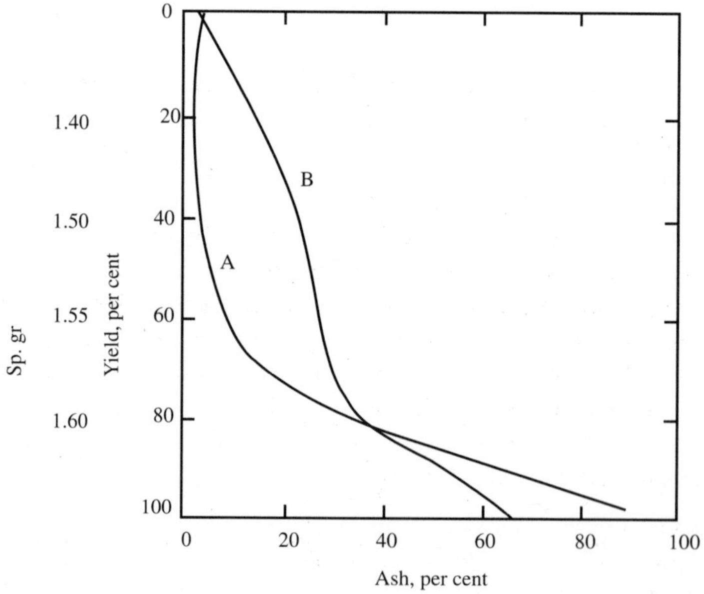

Figure 4.2 Washability characteristic curves

ash from particle to particle in the increasing order of higher specific gravity. Hence, this curve clearly shows how far it is possible to separate the dirt from the clean coal by mechanical methods. For an easily washable coal, this curve should have a sharp bend. On the other hand, in the case of a very difficult-to-wash coal, it tends towards a straight line joining the cleanest and the dirtiest particles (Fig. 4.2).

Table 4.2 Estimate of coal washing problem from the amount of near-gravity material

Amount of near-gravity material, per cent		**Estimate of coal washing problem**
greater than	less than	
0	7	Simple
7	10	Moderately difficult
10	15	Difficult
15	20	Very difficult
20	25	Exceedingly difficult
25	–	Formidable

Preliminary information about the washability of a coal may be obtained from the numerical data of the float-and-sink test. If the yield of fractions of specific gravity between 1.40 and 1.60 is very small, it indicates that the mineral particles are not intimately associated with organic mass and the coal is easily washable. If the yields of these fractions are considerable, it means that much of the mineral matter is intimately associated with the pure coal and hence difficult to separate.

The specific gravity at which a coal is to be cleaned is determined from the washability data and economic considerations. The ease of washing at this specific gravity may be judged from the amount of *near-gravity material* (ngm) present in the coal. The amount of this material is defined as the percentage of the coal that will float in a range ±0.10 of the specific gravity of the separation value. The presence of ngm causes misplacement of sinks in floats and floats in sinks. The larger the amount of ngm, the more difficult the cleaning operation, and vice-versa (Table 4.2).

The commercial methods of coal cleaning are based on differences in some physical properties of pure coal and the minerals present. Most of the commercial cleaning processes make use of the difference in the specific gravity. Coefficients of friction, wettability and resilience are used in a few other commercial processes. Some methods of coal preparation have been suggested also on the basis of optical and electrical properties. Coal cleaning processes are influenced to a greater or lesser extent by the size and shape of the solid particles.

Depending upon the medium used for cleaning, the various processes are grouped as dry or wet. Table 4.3 gives the classification of the important coal cleaning processes. Currently the washeries in India can be broadly divided into two groups.

1. *Pit-head washeries:* They treat coal from the same or two or more adjoining collieries.
2. *Central washeries:* They recover their feed supplies from a large number of sources located in specified common areas.

Dry processes

Pneumatic table: It is essentially a perforated and riffled sloping deck which is rapidly reciprocated (Fig. 4.3). The coal is fed in a thin stream from the upper corner of the table which is transversely inclined. The table is given a reciprocating movement and the air blown through the deck. The coal spreads out over the deck and gets segregated into clean coal and refuse under the combined action of air flow, reciprocating movement of the table and the friction resistance of the table. The heavy refuse is trapped in the riffles and transported to the end of

Table 4.3 Classification of coal cleaning processes

Name	Basis of separation
Dry processes	
1. Pneumatic tables	density and friction
2. Spiral separation	density and friction
3. Berrisford process	resilience and density
4. Air–sand process	density
Wet processes	
1. Dense or heavy medium separators	density, size and shape
2. Jig washers	density, size and shape
3. Cyclone washers	density, size and shape
4. Trough washers or launders	density, size and shape
5. Upward-current classifiers	density, size and shape
6. Froth flotation process	wettability and density
7. Concentration tables	density and friction
8. Oil agglomeration process	wettability, size and density

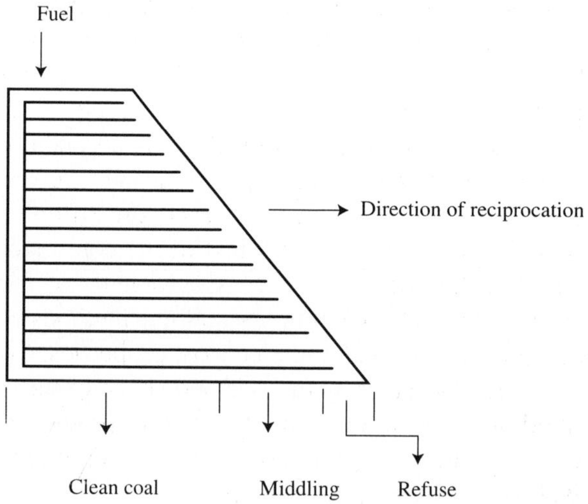

Figure 4.3 Pneumatic table

the table. The clean coal passes over the table. Therefore, middlings and clean coal are collected in separate receivers.

Spiral separator: Coal is fed to the top of a tall, vertical spiral. Owing to a lower coefficient of friction, the clean coal slides to the outside of the spiral while the refuse slides down the inside of the spiral.

Berrisford process: This is based upon the difference in the resilience of clean coal and dirt. The feed is dropped on an inclined plane containing a gap of chosen width. The clean coal bounces and falls on a receiver while the dirt falls through the gap.

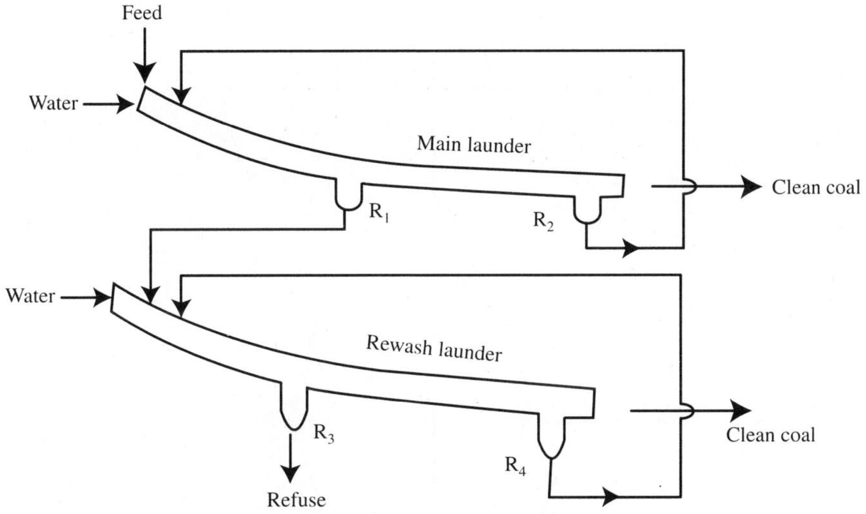

Figure 4.4 A two-trough Rheolaveur launder

Air–sand process: In this process a suspension of sand in air is used as the medium for coal washing. The mixture of air and sand acts as the dense medium of suitable specific gravity; the clean coal floats and the dirts sink.

The above mentioned processes are the dry processes of coal cleaning. The chief advantage of these over the wet processes is the elimination of the cost of drying wet coal and the difficulty of disposal of slurries of water and fine coal or dust. However, these processes suffer from a great disadvantage—the requirement of very close limits of screen sizes of the coal feed. Moreover, the dry processes cannot satisfactorily clean coals of difficult washability. These considerations greatly limit the use of the dry processes. Among these the tables are more widely used. There is no coal washery in India based on the dry cleaning principles.

Wet processes

Concentration tables: The working principle of the concentration tables is similar to that of the pneumatic tables. Water replaces air and the washing efficiency is higher, particularly with unsized feeds.

Trough washers: Trough washers or launders are of various designs. The Rheolaveur system consists essentially of one or more inclined troughs in which the clean coal and the refuse are separated by alluviation (Fig. 4.4). The coal feed and water enter at the top of the trough. The heavier dirts tend to settle at the lower strata of the bed. The lighter clean product remains in the upper strata and overflows at the lower end of the trough. The flowing refuse at the lower end is continuously removed through several special pockets called *rheo boxes* which are located along the length of the trough. The heavier material collected in the rheo boxes may be rewashed in another trough for recovering more clean coal.

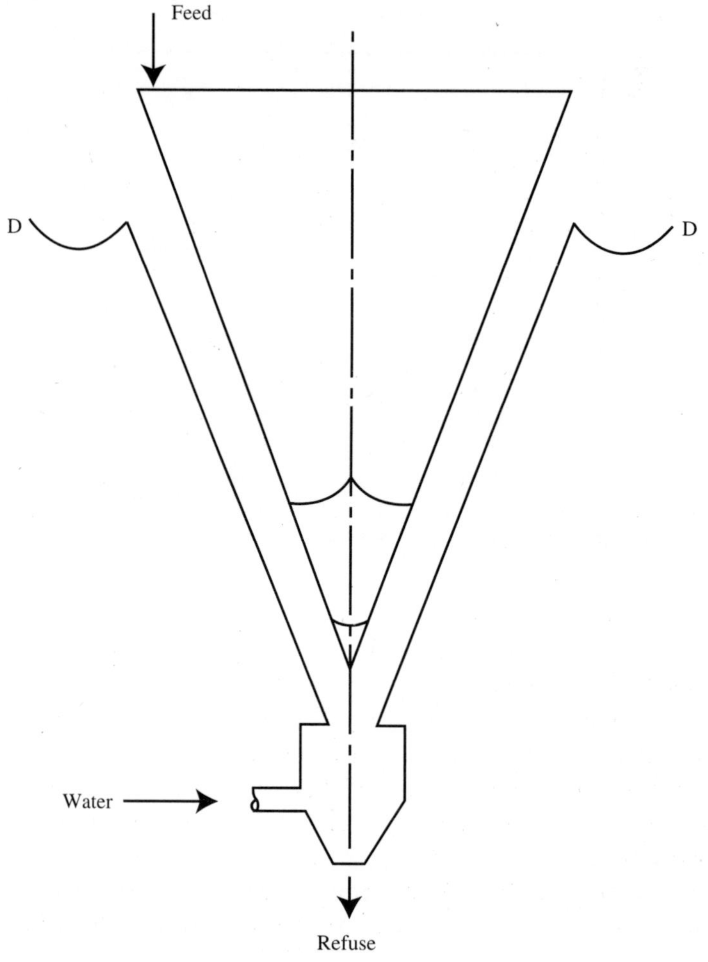

Figure 4.5 Double cone classifier

Upward-current classifiers: In the upward-current classifiers, the dirty coal is fed at the top of an inverted conical vessel and a continuous upward current of water enters through the base (Fig 4.5). The separation is effected by adjusting the velocity of water between the terminal velocity of fall of the clean coal particles and that of the dirts. The settling dirts impart to the bath some of the properties of a dense medium. Therefore, the size range of the feed can be greater than would be permitted under typical free settling conditions.

The use of concentration tables, trough washers and upward-current classifiers is limited to coals that are close in size and have good washability characteristics. The efficiency of the washers is low in the case of difficult to wash coals. They are not in use in India.

Jigs and dense medium separators are very widely used in coal washing plants. Most washeries in India are equipped with these two types of equipment. Jigs have been the most common type

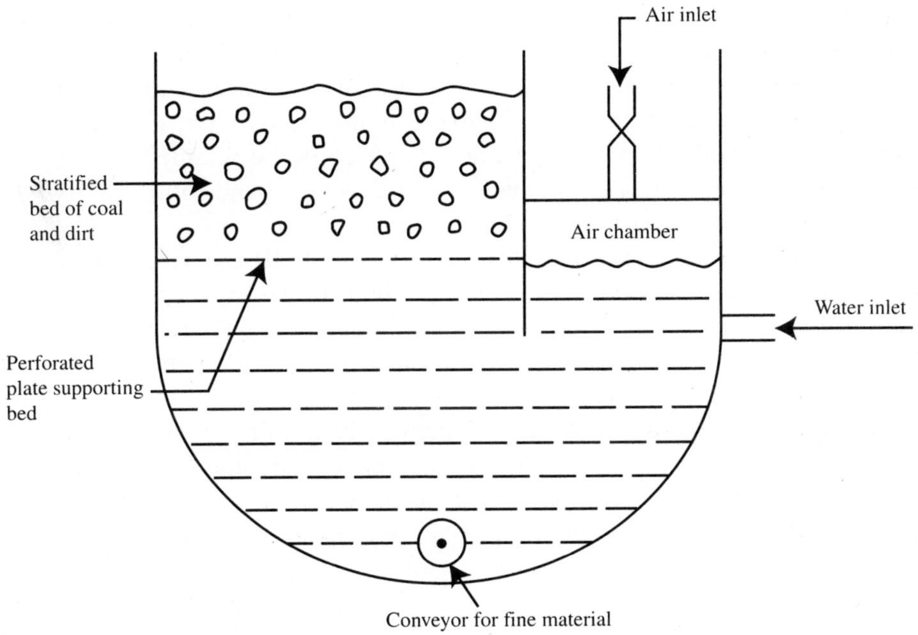

Figure 4.6 Baum jig

of coal washer the world over. Recently dense medium separators have become very popular owing to their better performance in the case of difficult coals. However, jigs are better adaptable to finer sizes than dense medium separators.

Jig washer: In a jig washer, a coal bed is maintained on a perforated plane and subjected to the action of periodic upward and downward current of water. This action stratifies the coal bed mainly according to the specific gravity of the particles. The clean coal is concentrated largely in the upper strata of the bed and the dirts accumulated in the lower strata. The pulsaing movement of water in the jig can be created by different means. The older types of jigs use reciprocating plungers or diaphragms.

The Baum jig produces pulsations in the water by alternately admitting and releasing compressed air. It consists of a U-shaped box divided longitudinally by a vertical partition into washing and air compartments (Fig. 4.6). The former is sub-divided into two sections by a transverse partition. Each section of the washing compartment is provided with a sieve. The water pulsation is caused at the rate of 30 to 60 cycles per minute in the air compartment and relayed to the adjoining washing compartment. The unclean coal is fed at one end of the washing compartment. Large pieces of the refuse collect at the bottom of the first section itself. The clean coal and the smaller sizes of the refuse are carried forward over the weir into the second section. While the refuse sinks to the bottom, the clean coal is carried by the flowing water over the discharge weir. The refuse is periodically removed from the two sections through valves. Fine dirts fall through the sieves into the bottom of the washer and are removed by a conveyer.

The size of coal that can be treated in jigs can be anything between 200 mm and 0.5 mm. However, jigs are now often used to wash small coal, say below 25 mm—the large sizes being treated in dense medium separators. Jigs cannot wash fines below 0.5 mm and their efficiency drops sharply when the content of sizes below 3 mm exceeds 25%. For such fines cyclone washers are quite suitable. Flotation cells are also used in the washing of coal fines. When the feed of 200 to 0 mm is jigged, the fines remain unwashed and form a slurry which needs further treatment for the recovery of combustibles.

The efficiency of jigs is more than 98% for coals with 10% near-gravity materials. The optimum capacity of jigs is 200 to 400 tonnes per hour (tph). However, commercial units of 20 as well as 600 tph capacity are known to exist. Baum jigs have been modified in several ways to increase their capacity for the same physical space, cope with feed fluctuations and treat fine coals with high efficiency.

The *feldspar jig* is an air-pulsated Baum-type jig which utilises a bed of feldspar (or some hard rock) through which all the waste materials migrate. With automatic control, this jig can cope with feed flucuations. It is usually used for the cleaning of –13 mm size coal without desliming. However, little ash reduction can be accomplished for the –48 mesh (0.29 mm) fraction of the feed.

The principle of cleaning coal by jigging is the same in the new *Batac jigs* as in conventional Baum jigs. However, the Batac jig has improved and automated methods of air distribution, pulsation and bed control. It provides a higher capacity in the same physical space. Baum jigs are mainly used for coarse coal washing, while Batac jigs are used for both coarse and fine coal cleaning. When cleaning fine coals, feldspar beds are used in some of the cells of the Batac jig. The performance characteristics of Batac jigs in fine coal cleaning are similar to those of Baum jigs in coarse coal cleaning.

Dense medium processes: The various dense medium processes are essentially large scale applications of the float-and-sink test. The separation can be made to almost correspond to the washability curves. These processes may be classified on the basis of the type of equipment used:

1. *Deep bath:* the Chance process and Barvoy's process,
2. *Shallow bath:* the Dutch State Mines process and Tromp process,
3. *Drum:* the Drewboy process and Link belt process.

The present tendency in the design of dense medium washers is to prefer a shallow bath to deep bath. The advantages of the shallow bath are reduction in the building height, reduction in the amount of excavation required for the collecting tank, requirement of much lower volume of liquid media, increase in the capacity for extracting middlings, and flexibility in flow arrangements. The drum wash boxes have also similar advantages and find use in newer installations.

The main point in favour of dense medium processes against jigs is that coals containing high proportions of near-gravity material (70% to 80% or even more) can be efficiently washed to about 99% efficiency in almost all types of dense medium separators. While the jigs are normally efficient in the gravity range of 1.55 to 1.80, the dense medium washers can efficiently separate at as low a specific gravity as 1.28 and at any high practical gravity level.

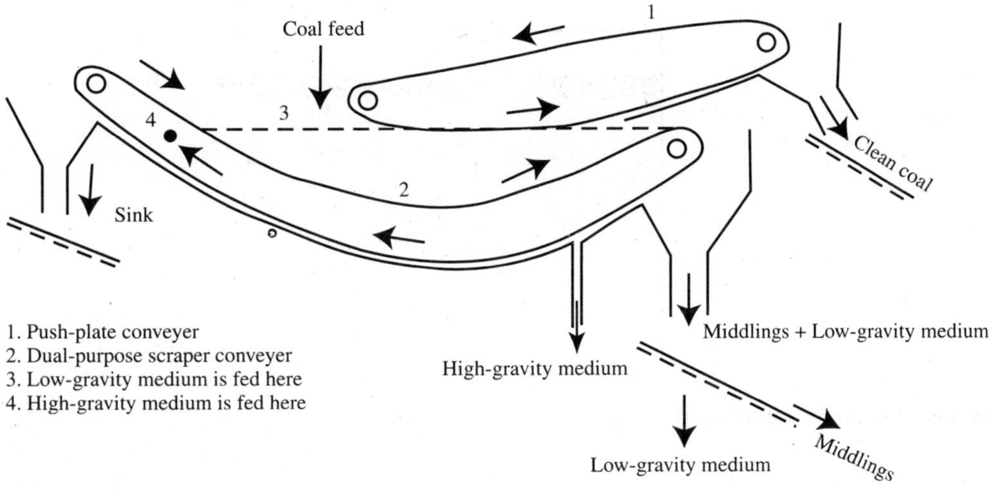

Figure 4.7 Heavy medium separator

The minimum and maximum capacities of commercial dense medium units are almost similar to those of jigs.

The dense medium processes are generally inefficient in cleaning small coals, say, below 6 mm because the effect of viscosity of the media greatly reduces the capacity of the plant, if an accurate separation is to be maintained. Moreover, the quantity of the solid media removed with the clean products is much greater in these sizes and it becomes necessary to regenerate a relatively greater proportion of the media, and that too with greater difficulty. The problem of viscosity has been successfully tackled by the addition of peptisers, such as sodium hexametaphosphate.

A shallow bath of the *Tromp process* is shown in Fig. 4.7. The bath is made of mild steel plate. It is provided with a push-plate conveyer (1) to discharge the cleaned coal and a scraper conveyer (2) to remove separately the middings and the refuse. A low gravity medium is fed at point (3) and a high gravity medium at point (4). Both media flow horizontally over the whole width and depth of the washer. The two conveyers are run in such a way that the swirl of the media is avoided. The charging of the feed and the collection of the products are clearly shown in the figure. The middlings may be crushed and recycled to the bath or used as a by-product fuel.

The dense media are prepared from the aqueous suspension of finely ground particles like magnetite, barytes, sand loess, sintered pyrites and froth flotation tailings. The washery products are separated from these particles by sieving followed by water spraying on sieves.

Froth flotation process: It is largely used in the beneficiation of mine dust, slack coal, washing rejects and slurries. It is especially suitable for producing ultra-pure coal for the manufacture of electrode carbon. The distinctive feature of this process is its suitability to clean fine coals., say 0.5 mm, which cannot be treated by other processes. The optimum particle size is

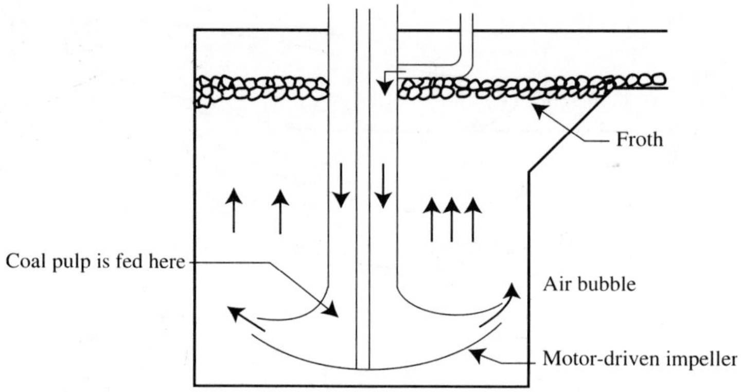

Figure 4.8 Flotation cell for coal

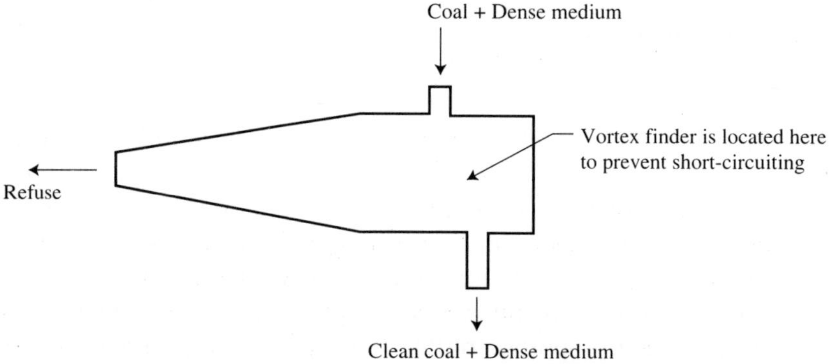

Figure 4.9 Horizontal cyclone washer

48 to 150 mesh, or 0.29 to 0.10 mm. However, 200 to 325 mesh or 0.074 to 0.43 mm, has also been processed. It is the only method commercially available for washing –200 mesh size coal. Froth flotation has not been popular in coal washing so far, owing to high capital and running costs. Its use has recently been encouraged as a modern trend in washing more and more fine coals.

A froth flotation plant consists of a battery of cells, each with a capacity of 5 to 10 tph. In each cell froth is made by bubbling air through water in the presence of frothing agents, such as, cresol, pine oil, kerosene and alcohols (Fig. 4.8). When fine coal is added to such a system, the pure coal particles adhere to the bubbles of the froth while the dirts are wetted by the water and sink to the bottom of the cell. The separation is improved by adding collecting agents, for example, spindle oil, creosote oil, xanthates and soaps to the system. The cleaned coal is recovered by the vacuum filtration of the froth.

Cyclone washers: These were developed during the Second World War in Holland and, within ten years, found wide application in washing small-sized coals (Fig. 4.9). The most distinctive features of these washers are a high quality of separation, and a high flexibility of operation with regard to percentage of near gravity material, coal particle size, feed rate and separation gravity. The modern cyclone washers compare favourably with jigs in capital cost. The slightly higher maintenance cost of the cyclone washer is more than compensated by the higher efficiency and greater flexibility.

Cyclone washers operate essentially on the principle of a settling chamber in which gravitational acceleration is replaced by centrifugal acceleration. Coal industries use low-pressure horizontal cyclones wherein the centrifugal force is about 40 times the gravitational force. This low value is sufficient to produce an accurate and rapid separation of raw coal as small as 0.5 mm in particle size, overcoming the effect of viscosity caused by 6 mm to 0.5 mm fines. On the other hand, the use of low feed pressure reduces energy consumption and simplifies the installation of the cyclone washer. Cyclone washers are usually built in the normal capacity of 70 to 150 tph.

A pulp of deslimed raw coal and heavy medium such as magnetite make the feed for a cyclone washer. The overflow contains the sinks. These are depulped and the products are rinsed with water before final disposal. The recovered slurry of heavy medium particles is reused after suitable treatment.

Oil agglomeration process: This process of coal cleaning is under development for early commercialisation. Particles are removed from the liquid suspension by selective wetting and agglomeration with a second immiscible liquid. The process relies on surface chemical effects rather than specific gravity differences as in conventional gravity separation processes. Since coal particles are essentially hydrophobic in nature, they can be readily agglomerated with many oils in an agitated aqueous slurry under high shear. On the other hand, the hydrophilic mineral particles are not affected and remain suspended in water. Furthermore, since the coal agglomerates of 0.7 to 2 mm are larger than the mineral particles, their removal can be readily achieved. Thus oil agglomeration is very efficient in handling extremely fine materials such as coal fines in the –200 mesh range or even –400 mesh, and materials with considerable amounts of clay slime.

Coal fines are agitated with water and middle distillate oil (5% to 9% of dry coal) in a flotation cell. No air is used. Heavier oils are avoided because they have functional oxygen and sulphur groups which can draw minerals also. Besides agglomeration of fines and partial de-ashing, dewatering also occurs. The collecting oil is adsorbed on the surface of the particles, displacing moisture. Since the surface area of agglomerates is considerably smaller than that of the starting material, the entrapped moisture content of the product is lower by 50% of the original coal.

Washer efficiency

The usefulness of a coal cleaning process for a particular coal, or a group of coals, depends essentially on the efficiency with which it separates the clean product from dirt. In the ideal case, the separation should be complete and there should not be any wrong migration of particles—heavier particles to the product and lighter particles to the tailings.

A number of formulae are available for calculating the washer efficiency, three of which are as follows:

*1. **Fraser and Yancey***

Efficiency, per cent

$$= \frac{\text{recovery of clean coal}}{\text{theoretical recovery at the given specific gravity}}$$

$$\times \frac{\text{per cent ash in feed coal} - \text{per cent ash in clean coal}}{\text{per cent ash in feed coal} - \text{per cent ash in theoretical clean coal}} \times 100$$

*2. **Yancey and Geer***

$$\text{Efficiency, per cent} = \frac{\text{recovery of clean coal}}{\text{theoretical recovery at the same ash level of the clean coal}} \times 100$$

*3. **Anderson***

$$\text{Efficiency, per cent} = 100 - (\text{sinks in clean coal} + \text{floats in refuse})$$

The recovery of the clean coal obtained in the washer is best calculated from the ash balance.

$$\text{Recovery, per cent} = \frac{\text{per cent ash in refuse} - \text{per cent ash in feed}}{\text{per cent ash in refuse} - \text{per cent ash in clean coal}} \times 100$$

The Fraser and Yancey formula is widely used. It does not, however, allow for any size degradation taking place in the actual washer operation which releases more dirt from the pure coal and alters the washability characteristics of the feed. Moreover, a high recovery of a low grade product may result in an efficiency greater than 100%. Similarly an exceptionally clean coal with low recovery will give misleading results.

The Yancey and Geer formula is an improvement upon the first formula. It is more simple and not likely to give misleading results as it is based upon the theoretical recovery at the actual ash level and not at the originally fixed specific gravity. In actual operation, the gravity of separation may differ from the set value. This formula, however, does not guard against the inaccuracy resulting from the size degradation of coal in the washer operation.

The demerits of the above two formulae are absent in the Anderson formula. The separation gravity and the quantity of wrongly migrated particles (sinks in clean coal and floats in refuse) are determined from the float-and-sink tests of the clean coal and the refuse. Any alteration in the set value of the specific gravity and the washability characteristics of the feed during washer operation is therefore taken into consideration. The Anderson formula however does not incorporate any term indicating the quantity of the noncombustible matter carried by the wrongly migrated particles. The total amount of migration cannot satisfactorily measure the quality of products.

4.2.2 Coal Washing in India

Although washability studies were first made in India as early as 1924, scientific coal preparation developed only recently. The problems in washing Indian coals are:

- large content of intimately associated mineral matter giving rise to difficult washability characteristics;
- presence of high percentage of near-gravity material at the desired density of separation;
- wide variation in washability characteristics of coals for different horizontal and vertical sections of the same seam;
- low average annual production of individual mines.

These problems are solved by using controlled coal crushing for better release of dirts, employing techniques of coal washing suitable for difficult-to-wash coal and establishing central washeries fed by a number of collieries. Jigs, heavy medium separators (shallow bath) and cyclones are the most widely used coal washers in India. Owing to high ash and difficult washability of coals, the washeries in India generally produce middlings in addition to the clean product and the rejects. The middlings have around 30% ash and are used in thermal power stations. The rejects are sometimes used in thermal power stations and in the back filling of mines. However, two product schemes producing a clean coal and a high ash by-product fuel (ash around 45%) are also recommended. No rejects are obtained in the scheme and the by-product fuel finds use in modern big thermal power stations and in the production of soft coke.

Table 4.4 gives the performance of washeries/washed coking coal production during the last five years, and the location and production figures as reported in the *Annual Report (2006–2007)*, of the Ministry of Coal, Govt. of India. The details of washeries in operation in the country as reported by the same source as above are given in Table 4.5. Both coking and non-coking coal sectors are covered.

The provisions as well as problems of central washeries are in many respects different from those faced by the pit-head washeries. It is logical to expect some appreciable variations in the economies of washing for these two different groups of washeries. For equivalent capacity level the capital cost of a central washery is obviously higher than that of a pit-head washery. This is mainly because in a central washery, 20%–30% of the capital cost is incurred on building elaborate marshalling yard and township, while in a pit-head washer, only 5%–10% of the capital cost goes towards railway siding and township.

The operating cost per tonne of throughput in a pit-head washery is about 70% of that incurred in a central washery. The higher cost of washed coal (cleans) in a central washery is mainly due to the higher freight handling charge on raw coals drawn from a large number of mines, many of which are located at a distance from the washery.

Table 4.6 gives the figures of production from more recent washeries in India in the years 1990–91, 1996–97 and 2003–04. The total raw coal feed for prime coking coal was 12.47 Mt in 1990–91 and only 4.64 Mt in 2003–04. The corresponding clean coal production was 6.44 Mt, 5.90 Mt and 2.86 Mt, and middlings productioin was 3.08 Mt, 3.70 Mt and

Table 4.4 Performance of washeries/washed coking coal production during the last five years (in million tonnes)

Source/Washery		2001–02	2002–03	2003–04	2004–05	2005–06	2006–07 (April–Dec. 2006)
BCCL							
2	Dugda-II	0.25	0.24	0.19	0.21	0.55	0.31
3	Bhojudih	0.68	0.66	0.58	0.50	0.67	0.32
4	Patherdih	0.17	0.09	0.08	0.07	0.07	0.05
5	Lodna	0.07	0	0	0	0.00	0.00
6	Sudamdih	0.28	0.25	0.22	0.22	0.27	0.15
7	Barora	0.07	0.04	0.01	0.00	0.00	0.00
8	Moonidih	0.42	0.41	0.46	0.54	0.49	0.24
9	Mohuda	0.25	0.31	0.30	0.33	0.24	0.10
10	Madhuband	0.14	0.16	0.02	0.00	0.00	0.00
Total	BCCL	2.33	2.16	1.86	1.87	2.29	1.17
CCL							
12	Kathara	0.65	0.54	0.57	0.58	0.54	0.28
13	Swang	0.35	0.37	0.40	0.38	0.31	0.20
15	Rajrappa	0.68	0.89	0.99	1.09	0.84	0.45
16	Kedla	0.33	0.38	0.40	0.59	0.61	0.42
Total	CCL	2.01	2.18	2.36	2.64	2.30	1.35
WCL							
17	Nandan	0.26	0.29	0.32	0.34	0.40	0.24
Total	CIL	4.60	4.63	4.54	4.85	4.98	2.76

•Provisional

Source: Annual Report 2006–2007, Ministry of Coal, Govt. of India, http://coal.nic.in

1.01 Mt. Including the medium coking coal, the total raw coal feed was 20.50 Mt, 22.20 Mt and 15.82 Mt while the data for clean coal production was 11.10 Mt, 10.90 Mt and 8.20 Mt and middling production was 5.60 Mt, 6.80 Mt and 4.85 Mt.

All-India production of some coal derivatives is given in Table 4.7. The total hard coke production was increased from 0.6 Mt in 1985–86 to 11.0 Mt in 1995–96 and to as high as 12.39 Mt in 2004–05 in the public sector alone. The company-wise production of coal is given in Table 4.8.

The National Steel Policy has an ambitions growth plan of 100 Mt coke per year by 2020, and to convert the country from a coke-importer to a coke-exporter. As a step towards ensuring the same, it has established long-term coal supply understanding with a number of reputed Australian coal suppliers. The coke demand for countries like India will have a continuous upward trend as the steel and automotive sectors are poised to grow.

The world coke production of the top ten hard coal producers is given in Table 4.9. Out of them India produced 295.2 million tonnes hard coke in 2006–2007.

The sector-wise demand of coal in India is given in Table 4.10 for the years 2001–07 and obviously the power utilities accounted for the lion's share of the demand: 216.472 Mt in 2006–07.

Table 4.5 Details of washeries in operation in the country

Sl. no.	Washery and Operator	Capacity (Mty)	Location	Consumer
Coking coal				
1	Dudga-II, CIL	2.00	Jharkhand	Steel plants
2	Bhojudih, CIL	1.70	West Bengal	-do-
3	Patherdih, CIL	1.60	Jharkhand	-do-
4	Moonidih, CIL	1.60	-do-	-do-
5	Sudamdih, CIL	1.60	-do-	-do-
6	Mahuda, CIL	0.63	-do-	-do-
7	Barora *, CIL	0.42	-do-	-do
8	Kathara, CIL	3.00	Jharkhand	-do-
9	Swang, CIL	0.75	-do-	-do-
10	Rajrappa, CIL	3.00	-do-	-do-
11	Kedla, CIL	2.60	-do-	-do-
12	Nandan, CIL	1.20	-MP-	-do-
	(A) CIL	**20.10**		
13	Durgapur, SAIL	1.5	WB	-do-
14	DCOP, DPL	1.35	WB	-do-
15	Chasnala, IISCO	2.0	Jharkhand	-do-
16	Jamadoba, TISCO	1.72	-do-	-do-
17	West Bokaro-II, TISCO	1.8	-do-	-do-
18	West Boakaro-III,TISCO	2.1	-do-	-do
19	Bhelatand	0.8	-do-	-do-
20	Govindpur, Auroma Coke Ltd.	1.0	-do-	-do-
	(B) PSU and Private	**12.27**		
	TOTAL (A + B)	**32.37**		
Non-coking coal				
1	Dugda-I,CIL	1.00	Jharkhand	TPS
2	Lodna,CIL	0.48	-do-	TPS
3	Madhuban,CIL	2.50	-do-	TPS
4	Gidi,CIL	2.50	-do-	TPS
5	Piparwar,CIL	6.50	-do-	TPS
6	Kargali,CIL	2.72	-do-	TPS
7	Bina,CIL	4.50	UP	TPS
	(A) CIL	**20.20**		
8	Dipka, Aryan Coal Beneficiation Pvt. Ltd.	5.0	Chattisgarh	RSEB, GEB, PSEB, KPCL, MSEB, cement.
9	Gevra, -do-	6.0	-do-	-do-
10	Chandrapur, -do-	2.0	Maharastra	MSEB, KPCL
11	Adilabad, Aryan Energy Pvt. Ltd.	2.0	AP	KPCL
12	Talcher, Aryan Energy Pvt. Ltd.	3.0	Orissa	Sponge iron
13	Wani, Kartikay Coal Washeries Pvt. Ltd. (Aryan)	2.0	Maharastra	TPS
14	Korba, ST-CLI Coal Washeries Ltd.	5.0	Chattisgarh	BSES, GEB, cement
15	Ramagundam, Gupta Coalfield & Washeries Ltd.	2.5	AP	KPCL, cement, small industries

Table 4.5 *Continued*

Sl. no.	Washery and Operator	Capacity (Mty)	Location	Consumer
16	Sasti, Gupta Coalfield & Washeries Ltd.	2.5	Maharastra	RPTS, KPCL
17	Wani, Gupta Coalfield & Washeries Ltd.	2.5	Maharastra	MSEB
18	Umrer, Gupta Coalfield & Washeries Ltd.	0.75	-do-	Cement
19	Bhandara, Gupta Coalfield & Washeries Ltd.	0.75	-do-	Sunflag Iron & Steel Co. Ltd.
20	Parasia, Gupta Coalfield & Washeries Ltd.	0.75	-do-	Cement
21	Bilaspur, Gupta Coalfield & Washeries Ltd.	1.2	Chattisgarh	Sponge iron
22	Ghugus, Gupta Coalfield & Washeries Ltd.	4.0	Maharastra	MSEB
23	Talcher, Global Coal Mining (P) Ltd.	2.0	Orissa	Sponge iron
24	Wani, Bhatia International Ltd.	3.0	Maharastra	MSEB
25	Chandrapur, Bhatia International Ltd.	1.0	Maharastra	Industries
26	Raigarh, Jindal	2.5	Orissa	Steel
27	Wani, Indo Unique Flame Ltd.	0.5	Maharastra	Sponge iron
28	Chhattishgarh Power & Coal Beneficiation Ltd.	1.2	Chattisgarh	Power and cement
	(B) Private	50.15		
	TOTAL (A+B)	70.35		

Source: Annual Report 2006–2007, Ministry of Coal, Govt. Of India, http://coal.nic.in

The requirements for industrial and domestic cokes are given in Tables 4.11 and 4.12, respectively.

4.3 STORAGE OF COAL

Coal consuming industries usually store large quantities of coal at the plant as a resource to offset interference in supply. During storage coal undergoes a series of changes on exposure to the atmosphere. This process is known as the *weathering* of coal. The extent of changes depends upon the method of storage and the nature of coal. Weathering is more pronounced with lower rank coals and greater exposure to air. By storing coal under water and in trenches, weathering may be minimised. The properties affected by the storage of coal are (i) size, (ii) friability, (iii) caking capacity, (iv) ultimate analysis, (v) calorific value, and (vi) yields of carbonisation products.

Weathering tends to reduce the coal size and increase its friability. The formation of slacks or small sizes of coal is particularly pronounced with sub-bituminous coal, lignite and peat. This imposes a serious problem in the handling, transport and storage of these fuels. Anthracite and mature bituminous coal do not slack.

Table 4.6 Production from coal washeries in India (million tonnes)

Feed and production	1990–91	1996–97	2003–04
Prime coking coal			
Raw coal feed	12.47	12.80	4.64
Clean coal production	6.44	5.90	2.86
Middling production	3.08	3.70	1.01
Medium Coking Coal			
Raw coal feed	8.03	9.40	11.18
Clean coal production	4.73	5.00	5.34
Middling production	2.46	3.10	3.84
Total			
Raw coal feed	20.50	22.20	15.82
Clean coal production	11.17	10.90	8.20
Middling production	5.54	6.80	4.85

Sources: (1) *Coal Directory of India*, Government of India, various years.
(2) *Annual Report 1997–98*, Ministry of Coal, Govt. of India, New Delhi

Table 4.7 All-India production of some coal derivatives (million tonnes)

Item	1985–86	1995–96	2004–05
Soft coke	1.7	0.2	–
Beehive hard coke	0.3	0.1	12.39
By-product hard coke	0.3	10.9	in public sector
Washed coal	10.9	11.9	8.7
Middlings	5.1	6.8	5.2
Total exports	0.2	0.1	1.3
Total imports	2.0	12.5	25.3

Source: Coal Directory of India (various years), Coal Controller's Organisation, Govt. of India.

Table 4.8 Company-wise coal production (million tonnes)

Company	Target 2006–2007 (April–December)	Actual Production* (2006–2007) (April–December)	Actual Production (2005–2006) (April–December)
CIL	257.367	250.073	241.721
SCCL	27.734	26.097	24.537
OTHERS	17.857	18.978	16.166
TOTAL	302.958	295.148	282.424

*Provisional (figures excluding Meghalaya)
Source: Annual Report: 2006–07, Ministry of Coal, Govt. of India, New Delhi

Table 4.9 Top ten hard coal producers (2006)

PR China	2482 Mt	Russia	233 Mt
USA	990 Mt	Indonesia	169 Mt
India	427 Mt	Poland	95 Mt
Australia	309 Mt	Kazakhstan	92 Mt
South Africa	244 Mt	Colombia	64 Mt

Source: World Coal Institute, Coal Facts, http://www.world.org

Table 4.10 Consumer-wise offtake of coal from CIL. SCCL and others for the last five years (million tonnes)

S.No	Name of Consumer	2001–02	2002–03	2003–04	2004–05	2005–06	2006–07 (April–Dec. 06)
1.	Power houses (middlings)	242.76 (1.80)	250.42 (1.71)	265.53 (1.67)	303.89 (1.83)	312.540 (1.370)	216.472 (1.450)
2.	Steel Plants and cokeries (raw coking coal)	20.28	18.93	18.91	17.51	18.59	12.109
3.	Loco	0.00	0.00	0.00	0.00	0.00	0.000
4.	Cement plants	11.85	12.56	13.65	14.70	14.970	10.942
5.	Fertiliser plants	3.20	2.49	5.21	2.18	2.260	2.307
6.	Soft coke manufacturing	0.00	0.00	0.00	0.00	0.00	0.00
7.	Brick kilns, textiles, chemicals paper and other industries	45.90	49.79	49.12	35.04	47.230	58.024
8.	Colliery consumption	1.80	1.48	1.33	1.18	1.070	0.745
9.	Total offtake	325.79 (1.80)	335.67 (1.71)	353.75 (1.67)	374.50 (1.83)	396.660 (1.370)	300.599 (1.45)

Note: Does not include Meghalaya Coal
Source: Annual Report: 2006–07, Ministry of Coal, Govt. of India, New Delhi, http://coal.nic.in

The caking capacity is very sensitive to aerial oxidation and always decreases on weathering. A slight decrease in the caking capacity may improve the coke size in some cases but generally the coking properties deteriorate during the storage of coal. Other parameters which fall on weathering are carbon content, hydrogen content, calorific value and tar yield. The oxygen content and yields of gas and liquor increase by atmospheric oxidation.

Weathering of coal is essentially a result of mild oxidation and hence, an exothermic process. If the heat liberated is not completely dissipated, the temperature of the coal rises. Since the rate of chemical reaction is approximately doubled by a rise of 10°C, the extent of atmospheric oxidation continues at an accelerated rate with a greater quantity of heat liberated. Ultimately the coal may reach its *ignition point* and burst into flame. This phenomenon is known as the *spontaneous inflammation* of coal and is associated with the storage of coal in bulk, say, 200 tonnes for bituminous coal and 50 tonnes for lignite.

Table 4.11 Requirements for industrial coke (IS: 439-1989)

Sl. No.	Coke designation	Grades	Ash, per cent, Max	Moisture, per cent, Max	Volatile matter, per cent, Max	Sulphur, per cent, Max	Phosphorus, per cent, Max	Shatter Index Min			Micum Index		Porosity, per cent	Temperature of ash fusion, °C (Under mildly reducing atmosphere)		Reactivity to CO_2	Size range
								Over 75 mm	Over 50 mm	Over 12.5 mm	+40 mm Min	−10 mm Max		Initial deformation	Fusion		
1	Blast furnace coke (BFC)	1	22	4	1.5	0.7	0.25	—	—	—	78	10	28 to 45	—	—	—	The material shall be in the size range between 100 and 25 mm with a size tolerance of 15 per cent by mass for oversize and 10 per cent by mass for undersize
		2	25	4	1.5	0.7	0.25	—	—	—	76	12	38 to 45	—	—	—	
2	Foundry	Special	20	4	2.0	0.7	0.15	30	70	92	—	—	35 to 45	—	—	—	Size of coke shall be as agreed to between the purchaser and the supplier
		1	22	4	2.0	0.7	0.3	30	70	92	—	—	35 to 45	—	—	—	
		2	25	4	2.0	0.7	0.3	25	70	92	—	—	35 to 45	—	—	—	
		3	30	4	2.0	0.7	0.3	25	70	92	—	—	35 to 45	—	—	—	
3	Coke for gas making (GC)	1	22	6	3.0	0.7	0.3	—	—	—	—	—	40 Min	>1240	>1350	—	Coke for producer and water gas manufacture shall be of the size range between 50 and 6.3 mm with a size tolerance of 5 per cent by mass on oversize and 10 per cent by mass on undersize
		2	25	6	3.0	0.7	0.3	—	—	—	—	—	40 Min	>1240	>1350	—	
4	Coke for ferro alloys industry (FAC)	Special	20	5	2.0	0.7	0.04	—	—	—	—	—	—	—	—	180	The material shall be in two size ranges, namely, between 50 and 20 mm and between 20 and 6.3 mm
		1	22	5	3.0	0.7	0.10	—	—	—	—	—	—	—	—	150	
		2	25	5	3.0	0.7	0.16	—	—	—	—	—	—	—	—	120	

Table 4.12 Requirements for domestic coke (IS:4286-1976)

Sl.No.	Characteristics	Requirements		
		Class 1	Class 2	Class 3
1.	Moisture, per cent max	8	8	8
2.	Volatile matter, per cent dry	3–10	3–10	3–10
3.	Ash, per cent dry max	30	40	50
4.	Size range, mm*	100–100	10–100	10–100
5.	Shatter index			
	+ 25 mm per cent min			
	+ 12.5 mm per cent min	65	65	65

Notes: (1)* Not more than 10% should pass through 100 mm and be retained on 10 mm sieves.

(2) In the case of domestic coke made from lignite, moisture and volatile matter, not exceeding 15% each, may be permitted.

This hazard may be avoided by decreasing the exposed surface of coal lumps and suppressing the ventilation. The rate of oxidation appears to increase proportionally to about the root of the surface area per unit weight of the coal. The exposed surface area is reduced by avoiding segregation and by packing the coal tightly and uniformly. Blanketing the pile with fine coal is also helpful in decreasing the ventilation. Another useful measure is to surround the coal pile by a wall. All of these constitute a positive measure in minimising the extent of oxidation—the root cause of spontaneous inflammation. Another approach is to enhance the ventilation in such a way that the liberated heat is fast dissipated. This method is not recommended in practice.

Whenever coal is stored in bulk for a long time, its temperature should be measured at various points of the pile from time to time. Above 70°C, the risk of spontaneous inflammation is great. Therefore, the coal should be consumed before this critical temperature is reached. Water spraying helps only in the initial stages and not after crossing the danger point. As a rule, coal should not be stored near a source of heat. The problems connected with the storage of coal also apply to coal in ships' bunkers and cargo. Here the conditions of ventilation and proximity of heat source are such that spontaneous inflammation may occur in piles of smaller quantities, that is, less than 200 tonnes. Therefore greater care must be taken to prevent any rise in temperature of coal beyond the critical value. Freshly mined coals are most susceptiable to spontaneous combustion. This hazard usually occurs within four to five weeks of mining. If coal has been safely stored for more than six months, it is not usually prone to spontaneous combustion.

4.4 CARBONISATION

The process of carbonisation may be of three categories, namely (i) low temperature carbonisation (LTC) at 600°C, (ii) medium temperature carbonisation (MTC) at 800°C, and (iii) high temperature carbonisation (HTC) at 1000°C. The low temperature process is commercially practiced for the production of semicoke, which is also called soft coke or char. The high temperature process is carried out in the production of hard coke and coal gas. The medium temperature process does not have commercial application. However, many commercial low temperature carbonisation processes approach the conditions of MTC. Because of the important use of hard coke in the blast furnace and cupola, HTC is practised on a much bigger scale than LTC. Soft coke is mainly used as a smokeless domestic fuel.

4.4.1 Low Temperature Carbonisation

Coal and semicoke are poor conductors of heat. In a static bed of externally heated coal, it is necessary to employ a high temperature gradient to achieve carbonisation in a reasonable time. This method may seriously affect the product by over-carbonising the surface layer. Therefore, other methods have been developed with a view to overcoming this difficulty. The heat transfer rate is enhanced in these methods by using one or more of the following steps:

- direct heating of the charge by hot gases,
- movement of the charge by vertical fall,
- rotating the carboniser,
- using fluidised bed technique.

There are two commercial LTC plants in India, situated at Neyveli (Tamil Nadu) and Naspur (Andhra Pradesh). Commissioned in 1968, the Neyveli plant produces 900 tonnes per day (tpd) of carbonised lignite briquettes sold under the trade name, Leco. The Naspur plant was commissioned in 1979 and has a capacity of 900 tpd of soft coke. Both these plants are based on internally heated Lurgi–Spül gas process (Fig. 4.10).

The heart of the Naspur plant is a vertical shaft consisting of three sections serving as the upper drying zone, the middle carbonising zone and the bottom coke cooling zone. Coal of size 25 to 125 mm is fed through a hopper. While falling through the vertical shaft it is successively dried and carbonised. The drying is effected by the combustion gases from the burner fitted to the drying zone. The carbonisation is carried out by a mixture of hot gases entering from the carboniser burner chamber and gases drawn from the coke cooling zone. The bulk of the gases in the drying zone is kept circulating by a fan while the rest escape through the chimney at the top. The rich gases from the carboniser are treated for the recovery of by-products and then sent to the coke cooling zone and also to various burners. Care is taken to prevent the entry of the carboniser gases to the dryer and the dryer gases to the carboniser.

The various products of low temperature carbonisation are semicoke, low temperature tar, liquor, crude low temperature spirit and gas. The quality and yield of these products depend upon the process and the coal used. LTC plants normally use low rank coals and lignite. These fuels lead to excessive smoke formation during burning, but the semicoke obtained from them is a clean fuel. Some typical yields (dry basis) by the Gray–King assay at 600°C are: semicoke 75%–80%, tar 9%–12%, liquor 5%–7% and gas 100–120 Nm^3/tonne. In commercial plants the yields of semicoke and tar are less and that of gas is higher. The crude spirit is usually recovered in 0.5%–1% yields.

Semicoke has volatile matter of 8% to 12%, is highly reactive and can be easily ignited into a smokeless flame. The screened product is an ideal domestic solid fuel (Table 4.12). The char produced by LTC in a fluidised bed is converted into suitable briquettes for domestic uses. Other uses of semicoke or char are production of iron in low shaft furnaces, gasification into

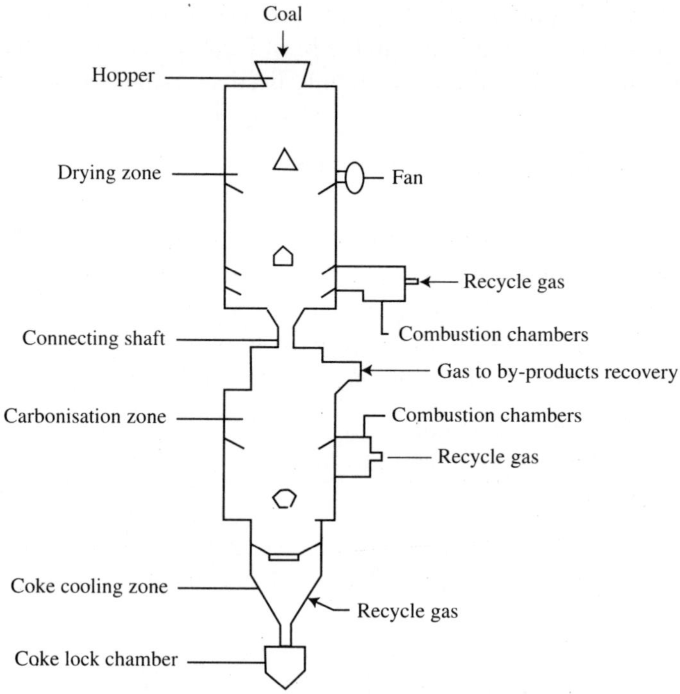

Figure 4.10 Lurgi–Spül gas low temperature carboniser

fuel gas or synthesis gas for fertilisers and chemicals, and preparation of coke oven blends. It may also be used as an industrial fuel.

Low temperature (LT) tar is often known as primary tar since the alteration by secondary cracking reactions is small due to relatively low temperature conditions. It is a very complex mixture of higher phenols, substituted aromatic hydrocarbons and various nitrogenous and oxygenated compounds of carbon and hydrogen. Many disinfectant preparations are made from this tar. Chemicals and liquid fuels may be produced from it by employing special processes. The LT tar has at present a much lower market value than the HT (high temperature) tar.

The crude spirit corresponds to the crude benzol of HTC but contains larger proportions of paraffins, naphthenes and olefins. Motor fuel may be produced by processing the crude spirit.

The properties of the gas obtained by low temperature carbonisation are:

- specific gravity (referred to air) 0.6,
- calorific value 6500 kcal/Nm^3,
- composition (in per cent): CO_2 4.0, C_nH_m 4.0, CO 7.0, H_2 33.0, C_nH_{2n+2} 45.0 and N_2 7.0.

This gas is richer in hydrocarbons and poorer in hydrogen compared to the gas produced by HTC. Its yield is lower but calorific value is higher. The low temperature gas is used up in the plant. The gas obtained from a direct heated LTC plant is a mixture of carbonisation and combustion gases and hence has a low calorific value of about 1500 kcal/Nm^3.

The low temperature carbonisation of coal has a bright future in India as it can supply the substitute fuel required to replace the large quantities of cattle dung and other biomass used for domestic hearths. However, the high capital cost and present poor market value of LT tar hinder the establishment of this industry. But the so-called *bhatta* coke is being produced in large quantities in the Bengal–Bihar coalfield area by a wasteful process. A heap of coal, 15 to 20 te, is set on fire and the volatile products are burnt off. After a suitable time, say 7 to 10 days, the fire is extinguished. The solid residue is essentially a reactive semicoke with pleasant free-burning characteristics. The Central Fuel Research Institute (CFRI) at Jealgora near Dhanbad in Bihar has developed a coking stoker which also produces soft coke by burning off a part of the volatile matter of coal. Its capital cost is small but no by-products are recovered.

4.4.2 High Temperature Carbonisation

This process is commercially practised in coke ovens and to a much lesser extent in gas retorts. The chief products from these two plants are coke and gas, respectively. The coke ovens may be of either the beehive type or the by-product slot-type. The by-products are recovered only in the latter type of coke ovens while they are burnt off in the former.

The gas obtained by HTC in gas retorts is know as coal gas. Its principal use is as a town gas. With the advent of natural gas, liquefied petroleum gas and electricity, the production of town gas by HTC in gas retorts is now more or less an obsolete process. Further, this retort coke is unsuitable in iron making, but is reactive and hence a good domestic solid fuel. Therefore, the two-fuel idea from coal is now staging a come back.

Three types of gas retorts are used in carbonising coal: horizontal retorts, intermittent vertical retorts and continuous vertical retorts. The presently defunet town gas plants at Mumbai and Kolkata were using horizontal retorts. A two-fuel complex using continuous vertical retorts is being installed at Dankuni, 20 km from Kolkata. The Dankuni coal complex may pave the way for many such projects in other viable areas in the country.

Continuous vertical retorts

Continuous vertical retorts were developed mainly in the UK during the early twentieth century. By 1966, there were at least 600 plants throughout the world, carbonising more than 25 million tonnes of coal per year.

These retorts are narrow vertical brick chambers, rectangular or elliptical in cross section. Indirect heating is achieved from both sides by burning producer gas in flues, which may be either horizontal or vertical. The chambers are wider at the bottom so that the charge can gravitate freely through it. The coal should ideally be a high-volatile, medium-caking coal or its blend with high volatile non-caking coal or a non-caking coal alone. It is charged at the top through a hopper. In its downward movement it is successively dried and carbonised. The coke is discharged by a mechanical device (coke extractor) into a sealed coke chamber. It is cooled here by quenching with steam and finally removed from the retort house. Steaming helps to recover the sensible heat of coke, augment gas yield by producing water gas and also improve the reactivity of coke by burning off a part of the coke and opening up its fine pore structure. Since the water gas so produced has a lower calorific value than the coal gas, it is obvious that the mixed gas obtained by steaming the coke is poorer than the coal gas. However, the total heat content of the product gas is increased by steaming. In other words the thermal efficiency of the plant is greatly enhanced by steaming due to the conversion of the sensible heat of coke into the potential heat of water gas.

The gas leaving the retorts is treated for purification from undesirable components, and for the recovery of valuable by-products. These treatments constitute the by-product recovery section of the plant. The basic principles of operation of the by-product recovery sections of the gas and coke plants are similar and will be discussed under the coke plant. The gas used as town gas may need an additional step, that is, partial drying because excessive water vapour causes corrosion and blockages in the distribution pipes. The drying may be achieved by chilling, compression or treating with dehydrating agents like concentrated calcium chloride solution. Sometimes crude benzol is not recovered so that the calorific value of the gas is higher due to the presence of the benzol hydrocarbons. These considerations also apply to coke oven gas meant for use through pipelines in places outside the coke plant. The coke obtained from the gas retorts is known as *gas coke*. Although a result of high temperature carbonisation, it is much more reactive than the hard coke obtained from coke ovens, partly due to the lesser caking properties and lower rank of the coal carbonised, and partly due to the effect of steaming. The gas coke is largely used as a fuel in miscellaneous industries and domestic ovens. Owing to its poorer strength it is not suitable in blast furnace operation.

Beehive coke ovens

These are firebrick chambers with an arched roof and derive their name from the shape of a beehive. The dimensions of a typical oven are: height 2 m and base diameter 3.5 m. There is a hole at the centre of the roof and a door at one side. A number of ovens are built in a row with common walls between neighbouring ovens. About 5 to 7 tonnes of coal are charged from the top hole into the hot oven and allowed to form a layer of about 60 to 90 cm depth by means of a rake. Carbonisation immediately starts and the evolved gases burn off with the air coming through an opening at the top of the closed door. The carbonisation proceeds from the top to the bottom of the charge and is complete in two to three days. Heat transfer is mostly by radiation from the hot roof to the surface of the charge and then by conduction through the layer of the charge. Heat is supplied by the burning of the volatile matter and hence no by-products are recovered. The exhaust gases are allowed to escape to the atmosphere. In some cases they are passed through flues built in the walls between the ovens and then released to the surroundings through a chimey. The thermal efficiency is thus increased and pollution also lessoned. Sometimes waste heat boilers are also run on the heat produced by the exhaust gases. When carbonisation is completed, the door is opened and the coke is quenched with water. The quenched coke is discharged manually or by a machine. The walls and roof of the emptied oven retain enough heat to initiate coking of the next charge.

The demerits of beehive coke ovens are:

1. There is no recovery of by-products,
2. There is lower coke yield due to partial combustion,

3. Quality coke can be produced only from good coking coals (blends of inferior coals cannot be used),
4. There is lack of flexibility of operation.

In spite of these demerits, beehive ovens are still in considerable use in many countries such as the USA and India. The advantages are:

1. Extremely good quality, strong and blocky coke of the required reactivity is obtained if the starting coals are of good coking quality.
2. The capital and running costs are low.
3. Coke production is not connected with the availability of a market for the by-products.

By-product slot-type coke ovens

Coal is indirectly heated in these ovens by combustion chambers separate from the carbonisation chambers. The ovens are built in a battery of 10–100 ovens, in which the combustion chambers alternate with the carbonisation chambers. A gaseous fuel is burnt in the combustion chambers for supplying the heat to the process. Coal is coked in the carbonisation chambers. The gases and vapours evolved on carbonisation are not permitted to mix with the combustion gases. The carbonisation gases are treated for the recovery of the valuable by-products, namely, tar ammonia, crude benzol and coke oven gas.

The carbonisation chamber is a long, narrow parallel piped of usual dimensions having length 12–14 m, height 4–5 m and width 400–500 mm. The modern trend is to build very large ovens with a useful capacity of around 40–50 m^3 leading to about 33–40 tonnes of coal per charge and giving about 50–60 tonnes per day throughout. This has been possible by increasing the height to 6–7.65 m and length to 15–17.25 m. The width of the oven is not uniform. A tapering of about 50 mm is maintained for easy pushing of the coke. The two ends of the chamber are provided with removable doors and the roof is provided with three to four holes covered with lids for charging coal from the top. The roof has an additional hole at one end for the escape of volatile products of carbonisation through a vertical pipe known as the standpipe.

The combustion chamber essentially consists of flues built in the dividing wall between two neighbouring carbonisation chambers and also in the two end walls of the battery. Gaseous fuels (usually coke oven gas or blast furnace gas) are supplied to the heating flues. The hot combustion gases are passed through regenerators or waste heat boilers and then let out into the atmosphere through the chimney. Modern coke ovens are always provided with regenerators which are built underneath the ovens. The heat recovered in the regenerators is utilised in preheating the air and blast furnace gas. Coke oven gas cannot be preheated because its hydrocarbons get cracked in the checker brickwork of the regenerators, resulting in the loss of its heating value and the blockage of the regenerators by carbon deposits. Regenerators work on the principle of alternate heating and cooling cycles. This is achieved by periodically changing the direction of flow of gases through the vertical flues (reversal) every 30 minutes or so.

The coke ovens are made of silica bricks which are preferred to firebricks mainly because of the higher stability required at the temperature of the combustion chamber of about 1350°C to 1450°C. On the other hand, regenerators are made of firebricks because they can withstand the stress caused by fluctuating temperatures in the operation cycle.

Each oven cokes a batch of about 20 tonnes or more. The batchwise carbonisation process is made continuous by building the ovens in a battery. In a modern coke plant there may be two to three batteries built in one row, with a total capacity of more than 1 Mt of coke per year. These big plants are attached to integrated iron and steel mills and are primarily meant for captive consumption of coke by their own blast furnaces. There are also *merchant* coke plants which sell coke to various agencies. These plants may vary widely in the annual production, the number of batteries and the number and size of ovens in a battery.

The by-product ovens may be of varying design. While simple ovens burn only rich gas (coke oven gas), there is provision for burning both rich gas and lean gas (blast furnace gas) in compound ovens. The coke oven battery may be supported either directly on the foundation, or on a set of tall columns erected on the foundation. The latter system constitutes the *underjet* ovens.

The method of control of gas and air supply differ in the two systems. In the former type, *gun flues* are required for gas supply. The heating flues in the combustion chamber may be horizontal (Semet–Solvay), or vertical (Koppers cross-regenerative, Simon Carves, Wilputte, Otto, Soviet PVR, Didier, Koppers–Becker and Soviet PK). The vertical flue ovens again differ in the manner of grouping of the flues. The most important consideration in the choice of heating flues is that horizontal flues are points of weakness in the brick structure. The situation gets worse with the large ovens and particularly with the compound ovens which need larger diameter flues for the passage of larger volume of combustion products from the blast furnace gas. Horizontal flues cannot be completely avoided even in the vertical flue systems because the combustion gases need a horizontal passage (bus flue) for movement. Koppers cross-regenerative ovens, Simon Carves (old), Carl Still and Wilputte ovens suffer from this disadvantage. Twin-flues in Otto, Soviet PVR ovens and Simon Carves (new) and crossover-flues in Koppers–Becker as well as the Soviet PK ovens have minimised the weakness of the brick structure by restructuring the horizontal flue length to the minimum. In the twin-flue system the combustion gases of the vertical flue leave the combustion chamber through the neighboring vertical flue (Fig. 4.11). Thus only two neighbouring flues are connected through a very short horizontal passage at the top. In the crossover-flue system the vertical flues are combined into six to seven groups by short horizontal flues at the top and the horizontal flues are each connected with one crossover flue (Fig. 4.12). The crossover-flues pass through the oven roof and serve as the connecting links between the two combustion chambers at the two sides of an oven. All the vertical flues of one of these combustion chambers serve as the heating flues at one time while those of the other combustion chamber act as the passage for the flue gases into the regenerator below. After the reversal, the vertical flues of the two combustion chambers interchange their function.

The operation of by-product ovens is as follows. The charge for the ovens is first prepared by an elaborate procedure based on careful considerations (Fig. 4.13). It consists of a blend of finely ground, say, 100% below 3 mm, and well mixed coals of different caking and coking properties. The coke oven charge is stored in service bunkers attached to the battery. Periodically a charging car receives the required quantity of the coal blend, moves over the rails laid down on the roof of the battery and releases the charge into an empty closed-door oven through the charging holes. The charge is levelled at the top by moving a bar through a hole at the front door.

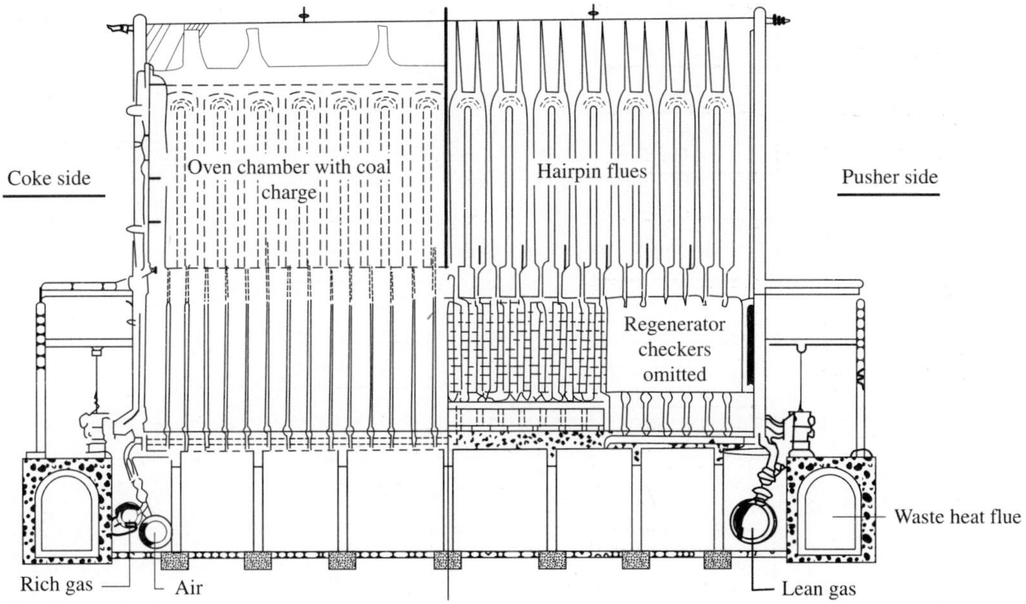

Figure 4.11 The Otto underjet hairpin flue oven

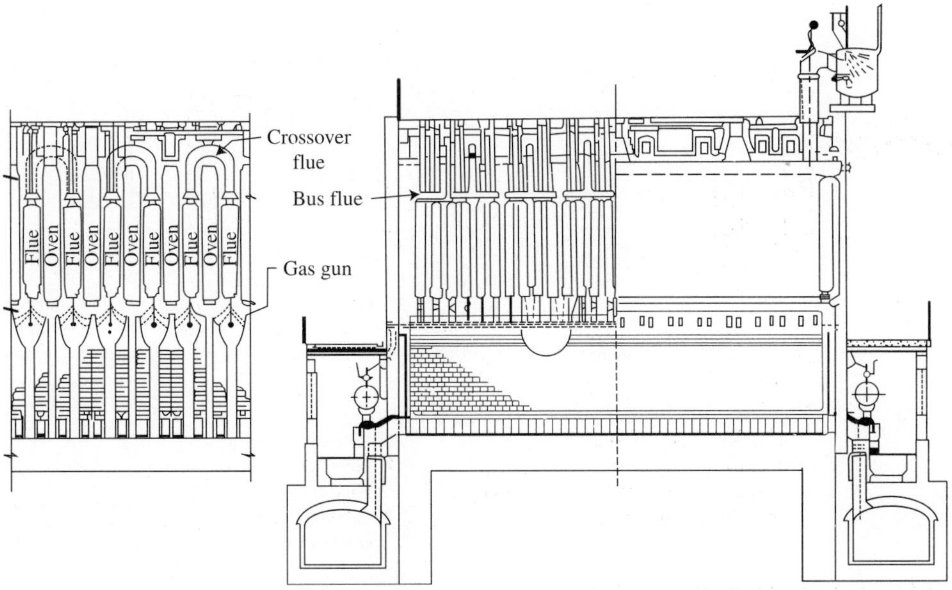

Figure 4.12 Coke oven with gas gun and crossover flue heating system

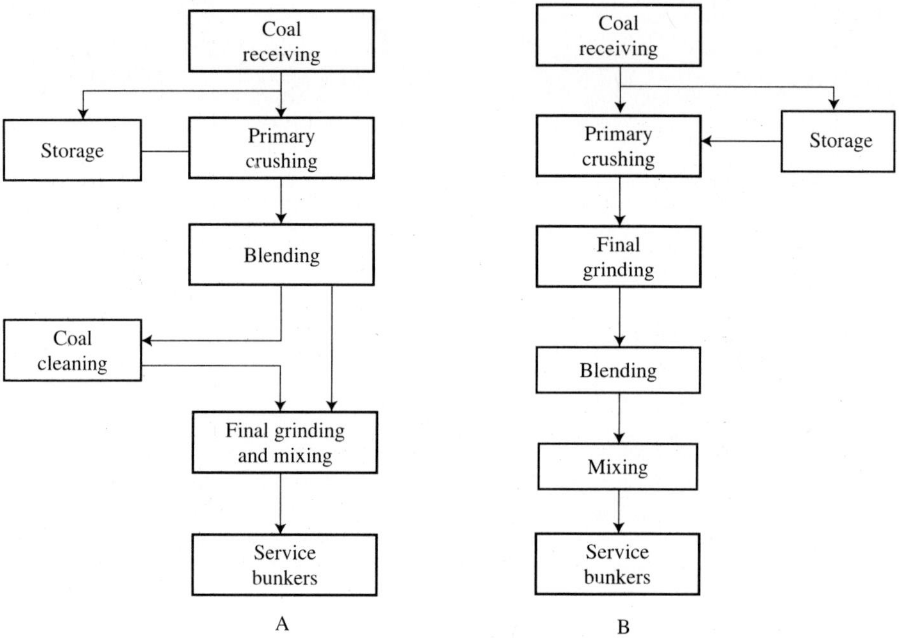

Figure 4.13 Two alternative flow-sheets, A and B, for coal preparation

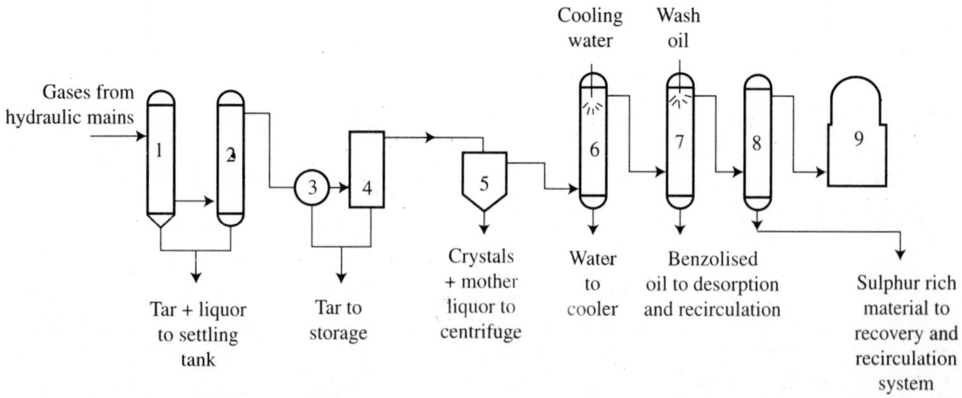

1. Down-comer (Separator)
2. Primary cooler
3. Exhauster (Blower)
4. Electrostatic precipitator
5. Saturator (Acid bath for NH_3 absorption)
6. Final cooler
7. Benzol scrubber
8. Sulphur remover
9. Gas holder

Figure 4.14 Flow diagram of semi-direct process

All holes are then closed. Carbonisation proceeds and the resultant gases and vapours escape through the standpipe for further treatment. After the coking period, usually 14 to 16 hours, the doors are opened and the glowing coke mass is discharged by the machine-driven coke pusher. into the coke-quenching car. The emptied oven is made ready for a fresh charge according to a properly maintained schedule. The hot coke is quickly quenched by a voluminous water spray and then dropped on the coke wharf. Dry quenching is also practised by circulating flue gases over the hot coke. In this case the sensible heat of the coke is partly recoverable in the form of steam raised in waste heat boilers which are run on the heated flue gases. From the wharf the coke is sent for screening and then the screened sizes are properly disposed of.

In modern coke ovens with a flue temperatures of 1,300–1,350°C, the productivity is 40 to 50 kg/m^3 per hour of oven volume and 9–20 kg/m^2 per hour of oven heating wall surface.

4.4.3 Recovery of By-Products

The gases and vapours leave the ovens at 600—700°C and pass a spray of flushing liquor at the goose-neck of the standpipe. The gas temperature drops sharply to about 80°C. Tar and steam are largely condensed here. The liquid and gas streams enter into a big horizontal pipe (hydraulic main or foul main) which is connected with all the ovens by the standpipes. The foul main delivers the streams to the by-product recovery section by a crossover-pipe.

Depending upon how ammonia is recovered, the process of by-product recovery is called *direct*, *indirect* or *semidirect*. In the direct process ammonia is fully recovered in the form of ammonium sulphate. In the indirect process it is recovered fully as an aqueous ammonia solution which may be later converted into the sulphate. In the semi-direct process, ammonia is recovered partly as the aqueous solution and partly as the sulphate. Owing to practical difficulties and production of inferior products, the direct process is not used in practice. Both the indirect and semi-direct processes are widely used. Figure 4.14 gives a flow chart of the semi-direct process.

The crossover pipes deliver the liquid and gaseous streams to a *down-comer* which separates the two streams. The liquid stream of liquor and condensed tar is sent to a *decanter* (settling tank) where they separate into two layers. The heavier layer of liquor goes to the flushing liquor circulation tank. The gases are cooled down to about 30°C in the primary cooler which may work on *direct (packed)* or *indirect (tubular)* principle. Almost all the residual tar vapours in the gas stream are completely condensed here. The water vapours are also largely condensed during the cooling process. A part of the ammonia gets dissolved in the liquor. The condensed tar and liquor are sent to a decanter. The cooled gases enter the *exhausters*.

The exhausters are essentially turbo-blowers ensuring smooth flow of gases in the entire by-product recovery section of the plant. The gases from the exhauster are admitted to the *electrostatic preciptators* which completely remove the minute droplets of tar (tar mist or tar fog) remaining suspended in the gas. The tar-free gas is heated to about 50°C to 60°C in the reheater and then bubbled through a weak sulphuric acid solution in the *saturator*. Reheating of the gas serves to increase its water vapour carrying capacity and keeps the volume of the saturator solution down to a minimum. Ammonium sulphate is formed in the saturator and recovered in the form of crystals. The gases leaving the saturator pass through an acid trap for the removal of the entrained acid droplets and are then cooled in the final cooler (direct type) to about 20–25°C before they are ready for the recovery of crude benzol. The cooled gases are subjected to an absorption treatement in a solvent (*wash oil*) in packed towers (*benzol scrubbers*). The *benzolised* oil is steam-distilled for the recovery of crude benzol. The *debenzolised* oil is recycled.

The gases leaving the benzol scrubbers contain sulphur compounds which are removed by one of the many processes available for the purpose. The *clean* coke oven gas is finally stored in gas holders. If the gas is to be transmitted over long distances through distribution pipes it should be carefully treated for minimising the content of naphthalene and water vapour which can choke and corrode the pipe lines.

The ammonia-containing liquor poses a big handling problem for a coke plant. The problem is lessened to some extent by using the indirect primary cooler in place of the direct type. The liquor may be treated in an ammonia-still for the recovery of ammonia vapours which are either admitted to the main gas stream before entering the saturator for conversion into ammonium sulphate, or condensed into concentrated ammonia solutions.

4.4.4 Coke Ovens in India

Coke was first manufactured in India towards the end of the nineteenth century in beehive ovens for the iron and steel works at Kulti near Raniganj. Since then the number of beehive ovens has been on the increase.

About 5000 beehive coke ovens are located in Jharia, Bokaro–Ramgarh–Karanpura and Raniganj coalfields. The larger batteries are of 20 to 30 beehive ovens each. The coke yield from a Jharia battery of 20 ovens is about 63%. By better control of the combustion conditions the yield may be raised to 72% and the coking time reduced from 72 to 60 hours.

In view of the low capital cost and simplicity of construction and the fact that beehive ovens can be shut down without any damage to the refractories, further installation of beehive ovens may be undertaken for the supply of hard coke to foundries and other consumers.

The first by-product coke plant was built in India in 1909 at Giridih for the then East India Railway. The first crude benzol recovery unit, however, was installed in 1920 at the Bararee coke plant. The others are merchant plants for supplying coke to foundries and to various other consumers. Diverse types of ovens have been built in the country, namely waste-heat and regenerative, horizontal-flue and vertical-flue, gun-flue and underjet, twin-flue and cross-regenerative, compound and rich-gas-fired. Currently the trend is to build twin-flue, underjet, compound ovens in the big modern coke plants of iron and steel plants. The total installed coking capacity is about 20 Mt while the coal throughput is about 14 Mt.

4.4.5 Products of HTC in Coke Ovens

Coke, tar, crude benzol, ammonia (as concentrated aqueous solution or ammonium sulphate crystals) and coke oven gas are usually produced by coal carbonisation in by-product ovens.

The typical yields on a dry basis should be generally, coke 78%–80%, tar 3%, crude benzol 0.8%, ammonia (as sulphate) 1% and coke oven gas 330 Nm^3/t. Other recoverable substances include sulphur (elemental or as sulphuric acid), pyridine, picoline, thiocyanates and ferrocyanides.

Typical yields from a steel plant coke oven in India are (on a dry charge): coke 77.3%, tar 2.8%, crude benzol 0.24% and gas 270 Nm^3/t.

Given the right type of coals, HT coke is much harder than LT coke, hence, the former is termed *hard coke* while the latter is known as *soft coke*. The hard coke produced in the coke plant is screened to various size grades, the nominal sizes being: extra large >100 mm, large 100–25 mm, medium 40–25 mm, small 25–10 mm and fine 10–0 mm. The respective trade names are foundry coke, blast furnace coke, nut coke, pearl coke and coke breeze. The large sizes are used in blast furnaces, foundries and water gas generators. Medium size is also now used in blast furnaces. Excepting coke breeze the other size grades are used as domestic or industrial fuel. The disposal of coke breeze is a problem. It may be used in boilers, sintering furnaces, and also in coke oven charges in small quantities. Typical yields of various size grades of coke on screening in a coke plant in India are: up to medium 82%, small 6% and breeze 12%.

Coal tar forms as a result of the secondary cracking reactions of the primary tar vapours in the high temperature conditions prevailing in the empty space over the charge and on the surface of coke oven walls. It is a complex mixture of aromatic and heterocyclic compounds together with small amounts (less than 5%) of non-aromatic hydrocarbons. Over 500 individual compounds have been identified and many more minor components await identification. The composition varies from tar to tar, depending on the nature of the coal and the coke oven conditions. The concentrations (per cent by weight, dry basis) of the important components in a typical tar are: benzene 0.25, toluene 0.2, xylenes 0.2, phenol 0.6, cresols 1.0, naphthalene 8.0, anthracene 1.0, phenanthrene 6.0, carbazole 1.5 and diphenylene oxide 1.5. Coal tar is rightly called a storehouse of chemicals.

Direct extraction of the components of coal tar is extremely difficult, if not impossible. The commercial practice is to distil the tar into several fractions and then recover the various components either in pure form or as concentrates.

The pure compounds and concentrates obtained fron coal tar are widely used in the manufacture of dye-stuffs, pharmaceuticals, plastics, explosives, motor spirits, solvents, disinfectants, absorbants and timber preservatives. Coal tar and some of its oily fractions act as excellent furnace oils. Pitch is used in surfacing roads and roofs and also in the production of carbon electrodes and other carbon articles.

Like coal tar, crude benzol also differs in composition from plant to plant. It is a mixture of a large number of compounds, of which the following are important: benzene 65, toluene 15, xylenes 5, naphthalene 1.5, carbon disulphide 0.5, and unsaturated compounds (such as cyclopentadiene) 10, all as per cent by weight. These compounds are recovered in various grades of purity and widely used in diverse fields of the chemical industry. The methods of recovery involve distillation and subsequent treatment of fractions.

The coke oven gas is also the result of secondary cracking reactions to which the gases and vapours of carbonisation are subjected and hence hydrogen forms its principal component. A typical composition is H_2 54.0, CH_4 28.0, CO 7.4, N_2 5.6, CO_2 2.0, C_nH_m 2.6, and O_2 0.4%.

The corresponding gross calorific value is 5020 kcal/Nm^3 (dry) and 4450 kcal/Nm^3 (wet). This gas is a good source of hydrogen for the production of ammonia, for example, in the fertiliser plant at Rourkela. Fractions of hydrogen and hydrocarbons are obtained by compression and liquefaction. The hydrocarbons give additional hydrogen by partial combustion or reaction with steam. Methanol industries may also use the hydrogen from coke oven gas. Owing to its high calorific value, the coke oven gas is widely used as a rich gaseous fuel by itself or in admixture with lean gases like blast furnace gas in steel mills and other industries.

4.4.6 Properties of Coke

Coke is a hard, porous, carbonaceous material used mainly as a source of reactive carbon in blast furnaces, foundries and gasifiers and also as a fuel in boilers and heating appliances. Apart from its size, three important properties of coke are its purity, strength and reactivity. The purity is ascertained by determining its moisture, ash, sulphur and phosphorus content. The percentage of volatile matter is essentially a rough guide of the extent of carbonisation. The ash and phosphorus are concentrated in coke during carbonisation while the percentage of sulphur in the fuel may remain practically the same due to the removal of volatile sulphur together with volatile matter of coal. The purity of coke is ascertained only by the purity of the coal carbonised.

There are many tests for evaluating the strength and hardness of coke, of which the Shatter and the Micum tests are the standards in India for evaluating metallurgical coke. In the Shatter test, a coke sample is dropped from a height on to a metal-plate and the hardness is reported as a percentage of oversizes on specified sieves. In the Micum test, the coke samples are put into cylindrical metal drums which are then rotated about the horizonal axis at a definite rate for a definite period; angle irons attached to the inner periphery of the drums carry up some of the coke pieces and drop them from a certain height thus giving a combined effect of abrasion and shatter. The results of screening degraded coke are expressed as indices for the strength and hardness of the original coke. An efficient operation of blast furnace needs minimum voidage of its charge for the heterogeneous reactions between the solid and gaseous materials. Out of the three components of the charge, coke has to play the dominant role in preserving the voidage which is disturbed by the formation of fines during the downward movement of the charge.

There is no standard test for evaluating the reactivity of coke for metallurgical uses. However, a specified minimum porosity indirectly ensures its reasonable reactivity. The porosity is calculated from the true and apparent specific gravities of the sample.

The reactivity of coke for general uses may be determined with respect to oxygen or air, carbon dioxide, steam and sulphur. The critical air blast (CAB) test measures the reactivity to air by determining the minimum rate of air blast which will maintain combustion of closely graded fuel in an ignited bed of specified dimensions. The more reactive the coke the lower the CAB value. This test has been standardised in India together with another test which indirectly gives a measure of the reactivity from the thermal value of the volatile matter of the coke. The more reactive cokes have higher thermal values of their volatile matter.

Table 4.13 Specification for domestic coke (IS 4286-1976)

Characteristic	Class 1	Class 2	Class 3
Volatile matter, per cent dry	3–10	3–10	3–10
Moisture, per cent max	8	8	8
Ash, per cent max	30	40	50

Note: In the case of processed fuel made from lignite, VM not exceeding 15% may be permitted.

The IS specifications for domestic coke and industrial coke are given in Tables 4.11 and 4.13 respectively.

An important property of coke for its storage and transport is its bulk density. The values expressed in kg/m^3 for different types of cokes are given below.

Source	Sized (18–40 mm)	Unsized
Coke oven	417–480	465–513
Horizontal retort	368–417	400–448
Continuous vertical report	337–453	368
Low temperature retort	305	320

It is noteworthy that the bulk density of coal lumps is 640–800 kg/m^3. Thus coke is much lighter than coal and needs much higher capacity in storage and handling for the same weight. The true density of coke is, however, quite high at 1750–2000 kg/m^3.

4.5 BRIQUETTING OF SOLID FUELS

The process of briquetting consists in applying pressure to a mass of particles with or without the addition of a binder and converting it into a compact or agglomerate. The products are called either briquettes containing binders or binderless briquettes, as the case may be. The object of briquetting is to convert a low-grade solid fuel into one of higher quality. Sawdust, peat, lignite, and coal fines, can all be briquetted into lumps of suitable size, shape and strength and used as convenient domestic and industrial fuels. Fuel briquettes in some form or other are almost universally used. Countries manufacturing fuel briquettes on a commercial scale include East Germany, West Germany, the old USSR, France, Holland, the UK, Spain, Belgium, the USA, Hungary, Poland, Czechoslovakia and India. India produces 13.97 Mt per year of lignite at Neyveli and 6.68 Mt per year in Gujarat.

4.5.1 BRIQUETTING OF LIGNITES

The process of briquetting is most widely used in the case of lignites or brown coals. Owing to the low structural quality of lignite, considerable amounts of the fuel are recovered as fines during mining; the coarser pieces disintegrate into fines after air drying. Briquetting is therefore an essential prerequisite for burning lignite in domestic ovens and stoker furnaces. However pulverised fuel combustion as in thermal power stations does not need this preparative step. Binders are not required in the briquetting of lignite and comparatively low pressure suffices in achieving the compaction of its particles. These considerations have contributed to the early commercial success of the process.

The briquetting of lignite involves three steps—grinding, drying and pressing. The formation of a good briquette requires that the grains are of a certain uniform size and optimum moisture content. Under these conditions the particles can approach very close to each other and allow the cohesive forces to come into play when an external pressure is applied. The process variables depend essentially on the nature of the lignite. No universal conditions can be laid down. The outline of the briquetting plant at Neyveli is given below.

Raw lignite is crushed to below 250 mm size in a blade crusher and stored in a bunker. This is further ground to below 50 mm in a roller crusher, and finally to below 8 mm in a hammer mill. The ground, but wet, lignite with a moisture content of 50%–56% is dried in a steam-tube rotary drier to an average moisture content of about 9%–11%. The discharge from the rotary drier has grain size below 6 mm. Before undergoing final grinding, it needs pretreatment which involves screening and redrying of the coarse fraction. The pretreated material is ground to below 1 mm in a hammer mill and then converted into briquettes by passing through quadruple extrusion presses at a pressure of about 1000 kg/cm^2. No binder is used. The optimum temperature for briquetting is about 70°C. Cooling of the dried lignite occurs during its transport from the drier to the press by a conveyor. The temperature rises during the pressing operation, and therefore it is necessary to allow for cooling before the briquettes are disposed of.

The lignite briquettes at Neyveli are carbonised at 650°C to 700°C. After cooling, the carbonised briquettes are sieved over 10 mm screens. The oversize is marketed under the trade name Leco. The undersize char fines are also used as industrial fuel.

4.5.2 Briquetting of Bituminous Coal

Fine coals are produced in large quantities during mining and processing. Before the advent of pulverised fuel combustion, these fines could be burnt in ovens and furnaces only after briquetting. Even today the process of briquetting of coal fines is being increasingly used in many countries since mechanical mining results in enhanced production of fines and the technique of pulverised fuel firing is not suitable for domestic ovens and small industrial furnances.

Briquetting bituminous coal without a binder is difficult. Compared to lignites, much higher pressures and much finer grain sizes are required. Therefore, processes that are commercially successful use a suitable binder like coal tar, pitch or petroleum bitumen. The whole process consists of drying and grinding the coal, mixing it with the binder, beating the mixture, pressing it into moulds and cooling the briquettes before final disposal. The moisture content of the dried coal is maintained at 2%–4%. Pitch and dried coal are often ground together.The grain size of the ground coal is about 80% less than 0.5 mm and 93% less than 0.88 mm. The pitch content is about 7%–8% of the coal. The temperature of the mixture is kept at a few degrees within the softening point of the binder before it is charged into the press. Briquetting pressure is normally about 150 kg/cm^2 . The quality of the briquettes depends upon many factors like the properties of the coal, the properties of the binder, and the precise way in which the various operations are carried out.

Many binderless briquetting processes have been developed. Economic considerations stand in the way of their wide commercial exploitation. Hot briquetting technique consists of preheating a coal blend containing one or more caking coals to a temperature in the plastic range and then compacting the mass. The softened coal acts as the binder and the required briquetting pressure is much lower than in the other binderless briquetting processes for bituminous coals.

4.5.3 Carbonised Briquettes

Briquettes of lignites and bituminous coals are carbonised for producing smokeless lumpy fuel and, to a lesser extent, substitutes for metallurgical coke. The manufacture of carbonised briquettes requires an additional stage over and above that required for the conventional coking practice. However, the process enables the production of coke-like material from lignites, sub-bituminous coals and non-coking bituminous coals. These are called *formed coke*.

There are many processes for the production of carbonised briquettes in various stages of development. The Lurgi–Spülgas process is one of the most successful of them. In this process binderless lignite briquettes are first produced and then carbonised in a three-chamber unit as described in the section on low temperature carbonisation. The carbonisation temperature may vary between 500°C and 850°C depending upon the use of the product as domestic or metallurgical fuel. The higher temperature reduces the volatile matter and considerably increases the strength of the briquettes. Briquettes meant for metallurgical fuel production are made at high pressures, about 2000 kg/cm^2 in ring-roll presses.

Briquettes are susceptible to deformation, fissuration and crumbling during carbonisation. The rate of heating during carbonisation is extremely important. In the case of coal briquettes, the quantity of pitch must not be excessive and the highly coking coals need pretreatment like oxidation or partial carbonisation before briquettes are made from them.

The technique of briquetting has also been employed for the production of agglomerates of fuel with one or both of iron ore and lime. The products, *ferrocoke* and *fluxed coke* are meant for iron production, but are not yet commercial commodities.

4.6 GASIFICATION OF SOLID FUELS

Solid fuels can be gasified by reaction with air, oxygen, steam, carbon dioxide, or mixtures of these, into a product that is suitable for use either as a fuel or as a raw material for making chemicals, liquid fuels or other gaseous fuels. Carbonisation may be considered as a partial gasification process yielding solid, liquid and gaseous products from the initial solid fuel. Most of the gasification processes, however, strive for total gasification of the solid fuel.

The gasification of coal is at present the subject of intense development activity. It is believed that a considerable portion of the coal consumed in future will be first gasified. There are many considerations behind the importance given to gasification: (i) the general advantages of using gaseous fuels (see Chapter 6), (ii) the product gaseous fuel is cleaner and therefore, environmentally more acceptable, and (iii) overall rise in efficiency of use of coal if gasification route is involved.

Gasification processes may be classified in a number of ways:

1. By the method of supplying the heat required

(i) Internal heating
- Autothermic, for example, producer gas
- Cyclic, for example, water gas
- Heat carrying fluids or solids, for example, Koppers process

(ii) External heating or allothermic

2. By the method of contacting reactants

(i) Fixed bed, for example, producrer gas, water gas, Lurgi
(ii) Fluidised bed, for example, Winkler
(iii) Suspension of particles in gasifying medium, for example, Koppers–Totzek

3. By the flow of reactants

(i) Cocurrent, for example, Koppers–Totzek
(ii) Countercurrent, for example, producer gas

4. By the gasifying medium

(i) Air with steam, for example, producer gas
(ii) Steam with oxygen or air or oygen-enriched air, for example, water gas, Koppers–Totzek, Lurgi, Winkler
(iii) Hydrogen in hydrogasification

5. By the condition of the residue removed

(i) Dry ash in nonslagging operation (all the common gasification processes fall in this group)
(ii) Slag in slagging operation

Some of the more important gasification processes will be taken up later in Chapter 6.

4.7 LIQUEFACTION OF SOLID FUELS

In analogy to gasification, the term liquefaction of solid fuels is used to cover all the processes that convert solid fuels into liquid products which may be either used directly as fuel, or converted into chemicals or other liquid fuels. Owing to the recovery of crude benzol or coal tar, the carbonisation process may be described as partial liquefaction of coal. The production of shale oil from shales by heating in retorts also falls in this category.

Hydrocarbon-type liquid fuels are obtained from solid fuels by the following routes:

1. Hydrogeneration of coal, for example, Bergius–Pier process,
2. Solvent extraction of coal and hydrogenation of coal extracts,
3. Refining of tar and oil obtained by the carbonisation of various solid fuels and oil shales,
4. Gasification of solid fuels into synthesis gas and conversion of the latter into liquid fuels and chemicals, for example, Fischer–Tropsch process and methanol synthesis.

Much work has been going on to develop these processes into economical propositions. The world has only one Fischer–Tropsch plant, SASOL in South Africa, and a few bituminous and brown coal tar refining plants in East Germany, West Germany and Czechoslovakia. Chapter 5 includes a treatment of these and other processes for making liquid fuels from solid fuels.

It is a century-old concept that coal fines may be dispersed in fuel oil and the resultant *colloidal fuel* may be burnt like a liquid fuel with an added advantage of high flame emissivity. This dormant concept has been revived in the seventies to conserve petroleum fuel, and now the liquid medium is offered by water, oil, methanol or a mixture of any two or all of them. All technological problems appear to have been nearly solved for the coal–liquid mixture (CLM) to be a viable alternative fuel to partially displace fuel oil and gas.

Coal should be in a very fine state: for example, 80% below 75 μ and 100% below 200 μ. This is difficult for Indian coals with poor grindability. However, larger sizes may be used with suitable dispersants and other additives. Coal loading of 50%–80% has been achieved—the higher figure with coal–water mixtures (CWM) and the lower with coal–oil mixtures (COM). Coals of any rank may be used. CWM system may be easily integrated with a coal preparation plant and its high potential lies in the fact that the high ash middlings and tailings may be used as CWM in fluidised bed combustion systems. COM is already commercialised in many countries. CWM is also a mature technology approaching commercialisation rapidly. It has a potentially lower cost than coal liquefaction and gasification due to the lower capital and operating costs. The main application of COM is in boilers and furnaces. It has a high potential as injection fuel for blast furnaces and as feedstock for gasifiers. Heat engines like stationary diesel engines are a potential target; the coals have to be extra-clean at 2% ash.

Example 1

The float-and-sink test data for an Indian coal and European coal are given below.

	A European coal crushed below 4 mm		An Indian coal crushed below 3 mm	
Specific gravity	yield per cent	ash per cent	yield per cent	ash per cent
Below 1.30	37.7	3.2	1.2	2.1
1.30–1.40	31.3	5.8	6.0	9.6
1.40–1.50	2.9	24.9	38.4	19.6
1.50–1.60	2.8	32.3	43.0	25.2
1.60–1.70	1.2	44.1	6.6	34.2
1.70–1.80	0.7	57.5	2.5	41.6
Above 1.80	23.4	78.6	2.3	57.4
Total	100.0	23.6	100.0	23.6

Without drawing the washability curves, what information can you extract from these data on the ease or difficulty of washing the two coals?

Solution

Yields of fractions between 1.40 and 1.60 specific gravity

European coal: 2.9 + 2.8 = 5.7%

Indian coal: 38.4 + 43.0 = 81.4%

This shows that the Indian coal (even after crushing below 3 mm) contains large quantities of particles where organic mass and mineral matter are intimately associated. The European coal contains a very small portion of such particles. Therefore minerals can be easily separated from the European coal while this is not so in the case of the Indian coal. This is confirmed by the high yield of 69% with ash below 5.8%, obtainable from the European coal at a specific gravity of 1.40. The intimately associated nature of the mineral matter in the Indian coal is evident from the high ash value between 19.6% and 25.2% on the 1.40–1.60 fraction with a high yield of 81.4%.

If 1.50 is chosen as the specific gravity of washing, the fraction 1.40–1.60 forms the near gravity material. Table 4.2 shows that the coal washing problem will be *simple* for the European coal and *formidable* for the Indian coal. However, by tolerating high ash in the clean product if we choose 1.7 as the specific gravity of washing for the Indian coal,

ngm = 6.6 + 2.5 = 9.1%

Therefore the coal washing problem becomes *moderately difficult*.

Ash of the product at 1.70 specific gravity.

$$= \frac{1.2\times0.021+6.00\times0.096+38.4\times0.196+43\times0.252+6.6\times0.342}{1.2+6.0+38.4+43.0+6.6}\times100$$

$$= 22.3\%.$$

Example 2

In a central washery in India with a feed coal of 21.0% ash content, the clean product has an ash content of 18.7% and the sinks have an ash content of 41.2%. If the theoretical recovery is 91.6% for a product of the same ash level, estimate the performance of the washery.

Solution

Using the ash balance, the plant yield of clean product

$$= \frac{41.2-21.0}{41.2-18.7}\times100 = 89.8\%$$

Therefore, Yancey and Geer efficiency $= \dfrac{89.8}{91.6} \times 100$

$= 98.0\%$

PROBLEMS

1. The float-and-sink analysis of an Indian coal is given below. Construct the three important washability curves and comment on the washability characteristics of the coal.

Specific gravity	Yield per cent	Ash per cent
Below1.30	8.57	3.9
1.30–1.40	47.25	12.6
1.40–1.50	13.31	21.8
1.50–1.60	12.13	30.4
1.60–1.70	12.04	40.1
1.70–1.80	3.30	49.9
Above 1.80	3.40	61.3

2. The following table gives the float-and-sink analysis of a European coal. After constructing the washability curves, compare its washability characteristics with those of the Indian coal given in Problem 1.

Specific gravity	Yield per cent	Ash per cent
Below 1.30	66.88	3.75
1.30–1.40	7.22	11.70
1.40–1.50	5.71	21.75
1.50–1.60	2.55	30.70
1.60–1.80	3.44	41.40
Above 1.80	14.20	73.50

3. For the two coals in Problems 1 and 2, choose the suitable specific gravity for washing and estimate the ease/difficulty of washing from the amount of near-gravity material. (Use Table 4.2).

4. On washing in a plant, the coal of Problem 2 gave a product of 7.3% ash at a yield of 79.5%. Evaluate its performance.

   [Answer: Yancey and Geer η = 94.4%]

5. The feed to a coal washing plant contains 20.9% ash and its washability curve indicates a yield of 77.5% clean coal with 7.9% ash at the chosen specific gravity. However the actual product has an ash of 9.0% and the rejects have an ash of 67.2%. Evaluate the performance of the washery. (Calculate the plant yield from the ash balance and then the washery efficiency from the first two formulae. For calculating yield at the same ash level, assume linear relation between yield and ash).

   [Answer: Fraser and Yancey η = 94.0%,
   Yancey and Geer η = 90.2%]

6. The float-and-sink test data of an American coal are given below. Washing in a jig gave a char product of 13.9% ash and refuse of 53.4% ash. Rate the performance of the jig.

Specific gravity	<1.30	1.30–1.40	1.40–1.50	1.50–1.60	1.60–1.70	1.70–1.80	>1.80
Yield of fraction	16.8	31.2	12.0	3.8	5.0	4.6	22.1
Ash of fraction	4.3	11.2	23.5	32.6	43.1	51.0	71.4

[Answer: Yancey and Geer η = 88.4%]

7. An Indian coal gave the following float-and-sink data. A cyclone washer produced a product of 19.3% ash with a yield of 53.0%. How did the cyclone perform ?

Specific gravity	<1.30	1.30–1.40	1.40–1.50	1.50–1.60	1.60–1.70	1.70–1.80	>1.80
Yield of fraction	4.7	11.4	35.7	25.7	7.6	5.2	9.7
Ash of fraction	4.5	14.0	21.4	30.3	36.3	44.0	64.5

[Answer: Yancey and Geer η = 90.5%.]

Bibliography

1. Bureau of Indian Standards, New Delhi.
   a. IS: 437 (1979) Size grading of coal and coke for marketing
   b. IS: 1354 (1992) Test for coke – special tests.
   c. IS: 4023 (1991) Reactivity of coke.
   d. IS: 7929 (1975) Determination of electrical resistivity of chemical coke.
2. Coal Carbonisation, *The Wealth of India, Industrial Products*, Pt. II,128, 1951.
3. Coal Tar, *The Wealth of India, Industrial Products*, Pt. II,142, 1951.
4. Ghosh SR, Sarkar S, Das Gupta NN and Lahiri A. Properties of coal and coke for metallurgical uses, *Journal of Mines, Metals and Fuels*, Special Issue on Coal Carbonisation, 305, 1959.
5. Grainger L and Gibson G. *Coal Utilisation – Technology, Economics and Policy*, London: Graham and Trotman, 1981.
6. Lowry HH, (ed.). *Chemistry of Coal Utilisation*, Vol. 1, New York: Wiley,1945.
7. Ratnam CVS. Chemical engineering in the utilisation of lignite, *Chemical Age of India*, Vol..20 (2), 129, 1969.
8. Sarkar S. Kinetics of coal carbonisation reactions in the plastic state, *J. Sci. Ind. Res.*, 19B: 246–8. 1960
9. Sarkar S. Rational utilisation of India's coal resources, *Urja*, 143, Aug. 1987.
10. *Studies on Coal Preparation in India*. Central Fuel Research Institute, Dhanbad,1972
11. Tsai SC. *Fundamentals of Coal Beneficiation and Utilisation*. Amsterdam: Elsevier, 1982.
12. Whitaker JW and Sarkar GG. *Studies in Coal Washability*, Parts1, 2, 3 (1950), Parts 4, 5, 6 (1951), Parts 7, 8, 9 (1953). Central Fuel Research Institute, Dhanbad.
13. Wilson PJ and Wells JH. *Coal, Coke and Coal Chemicals*, New York: McGraw Hill, 1960.

5 Liquid Fuels

5.1 INTRODUCTION

The importance of liquid fuels is best exemplified by the fact that almost all internal combustion engines run on them. Life would come to a standstill if we are denied this commodity, either from nature or from synthetic sources. The use of liquid fuels in heat generation in ovens and furnaces is expanding rapidly. The development of highly efficient oil burners of widely varying burning rates has made this possible. Furthermore, we are increasingly dependent on petroleum as a source of carbon and hydrogen for the production of chemicals, fertilisers, dyestuffs, plastics, detergents and fibres.

Petroleum accounts for the bulk of the liquid fuels. The other liquid fuels in use are coal tar, crude benzol, synthetic liquid fuels made from coal, shale oil, alcohol and oil from tar sands.

5.2 PETROLEUM

The term *petroleum* means oil obtained from the earth. Mineral oil, crude petroleum, crude oil, crude or simply oil are other names for petroleum. Petroleum varies widely in appearance. It may be a straw-coloured liquid or even a black, viscous material. Most petroleums, however, are dark-coloured, highly mobile liquids in which gas is dissolved and solids are both dissolved and dispersed. Sometimes, the gas separates naturally and gives rise to reserves of natural gas. Once the liquid and gas are removed, the solids are separated on exposure to the surface of the earth and a natural asphalt deposit results. Harder types of natural asphalt are readily mistaken for a sapropelic coal.

Petroleum occurs in a reservoir rock which has fluid holding capacity, porosity and fluid transmitting capacity and permeability. Most reservoir rocks are sedimentary rocks consisting of coarse-grained minerals, such as, sand, sandstone, grit, limestone and dolomite. The upward migration of the fluid from the reservoir rock is prevented by an upper stratum called *cap rock* or sealing formation. A cap rock has very fine pores and low permeability. Typical cap rocks are clays and shales. The migration of the fluid in other directions is prevented by suitable geological formations of the reservoir rock and cap rock. These are known as *oil traps*. In principle, the boundary between the cap rock and the reservoir rock should be convex upward. The exact form varies widely. The simplest forms are flat-lying convex lens and the anticline and dome. The trapping of oil and gas is facilitated by the presence of water or brine seals (Fig. 5.1).

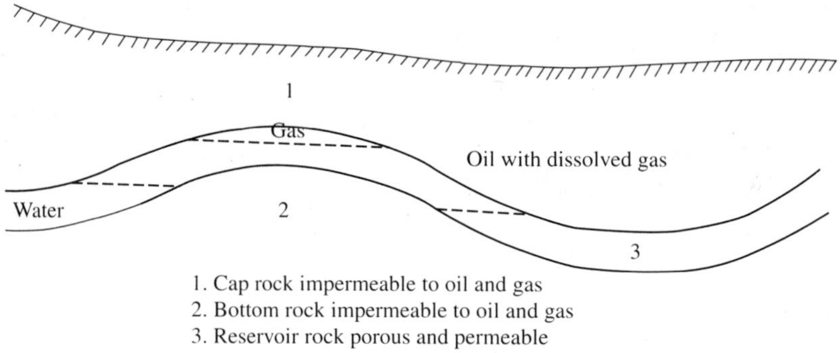

Figure 5.1 Vertical cross section of an anticlinal accumulation of oil and gas

Table 5.1 Reserves of crude oil in India (million tonnes)

Area	1990	2000	2002	2003	2004	2005	2006	2007
Onshore	307	317	332	339	357	376	387	357
Offshore	432	386	409	422	382	410	369	369
Total	739	703	741	761	739	786	756	726

Source: Ministry of Petroleum and Natural Gas, Govt. of India, Petroleum Statistics, http://petroleum.nic.in/petstat.pdf

Exploration for oil reservoirs is conducted by geological and geophysical techniques. Once promising rock structures are confirmed, the actual presence of oil is established by drilling. Usually, several deep bores are drilled to prove the existence or absence of a commercial reservoir. Petroleum deposits occur in most parts of the world but commercial quantities are produced mainly in a wide belt stretching from Indonesia through Myanmar (Burma), India, the Middle East and Central Europe to North and South America. The most widespread distribution is in the continent of North America, throughout the United States, Canada and Mexico. The major oil fields are located in the Middle East (mainly Saudi Arabia, Iran, Kuwait, Iraq and Abu Dhabi) accounting for about 56.3% of the world proved reserves of petroleum of about 1.317 billion barrels (Table 1.2 and Fig. 1.2). The other major oil-bearing countries are Russia 8.6%, Mexico 7%, USA 5%, Venezuela 3.6%, Libya 3% and Nigeria 2.3%.

Petroleum production in India

About 1.4 Mkm^2 or about 42% of total land area in India consists of oil yielding sedimentary rocks. The offshore sedimentary basin exists up to a water depth of 200 m with an area of 0.4 Mkm^2. India's first oil well, 31m deep, was drilled with a steam-powered drill near an oil seepage at Nahar Pong in Assam in 1866. It failed to strike oil. On 26 March 1867, oil was struck at 35.5 m at Makum in the Digboi area. A few more wells were dug and commercial production was successful by1889—only 30 years after the drilling of the Drake well in the

Table 5.2 Production of crude oil in India (thousand tonnes)

Item	2000–01	2001–02	2002–03	2003–04	2004–05	2005–06	2006–07* (April–Dec)
(a) Onshore							
Gujarat	5815	6002	6042	6131	6187	6251	4697
Assam	5199	5095	4660	4592	4703	4474	3329
Arunachal Pradesh	78	69	74	77	83	104	81
Tamil Nadu	436	440	395	375	391	385	267
Andhra Pradesh	263	283	300	281	226	216	185
Total (a)	**11791**	**11889**	**11471**	**11456**	**11590**	**11430**	**8559**
Of which							
OIL	3286	3182	2951	3002	3196	3234	2351
ONGC	8428	8636	8445	8380	8320	8095	6087
JVC/Private	77	71	75	74	74	101	121
(b) Offshore							
ONGC (Mumbai High)	16629	16074	17560	17677	18165	16309	13409
JVC/Private	4006	4069	4013	4240	4226	4451	3516
Total (b)	**20635**	**20143**	**21573**	**21917**	**22391**	**20760**	**16925**
Grand total (a+b)	**32426**	**32032**	**33044**	**33373**	**33981**	**32190**	**25484**

*Provisional

Source: *Annual Report 2006–2007*, Ministry of Petroleum and Natural Gas, Govt. of India

USA. Initially the crude was refined in a small refinery at Margherita. The first full-fledged refinery was commissioned at Digboi in 1899, with a capacity of about 90 kl per day. Through the exploration work of government and semi-government agencies, several more oil fields have been discovered—Ankleswar, Kalol, Sanand, Nawagam, North Kadi, and others in Gujarat, Moran-Nahorkatiya, Hugrijan, Rudrasagar, Lakwa and Galeki in Assam, Bombay High and Cambay in the western offshore, Palk Strait in the south and the Godavari offshore in the east. Exploration is going on to ascertain the proved and recoverable oil deposits in the country.

The story of Bombay High [ONGC (Mumbai High)] is fascinating. The first step in offshore exploration was taken by the Oil and Natural Gas Commission (ONGC) in 1962 in the Gulf of Cambay. The Bombay High was delineated by seismic surveys during 1964–67. Then the pace quickened and detailed seismic surveys were carried out. In view of the difficult bydrometeorological and sea-bed conditions, foreign experts were consulted. The jack up drilling rig, Sagar Samrat, arrived in Bombay in 1973.

Commercial production was started in May 1976 at the rate of 4300 barrels of oil per day (BOPD) or 0.2 million tpy. The contribution of Bombay High led to a sharp rise in the production figure. Table 5.1 gives the crude oil reserves up to 2007. The total reserves in 2007 stand up to 726 million tonnes, out of which 49% was in the onshore area and the remaining 51% was in the offshore area.

Table 5.3 Company-wise locations and capacity of refineries as on 1 July 2005 (million metric tonnes per annum)

Sl. No.	Name of the company	Location of the refinery	Capacity (mmtpa)
1.	Indian Oil Corporation Ltd. (IOCL)	Guwahati	1.00
2.	IOCL	Barauni	6.00
3.	IOCL	Koyali	13.70
4.	IOCL	Haldia	6.00
5.	IOCL	Mathura	8.00
6.	IOCL	Digboi	0.65
7.	IOCL	Panipat	6.00
8.	Hindustan Petroleum Corporation Ltd. (HPCL)	Mumbai	5.50
9.	HPCL	Visakhapatnam	7.50
10.	Bharat Petroleum Corporation Ltd. (BPCL)	Mumbai	6.90
11.	Chennai Petroleum Corporation Ltd. (CPCL)	Manali	9.50
12.	CPCL	Nagapattnam	1.00
13.	Kochi Refineries Ltd. (KRL)	Kochi	7.50
14.	Bongaigaon Refinery & Petrochemicals Ltd. (BRPL)	Bongaigaon	2.35
15.	Numaligarh Refinery Ltd.(NRL)	Numaligarh	3.00
16.	Mangalore Refinery & Petrochemicals Ltd. (MRPL)	Mangalore	9.69
17.	Tatipaka refinery (ONGC)	Andhra Pradesh	0.078
18.	Reliance Petroleum Ltd. (RPL) (Pvt. Sector)	Jamnagar	33.00
	Total		**127.37**

Source: Ministry of Petroleum and Natural Gas, Govt. of India, http://petroleum.nic.in/refi.htm

The state-wise crude oil production is shown in Table 5.2. The total production in 2005–06 was 32.190 Mt, out of which Bombay High alone accounted for 16.309 Mt (or 50.7%). Among the onshore companies, Oil India Limited accounted for about 3.2 Mt in 2005–06 and ONGC for about 8.1 Mt.

It is of interest to note from Table 5.2, compared to Table 5.1, that the crude oil production in the country is very low in comparison to total reserves. There is thus a high scope for expansion in the production figures by expanding the existing refineries/establishing new refineries. According to the report of the Ministry of Petroleum and Natural Gas, 2005, there were 18 refineries in the country with a total capacity of 127.4 Mt per year, the throughput being as high as 127.1 Mt so that the per cent utilisation of the capacity was very high (99.8). Table 5.3 gives the break-up of the refineries in India. Gujarat has the two largest refineries: Reliance Industries Limited at Jamnagar (33.0 Mt per year) and the India Oil Corporation at Koyali (13.7 Mt per year). Mention may also be made of the Mangalore refinery in Karnataka (9.7 Mt per year) and the Manali refinery in Tamil Nadu (9.5 Mt per year). The Haldia refinery in West Bengal is also mentionable (6.0 Mt per year).

5.2.1 Origin of Petroleum

The origin of petroleum is not satisfactorily proved. A large number of theories have been suggested from time to time. According to some, petroleum is of inorganic origin, being formed by the action of water on metallic carbides or by the interaction of water and carbon monoxide on alkali or alkaline earth metals. According to others, petroleum was formed from organic sources, animal or vegetable, by a process of destructive distillation. All these theories have been rejected because they require some improbable materials or processes.

The vast pool of information gathered by petroleum geologists, chemists and geophysicists has led to the emergence of some rational views on the genesis of petroleum. For the sake of convenience four phases of petroleum genesis are recognised:

- embeddment of organic matter in source beds,
- conversion of source material into a fluid,
- migration of oil into the reservoir rock,
- secondary transformation of oil in the reservoir rock.

It is generally accepted that crude has been derived from the organic matter originally present in marine sediments. The probable sources are plankton, marine algae, sea grass and larger marine animals. Some terrestrial and freshwater plants may also have taken part in petroleum formation. The absence of a reasonable amount of phosphorus in petroleum is an argument against the involvement of fish in the processes. The dead organic matter settles down to the bottom of shallow seas and lagoons. The settling process is facilitated by the presence of mineral particles transported by flowing water to places where the current is less strong. The settled debris is attacked by anaerobic bacteria. Most of the organic compounds are destroyed and lost. Unsaturated fatty oils and fatty acids which survive the bacterial attack undergo polymerisation into insoluable solids. These are buried under the steadily increasing cover of sediments, while anaerobic bacteria continue to act on them. In the process of compaction, the water content diminishes from 70%–80% to 10% or less. As a result of continued bacterial decomposition under anaerobic conditions (reducing atmosphere), the decarboxylation and other reduction processes take place and the composition of the organic matter becomes more and more petroleum-like.

The formation of oil from the consolidated organic material of the source bed needs a considerable degree of degradatioin. The presence of thermo-labile and optically active substances like porphyrines and some nitrogen and sulphur containing complexes in crudes rule out the involvement of high temperatures, such as, above 200°C, in the genesis of petroleum. The probable processes by which the degradation of the source material takes place are

- anaerobic bacterial action,
- low temperature cracking in presence of clay and mineral catalysts,
- irradiation by radioactive materials associated with sedimentary rocks.

Low temperature catalytic cracking is believed to be the major process by which oil was produced from consolidated source material.

Table 5.4 Installed capacity and refinery crude throughput (thousand tonnes)

Refinery/Location	Installed capacity as on		Refinery crude throughput						
	1.4.226	1.4.2007	1990–91	2000–01	2002–03	2003–04	2004–05	2005–06	2006–07*
(a) PUBLIC SECTOR	**99468**	**105468**	**51772**	**77411**	**82015**	**89495**	**93107**	**96946**	**108172**
IOC, Guwahati, Assam	1000	1000	783	707	458	891	1002	864	839
IOC, Barauni, Bihar	6000	6000	2416	3122	2994	4304	5082	5553	5469
IOC, Koyali, Gujarat	13700	13700	9334	12006	12434	12758	11698	11543	12953
IOC, Haldia, West Bengal	6000	6000	2835	3873	4513	4518	5418	5502	5836
IOC, Mathura, Uttar Pradesh	8000	8000	7808	7133	8207	8248	6387	7938	8883
IOC, Digboi, Assam	650	650	566	678	581	602	651	615	586
IOC, Panipat, Haryana	6000	12000 $	0	5707	6101	6338	6390	6507	9435
Total IOC	**41350**	**47350**	**23742**	**33226**	**35288**	**37659**	**36628**	**38522**	**44001**
BPCL Mumbai, Maharashtra	12000	12000	6957	8683	8711	8757	9138	10298	12030
HPCL, Mumbai, Maharashtra	5500	5500	5766	5575	6078	6108	6118	6249	7419
HPCL. Visakh, Andra Pradesh	7500	7500	3464	6405	6851	7591	8121	7980	9377
Total HPCL	**13000**	**13000**	**9230**	**11980**	**12929**	**13699**	**14239**	**14229**	**16796**
KRL, Kochi, Kerala	7500	7500	5006	7520	7580	7854	7924	6939	7742
CPCL, Manali, Tamil Nadu	9500	9500	5698	6046	6176	6387	8181	9680	9784
CPCL Narimanam, Tamil Nadu	1000	1000	0	579	643	653	742	682	618
Total CPCL	**10500**	**10500**	**5698**	**6625**	**6819**	**7040**	**8923**	**10362**	**10402**
BRPL, Bongaigaon, Assam	2350	2350	1139	1488	1463	2126	2311	2356	2067
NRL, Numaligarh, Assam	3000	3000	0	1451	1879	2200	2042	2133	2504
ONGC, Tatipaka, Andhra Pradesh	78	78	0	0	93	91	93	93	94

Table 5.4 *Continued*

Refinery/Location	Installed capacity as on		Refinery crude throughput						
	1.4.226	1.4.2007	1990–91	2000–01	2002–03	2003–04	2004–05	2005–06	2006–07*
MRPL, Mangalore, Kamataka	9690	9690	0	6438	7253	10069	11809	12014	12536
(b) PRIVATE SECTOR	**33000**	**43500**	**0**	**26033**	**30544**	**32345**	**34309**	**33163**	**38379**
RPL, Jamna-gar, Gujarat	33000	33000	0	26033	30544	32345	34309	33163	36616
ESSAR Oil Ltd, Vadinar	>	10500						1763	
Total (a+b)	**132468**	**148968**	**51772**	**103444**	**112559**	**121840**	**127416**	**130109**	**146551**

* Provisional

Source: Ministry of Petroleum and Natural Gas, Govt. of India, Petroleum Statistics, http://petroleum.nic.in/petstat.pdf

With the increase in the overburden, the source rocks are subjected to compression and cause the liquids to migrate to the reservoir rocks. During migration the composition of oil may change by filtration, solution or adsorption. Both vertical and lateral migrations are possible. Oil reserves were built up by migration of oil from a number of source beds. After reaching the reservoir rocks, oil may or may not have undergone significant changes in its composition, depending upon its early history and the nature of the reservoir rock. The agencies for secondary transformations are radioactivity, bacterial action and catalytic influence of rocks. An additional factor in the case of shallow layers that are imperfectly sealed at the top is evaporation.

The reservoir rocks have been found to belong to widely varying geological eras, namely palaeozoic, mesozoic, teritiary and quanemary, (see age of coal deposits, Table 3.16). The bulk of the oil fields are from the mesozoic (120–200 million years) and tertiary (oilgocenemiocene periods, 20–40 million years) eras. Although it has not been possible to establish the precise ranks or degree of maturity of crudes of different geological ages, it appears from general observation that crudes from older reservoir rocks are lighter and more simple in structure and yield larger amounts of distillates compared to younger crudes. There are, however, exceptions to this. It has also been generally found that oils become lighter in going basinward in any horizon (Hilt's rule).

5.2.2 Petroleum Production

Occasionally oil may be obtained from surface outcrops, from hand dug shallow pits or from seepage into mine galleries. However, the usual production is only from wells. The method that is mostly used for digging a well is the *rotary drilling* system which utilises the principle of torsional grinding or cutting. The equipment assembly is called a drilling rig and it consists of four main parts:

- *Derrick*, a tall steel structure equipped with hoisting gear for raising and lowering the drilling equipment;
- *Drill column* or *stem* consisting of a main section of steel pipes (*drill pipe*) with at its lower end some extra heavy pipe (*drill collars*), to which the bit is attached;
- *Rotating mechanism* at the upper end of the drill column, that is at the surface, which is a hollow bar of square or hexagonal cross section (the *kelly* or grief *stem*), and this slides through a square or hexagonal hole in the machine (rotary table) which provides the rotary movement;
- *Mud circulation system* at the top of the kelly is the so-called swivel through which a thin mud (drilling fluid) passes from a flexible hose; the bottom half of the swivel revolves with the kelly while the top does not, and the flexible hose is attached to it. The top part of the swivel is also attached to the hoisting mechanism.

The hoisting system is controlled from a *working platform* built within the derrick above ground level. A central power unit drives the various moving parts usually by mechanical power transmission. The heart of the fluid circulation system is the mud pump plant. The mud is pumped down the drill pipe through the holes of the bit and returned to the pump section via the annular space between the drill pipe and the walls of the bore hole. The mud brings the bit cuttings with it. It is thereafter passed through a vibrating screen and/or settling pits before it returns to the pump section tank.

An efficient mud circulation is perhaps the most important single factor required to drill an oil well rapidly without accidents and difficulties. The drilling fluid serves four principal functions, namely (i) cools the bit, (ii) flushes out the cuttings, (iii) supports the bore wall by hydraulic pressure, (iv) penetrates porous rock to form a seal of mud cake and (v) prevents the escape of oil or gas through pores and cracks. Drilling muds are essentially colloidal clay suspensions with additives to control viscosity.

The main boring is of 20–30 cm in diameter and the depth of a well may vary from 1.5–4.5 km. Wells of more than 7.5 km have also been drilled while quite shallow wells also exist. The entire well is provided with a casing of steel pipes right upto the productive work layer. This prevents the collapse of the bore walls and ensures that other extraneous fluids do not enter the well. A steel tubing is suspended from the well-head up to the oil-bearing rock. It is provided with control valves and outlet pipe at the well-head.

Oil is normally under pressure in the reservoir. The source of the pressure may be a body of compressed gas above the oil or a mass of water underlying the oil. These pressures cause the oil to flow towards the foot of the well and drive it up the tubing to the surface (natural flow). When oil pressure is low or it falls as the oil well is depleted, the lifting may be done by sending gas at high pressure through the annulus between the casing and the tubing, or by providing a suitable pumping device at the lower end of the tubing. *Water flooding* is another method of secondary recovery of oil; water is injected into the reservoir rock through separate wells on the periphery of the oil field.

The total production from each well is piped to a *gathering station* meant for collecting the production from a number of wells in the area. The gas is separated by directing the fluid against baffles in cylindrical pressure vessels. The water and solids are separated either by settling out in tanks or by centrifuging. The gas is called *wet natural gas*. It is processed for the recovery of LPG and natural gasoline and put to other uses. The crude is sent to the refinery. Rarely is crude oil directly used.

5.2.3 Composition of Petroleum

Unlike coal, the ultimate analysis of crude petroleum varies within a narrow range; carbon 83%–87%, hydrogen 11%–14%, sulphur 0.5%–3%, nitrogen 0.1% (rarely up to 1%) and oxygen 2%–3%. The ash of crudes is only 0.1% or less. Practically all metals have been found in the petroleum ash; the most widely occurring elements are silicon, iron, aluminium, calcium, magnesium, nickel and sodium. A few crude oils contain appreciable quantities of vanadium.

Crude petroleum is essentially a mixture of hydrocarbons. Paraffins, naphthenes (cycloparaffins) and aromatics are present in all crudes in widely varying proportions. Only olefins and acetylenes are absent. Olefins are formed in the processing of petroleum. Only the lower boiling fractions of crudes contain different types of hydrocarbon as such. The higher molecular weight hydrocarbons include complex structures containing aromatic and naphthenic rings with paraffinic side chains. Sulphur, nitrogen and oxygen are present in the form of derivatives of hydrocarbons, for example, aliphatic mercaptans and disulphides, aromatic sulphides, polyalkylated quinolines, isoquinolines and pyridines, fatty acids, naphthenic acids, phenols and complex asphaltenes and resins.

5.2.4 Classification of Petroleum

The great variations in the nature and concentration of petroleum components render it very difficult to evolve a simple but satisfactory classification system. However, a three group classification is sometimes used by petroleum refiners for commercial practices. According to this system, the following three types of crudes are recognised:

1. *Paraffin-base crude* which produces a residue containing both wax and asphalt,
2. *Asphalt-base crude* which yields a black and lustrons residue of asphalt,
3. *Mixed-base crude* which produces a residue containing both wax and asphalt.

A misnomer 'naphthene-base' is sometimes used in lieu of asphalt-base in the mistaken belief that the asphalt-base crudes consist largely of naphthemic hydrocarbons. About 90% of the crudes produced at present fall in the category of mixed-base.

The above classification is based on the principle that the nature of the residue determines the nature of the whole crude. This is only of limited validity. Moreover, the grouping of the bulk of the crudes under mixed-base also give an incomplete picture. These considerations have led to the development of a number of improved systems. Out of them, the classification of Lane and Garton is most widely used (Table 5.5). This is briefly discussed overleaf.

Table 5.5 Classification of crude petroleum according to Lane and Garton

Sl.No.		Key fraction 1 Sp.gr.60°/60°F 15.5°/15.5°C °API	Key fraction 2 Sp.gr. 60°/60°F 15.5°/15.5°C °API
1.	Parafffin	0.8251 or less (40 or more)	0.8762 or less (30 or more)
2.	Paraffin-intermediate	0.8251 or less (40 or more)	0.934–0.8762 (30 to 20)
3.	Paraffin–naphthene	0.8251 or less (40 or more)	0.934 or more (20 or less)
4.	Intermediate	0.8251–0.8601 (40 to 33)	0.934–0.8762 (30 to 20)
5	Intermediate-paraffin	0.8251–0.8602 (40 to 33)	0.8762 or less (30 or more)
6	Intermediate-naphthene	0.8251–0.8602 (40 to 33)	0.934 or more (20 or less)
7	Naphthene	0.8602 or more (33 or less)	0.934 or more (20 or less)
8	Naphthene–paraffin	0.8602 or more (33 or less)	0.8762 or less (30 or more)
9.	Naphthene-intermediate	0.8602 or more (33 or less)	0.934–0.8762 (30 to 20)

Note: Figures in brackets give density as degree API, vide section 5.2.7.

Two arbitrary fractions are recovered from the crudes on the basis of boiling range: (i) 250–275 °C at 1 atmosphere pressure, and (ii) 275–300°C at a pressure of 40 mm of mercury. The former is termed *key fraction* 1 and the latter, *key fraction 2*. Each fraction is classified as paraffin, intermediate or naphthene according to specific gravity, and the crude is classified on the nature of the key fractions (Table 5.5). Following this system, a crude oil is called wax-free or wax-bearing according to whether the cloud point of the key fraction 2 is below or above 5°F(-15°C), respectively. Lane and Garton have found that 85% of the crudes examined fell into the three simple namely paraffin, intermediate or naphthene. Paraffin-naphthene or intermediate-naphthene crudes were not found to exist.

The system of Lane and Garton is an improvement over the three-group classification in as much as it recognised that one type of hydrocarbon might predominate in one fraction and another might predominate in another fraction. However, it characterises the nature of only two fractions and that too by the approximate method of using specific gravity data. More detailed studies are necessary for the complete characterisation of crudes.

Not each crude can yield all the products of right quality and in commercially recoverable quantities. Typical paraffin-base and asphalt-base crudes differ widely in this respect (Table 5.6). Mixed-base crudes have properties intermediate between these two. Modern refinery processes, however, can overcome the deficiencies of crude to a large extent. In general paraffin-base crudes yield good quality kerosene, diesel fuel, lubricants and waxes, while asphalt-base crudes yield good quality gasolines and asphalts. The refinery processes are largely influenced by the type of crude to be treated.

Crude oil from Pennsylvania (USA) and certain fields in Rumania, Poland and the old USSR may be classified as paraffin-base. Crudes from California (USA) and some areas in Mexico, Venezuela and the old USSR serve as examples of asphalt-base oils. The Middle East crudes form a large group of the mixed-base type.

Some crudes need special mention. Borneo crudes yield gasoline fractions containing up to 40% simple aromatic hydrocarbons. Another Indonesian crude (Tarakan) may be used directly

Table 5.6 Comparison of paraffin-base and asphalt-base crudes

Properties	Paraffin-base	Asphalt-base
Specific gravity	low	high
Yield of straight-run gasoline	high	low
Octane number of gasoline	low	high
Odour of gasoline	sweet or sour	aromatic or sour
Sulphur content of fractions	low	high
Smoke point of kerosene	high	low
Cetane number of diesel	high	low
Pour point of diesel	high	low
Yield of lubricants	high	low
Viscosity index of lubricants	high	low

as diesel fuel or light furnace oil. Light Kettleman Hills (California) crude practically consists of gasoline only.

Nature of Indian crudes

The oil fields of India yield highly waxy and mixed-base crudes. Some of the characteristic properties of three Indian crudes are given in Table 5.7. The API gravity varies from 22.3–38.5, the lightest being Bombay High and the heaviest the Cambay crude. The sulphur content is very low in all Indian crudes. The crude from Bombay High is almost sulphur-free. The wax content is hightest for Cambay and lowest for Bombay High.

The yield of straight-run gasoline is more than 20% from Bombay High and Nahorkatiya and less than 10% from Cambay. The recovery of total distillates is the lowest from Cambay.

The Bombay High crude is very highly aromatic. Its straight-run gasoline contains more than 60% aromatics. The corresponding figures for Nahorkatiya and Cambay are 20–25 and 3%–5%, respectively. The high aromatic content is a common feature of Southeast Asian crudes.

5.2.5 Petroleum Processing

After removal of dirt, water and much of the associated natural gas, the crude is separated into fractions by distillation and the resultant fractions are further subject to simple purification treatment or more complex conversion processes for the manufacture of diverse petroleum products. All these processes are referred to as petroleum refining and are carried out in a refinery.

Refinery processes may be conveniently grouped under three main heads, namely

1. *Physical separation processes*, for example, distillation and solvent extraction,
2. *Breakdown processes*, for example, cracking, visbreaking and coking,
3. *Rebuilding processes*, for example reforming, alkylation, isomerisation and polymerisation.

Table 5.7 Characteristic properties of Indian crudes

	Cambay, Gujarat	Nahorkatiya, Assam	Bombay High offshore
Specific gravity, 15.5°/15.5°C °API	0.8638 22.3	0.8574 33.5	0.8324 38.5
Water content, per cent v/v	2.60	0.06	–
Salt content, per cent w/w	0.014	nil	–
Sulphur content, per cent w/w	0.13	0.17	0.05
Kinematic viscosity at 37.8°C, CS	35.6	4.91	5.1
Calorific value, kcal/kg	10800	–	–
Pour point, °C	35.0	30–32	27.0
Distillation characteristics			
IBP, °C	62.5	41.0	38.0
Up to 150°C, per cent v/v	9.2	23.0	22.0
150–300 °C, per cent v/v	26.5	32.5	33.0
300–465 °C, per cent v/v	35.25	32.2	–
Total up to 465 °C, per cent v/v	70.95	87.7	–
Fraction up to 150 °			
Sp.gr.(15.5°/15.5°C)	0.7195	0.7471	–
Refractive index, 20°C	1.4041	1.4222	
Aromatic content	3–5	20–25	
Fraction 150–300 °C			
Sp.gr. (15.5 °/15.5°C)	0.7920	0.8516	
Refractive index, 20°C	1.4404	1.4777	
Fraction 300–450°C			
Sp.gr. (15.5°/15.5°C)	0.8698	0.9273	
Sulphur, per cent w/w	0.05	0.24	
Wax content, per cent w/w	32.2	24.4	
Short residue (>465°C)			
Sp.gr. (15.5 °/15.5 °C)	0.9554	1.0050	
Penetration number	50	16	
Softening point (R and B), °C	55	67	
Wax (300–465°C), on the basis of crude, per cent w/w	16	11	12

Various product streams are further refined before marketable products are obtained. According to the processes used, refineries may be classified as simple, more complex, or fully integrated.

Simple refineries produce only a limited range of products, mainly gasolines, kerosenes and aviation turbine fuels, gas and diesel oils, and fuel oil. Besides a crude oil atmospheric distillation unit, the only other units are those for reforming and for purification treatment. The additional units in more complex refineries are usually a vacuum distillation unit and a cracking plant together with the units for the processing of cracked gases by polymerisation, alkylation and other methods. Premium motor and aviation gasolines and bitumen are the notable additional products.

The whole range of petroleum products is obtained in the fully integrated refineries. The additional units include a high vacuum fractionating unit for the preparation of lubricating oils, a de-asphalting unit, a solvent extraction unit, a dewaxing unit, and a coking unit. Since

1950, integrated refineries which are cheaper to build and operate in the long run have been preferred. There are export refineries located at, or near the crude oil sources and market-oriented refineries located near the markets for the petroleum products. The advent of huge oil tankers, the developments in the piping of oils and the oil strategy of government have led to the increasing installation of market-oriented refineries.

A refinery may draw its crude oil supply from a single or many oil fields. If the crudes are not very different they may be blended before processing. Those yielding superior distillate fuels may be kept separate from those producting inferior stocks. Similarly, the oils that yield superior lubricants or asphalts may be segregated. It is not necessary to extend such segregation to the cracking of heavier fraction for gasoline production.

Petroleum distillation

The refining of petroleum begins with the separation of the crude by distillation into a number of fractions. The distillation is carried out first under atmospheric pressure and the residue from this primary distillation (reduced crude or long residue) is then distilled under vacuum. The primary distillation is also known as the topping of the crude oil. The maximum temperature upto which a crude can be heated without cracking is about 400ºC. This sets the limit of the maximum amount of distillate that can be collected by atmospheric distillation. Steam is used to facilitate the vaporisation of the high boiling constitutents. The recovery of still heavier fractions, upto 600ºC boiling range, is possible in the vacuum units further assisted by the admission of steam.

Fractions into which crudes are commonly separated are named in Table 5.8 with their approximate boiling ranges and uses. Primary distillation usually recovers fractions upto gas oils, while vacuum distillation recovers vacuum gas oil, raw lubricating oil distillates and asphalt or bitumen. The actual practice in a plant considerably varies from plant to plant, depending upon the type of crude input and the demand of products. On the one hand the crudes themselves yield fractions of different quality and quantity, and on the other, the boiling ranges of petroleum products considerably overlap each other. The nomenclature of the fractions also changes accordingly.

While vacuum distillation is a one-stage operation, primary distillation may be run in one or two stages. Figure 5.2 shows a flow-sheet of a complete three-stage crude distillation unit. The first stage operates at 3 to 4 kgf/cm^2 pressure and produces gases and light distillate boiling upto 90ºC. The bottoms from the first stage are heated to about 400ºC and sent to the second stage distillation column which operates at atmospheric pressure. Naphtha is recovered as top stream while kerosene and gas oil are withdrawn as side streams. The reduced crude is reheated to about 400ºC and pumped to the vacuum column (50 to 100 mm Hg pressure). Here the top stream is a raw lubricating oil. The bottoms are used in making asphalt or bitumen and heavy furnace oil. Parafffin wax is manufactured from vacuum gas oil while microcrystalline wax is recovered from certain types of heavy lubricating oil distillates or residues from paraffinic crudes.

The side streams contain some lighter components as impurities. These are purified by passing through steam-strippers. Steam carries off the lighter components back to the main column while the purified fraction is obtained at the bottom of the stripper.

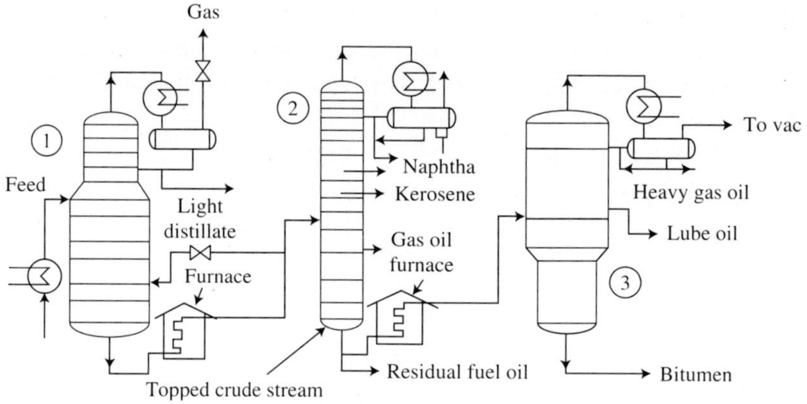

Figure 5.2 Three-stage distillation of crude oil

Table 5.8 Common fractions from crude petroleum

Fraction	Approximate boiling range, °C	Uses
Fuel gas	−162 to −42	containing methane, ethane and some propane; used as refinery fuel.
Propane	−42	liquefied petroleum gas (LPG)
Butane	−12 to −0.5	LPG component of gasolines in cold climate
Light naphtha	−1 to 150	straight run gasoline, solvents, depentanised added to heavy naphtha feed for reforming
Heavy naphtha	150 to 205	catalytic reformer feed, component of jet fuels and solvents
Kerosene	205 to 260	illuminant, fuel
Stove oil	205 to 290	fuel
Light gas oil	205 to 315	fuel oil, diesel fuel, absorbent
Heavy gas oil	315 to 430	catalytic cracking feed
Vacuum gas oil	430 to 540	catalytic cracking feed, suitable crudes yield lubricating oils
Residue (short residue)	595 and above	heavy fuel oil and asphalt or bitumen, petroleum coke making

Fractionating columns vary greatly in size. The usual dimensions are 3 m diameter and 30 to 35 m height. Plates of various designs are used by different refineries. Three to five plates, at 0.56 m spacing are usually sufficient between two neighbouring product streams in atmospheric columns. This number is two to three for vacuum columns where fractionation is more efficient owing to the increased difference of vapour pressure of components and also because of the larger spacing of plates.

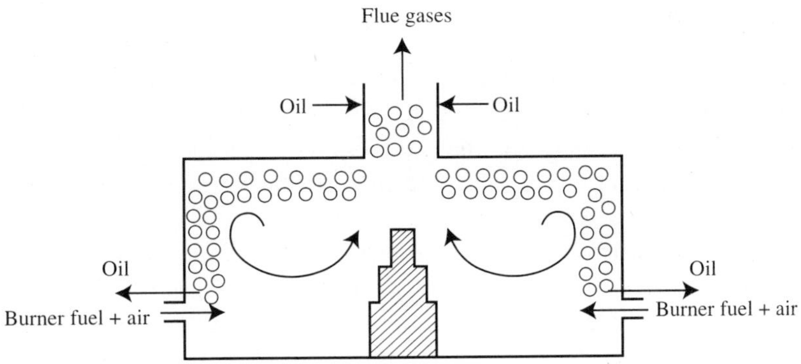

Figure 5.3 Box-type pipestill

In the distillation units, crudes are heated in pipestills. A pipestill is a special type of furnace with a relatively long but small diameter pipe. The pipe consists of a large number of tubes connected through bends. The tubes are housed inside the furnace chamber in single and multiple rows. The furnace is built in two sections, namely, radiant and convection. The tubes in the radiant section receive heat directly from the flame and those in the convection section recover heat from the hot gases travelling to the stack. There are different designs for pipestills depending upon the arrangement of tubes and direction of liquid flow. Box-type stills are the oldest and quite widely used (Fig. 5.3). The furnace is divided into two chambers by a bridge wall. Separate burners are put on the side walls. The convection section is separated from the radiant section by a row of tubes. The crude is pumped into the pipe inlet at the convection section. The final heating to about 400°C takes place when the preheated crude passes through the radiant section tubes. High rates of heat transfer are obtained by radiation and convection and by high liquid flow rates in the pipe. The high flow rate of the crude also ensures high productivity and prevents local overheating which might otherwise cause cracking and coke deposition.The box-type stills have input capacities of 5–20 million kcal/h. The heat flux is 15,000–40,000 kcal/m^2 of circumferential tube area per hour and the heat release is 20,000–45,000 kcal/m^3 furnace volume per hour. A high proportion of the total heat transfer between 55% and 60% is by radiation.

***Stabilisation*:** Natural gasoline, cracked gasoline and similar stocks contain dissolved hydrocarbons (gaseous and very highly volatile liquids) which tend to escape during storage. Stabilisation is the process of removing these lighter components and making such stocks stable and suitable for storage and handling. A typical cracked gasoline may contain more than 10% of butane and lighter hydrocarbons. Stabilisers are essentially high pressure, 8 to 15 kgf/cm^2, fractionating columns using 40 to 50 plates and high reflux rations. The escaping vapours on condensation separate into gaseous and liquid streams. The liquid is refluxed and also recovered as a speciality product. The stabilised gasoline is withdrawn from the bottom through a heat exchanger which preheats the feed. Extra heat is supplied at the bottom of the tower by means of a reboiler.

Breakdown processes

***Cracking*:** This term refers to the processes by which heavier fractions are converted into lighter fractions by the application of heat, with or without a catalyst. The chief application of commercial cracking in refineries is the production of gasolines from gas oils. Other fields of application are the production of olefins from naphthas and gas oils for the petrochemical industry, the reduction of viscosity of straight-run residues (visbreaking), for fuel oil production, and the conversion of heavy residual products into coke and lighter products (coking).

When cracking is carried out without any catalyst, the process is called thermal cracking. This is accomplished by heating the feed upto 450° to 750'C at pressures ranging from 1–70 atm. The important reactions are decomposition, dehydrogenation, isomerisation and polymerisation. The paraffinic molecules are first decomposed into lower molecular weight hydrocarbons, usually a paraffin and an olefin. For example,

$$\underset{n\text{-decane}}{CH_3(CH_2)_8CH_3} \longrightarrow \underset{n\text{-hexane}}{CH_3\,(CH_2)_4CH_3} + \underset{\text{butene-1}}{CH_3\,CH_2CH = CH_2}$$

While *n*-decane is a component of gas oil, *n*-hexane is one of gasoline. The olefins can then undergo isomerisation, dehydrogenation and polymerisation reactions. For example,

$$\underset{\text{butene-1}}{2CH_3CH_2-CH{=}CH_2} \quad \underset{\text{(polymerisation)}}{\longleftrightarrow}$$

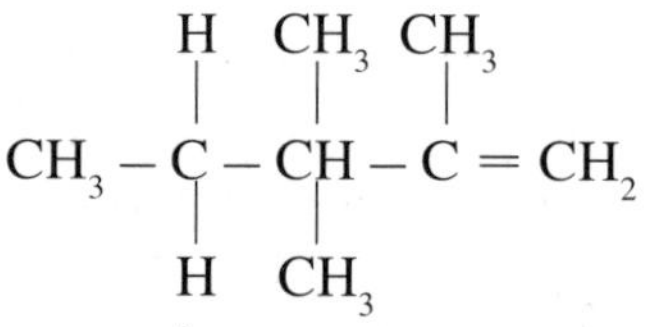

iso-octene

Iso-octene is again a gasoline component. Owing to various reactions during cracking, gases and solids (coke) are also produced together with the liquid products. The ease of cracking depends upon the boiling range and the chemical nature of the feed. Crackability is higher for the heavier fractions. Paraffinic feed is more crackable than aromatic.

The use of catalysts not only accelerates the above reactions but instead introduces new reactions which considerably modify the yield and nature of the products. Compared with the thermal process, the catalytic cracking yields less coke, less gas and more liquid products. Moreover, the gasoline obtained from the catalytic units is far superior in quality owing to higher content of iso-paraffins and aromatic hydrocarbons. All new cracking units for gasoline production are based on the catalytic process. Thermal cracking is now restricted to olefin production, visbreaking and coking. Conventional thermal cracking units for gasoline production, for example, the Dubbs process, may still be found in old refineries.

The usual catalysts generally are either synthetic composites of silica and alumina, or acid-treated natural clays of the montmorillonite group. The most recent advance is the introduction of zeolites as cracking catalysts with better activity. The cracking catalysts are used in the form of pellets, beads or powder. The pellets and beads may be used either as a fixed bed in a cyclic process, for example, the Houdry process, or as a moving-bed in a continuous process,

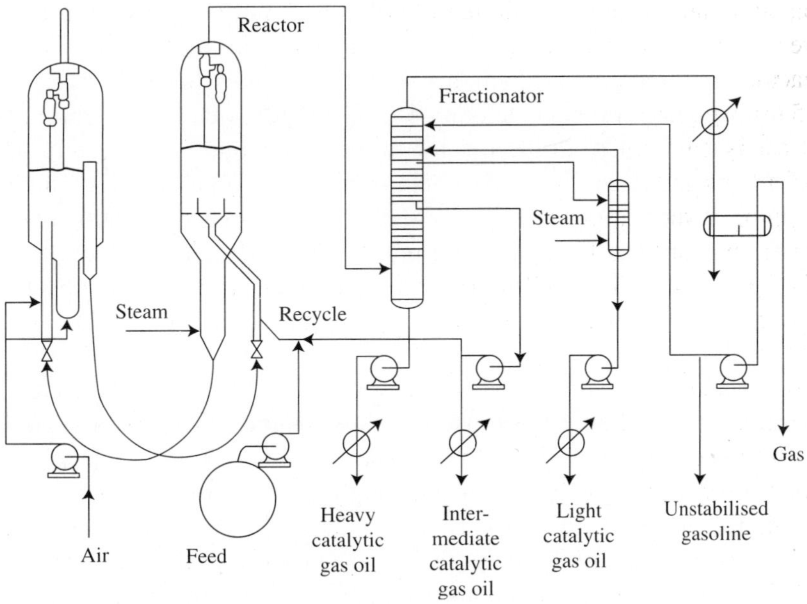

Figure 5.4 Fluidised bed (down flow) catalytic cracking process

for example, thermofor catalytic cracking or TCC process, Houdriflow process, and others. The power may be used either as a suspension in the liquid fuel, for example, the suspension process or as a fluid-bed with the vaporised feed as the fluidising medium. Out of the various processes now operating, fluid catalytic cracking is by far the most widely used. The use of catalysts minimises the formation of coke during cracking, but cannot completely avoid it. The coke particles settle on the catalyst surface and deactivate.

Catalyst regeneration is therefore vital to the economics of the process. This is accomplished by the burning of the deposited coke particles in air. The movement of catalysts from reactor to regenerator and back is much more easily achieved in the fluid cracking process than in the other continuous processes. In the fixed bed process, the catalyst is regenerated in the reactor itself by using a cycle of cracking followed by regeneration and continuing the cycle till further activation is not possible. The most obvious demerit of the fixed bed process is the variation of temperature and activity of the catalyst along the height of the bed and with the progress of the cycle, causing non-uniformity of the yield and quality of products. Besides easy catalyst movement, the fluid process has other merits as well, namely, the fluid bed is essentially an isothermal reactor owing to high heat transfer rates in the systems, and moreover, high mass transfer rates between the catalyst surface and gas phase help to increase reaction rates.

Figure 5.4 shows the flow-sheet of a typical *down flow* fluidised bed catalytic cracking process. Fresh and recycle gas oil (feed) are mixed with hot activated catalyst from the regenerator and admitted at the bottom of the reactor. The catalyst is fluidised in the reactor by the vapours issuing from the feed line. The cracker gases and vapours are passed through internal cyclones that remove the entrained catalyst powder and feed it back to the bed. The gases and vapours enter a large fractionating column (5 m diameter, 30 m height, 30 to 40 plates) where wet gas, gasoline (unstablised) and gas oils are recovered. A gas oil stream is recycled. Reactor conditions are: temperature 470– 520°C, pressure 8–15 kgf/cm^2, catalyst-to-oil ratio 5–15 (by weight) and space velocity 1–3 kg oil per hour per kilogram of catalyst.

The spent catalyst is sent from the reactor to the regenerator by a system of stand-pipes, U-bend transfer lines and risers. It always contains absorbed oil molecules which are removed by steam-stripping at the leg of the reactor. Air for catalyst regeneration is injected partly at the catalyst riser and partly at the regenerator bottom. Regenerator conditions are: temperature 550–660°C and pressures 1.3–1.7 kgf/cm^2. The regenerated catalyst is sent back to the reactor by a transfer system comprising an overflow weir, standpipe, U-bend and riser. The combustion gases are led to the stack.

Cracking is an endothermic process. The heat requirement is met by the combustion of coke in the regenerator, the heat being transferred as the sensible heat of the regenerated catalyst. The catalyst size varies and may consist of particles from 500 μ upper size to fines below 5 μ. Instead of powder form, catalysts are often used in the form of microspheres of required sizes. The size degradation due to attrition is very low in the case of microspheres.

The extent of cracking is deliberately kept at a low level so that coke and gas formation is minimised. By recycling cracked gas oil streams, the per cent conversion is raised. There are fluidised bed cracking units in the refineries at Bombay, Visakhapatnam, Koyali and Mathura.

***Visbreaking*:** Heavy furnace oils are made by blending residues with gas oils to meet the requirement of viscosity and pour point. The residues are highly viscous and need large quantities of the valuable blending oils. The main purpose of visbreaking is to reduce viscosity of the residues by thermal cracking and thus reduce the consumption of the blending oils. Visbreaking is carried out by passing the feed through a pipestill at 470 to 520°C at an outlet pressure of 4 to 20 kgf/cm^2. It is a one-pass process. The thermal cracking takes place in the liquid phase. The cracked material is led into a flash-distillation chamber. A gas, a low quality gasoline and a light gas oil are recovered from the overhead stream. The liquid products from the flash chamber are cooled and further fractionated under vacuum when a heavy gas oil and a tar of reduced viscosity are obtained. This tar is the main product of visbreaking. It is blended with an oil to give a furnace fuel while the other products join other corresponding streams of the refinery for further processing.

***Coking*:** The residual products of the refinery form the feed for coking. Under relatively severe conditions of thermal cracking, the heavy liquid feed is converted into gas, naphtha, heating oil, gas oil and coke. Coking is practised to get either gas oil or coke as the main product. The gas oil is primarily used as the feed for catalytic cracking while the coke (petroleum coke) is used as a fuel or in the production of carbon electrodes and chemicals. There are two methods of coking, namely delayed coking and fluid coking. Higher coke yield is obtained by the former method while the fluid coking yields less coke and more gas oil. In delayed coking, the feed is heated to about 500°C and charged into a chamber called the coke drum. Coking is continued in this chamber till the cracked products leave the top and the coke is left behind.

The conditions in the coke drum are: temperature around 425°C, pressure 2 to 6 kgf/cm². The overhead products are fractionated and the heavy oil is recycled. The coke is cooled and removed from the drum by hydraulic cutting.

In fluid coking a spray of the liquid (260° to 360°C) is injected into a fluidised bed of hot, fine, coke particles (480 to 560°C) at substantially atmospheric pressure. The material gets cracked. The gases and vapours are passed through a system of cyclone, scrubber and condenser for recovering entrained coke particles, gas oil and other products. The coke is deposited on the fluidised coke particles. It is recovered by withdrawal from the unit from time to time. The heat of the process is supplied by burning a part of the coke in a separate chamber (burner). A continuous movement of the coke is maintained from the reactor to the burner and back.

Reforming: This is the process of preparing high quality gasoline by heating, with or without a catalyst, the naphtha fractions of the gasoline boiling range. The upgrading is essentially a result of rearrangement or reforming of molecules without greatly disturbing the average molecular weight. The main reactions are the formation of aromatics by the dehydrogenation of naphthenes and the dehydrocyclisation of paraffins. For example,

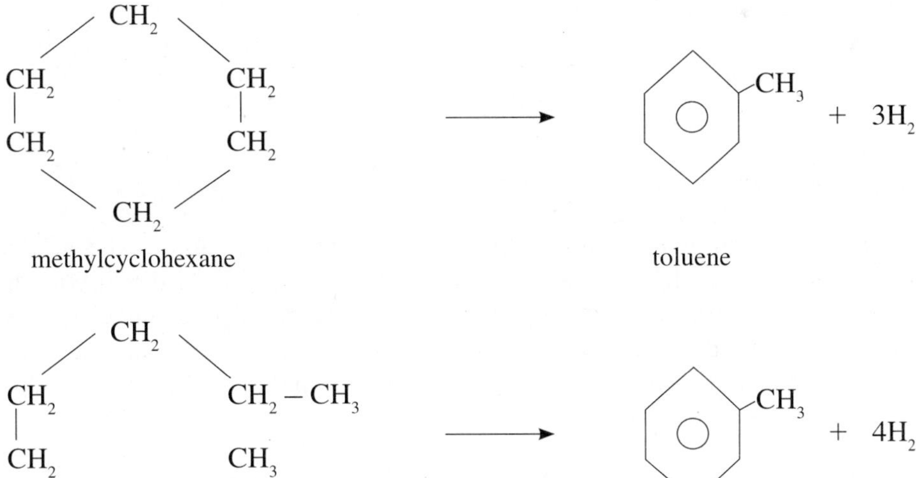

methylcyclohexane toluene

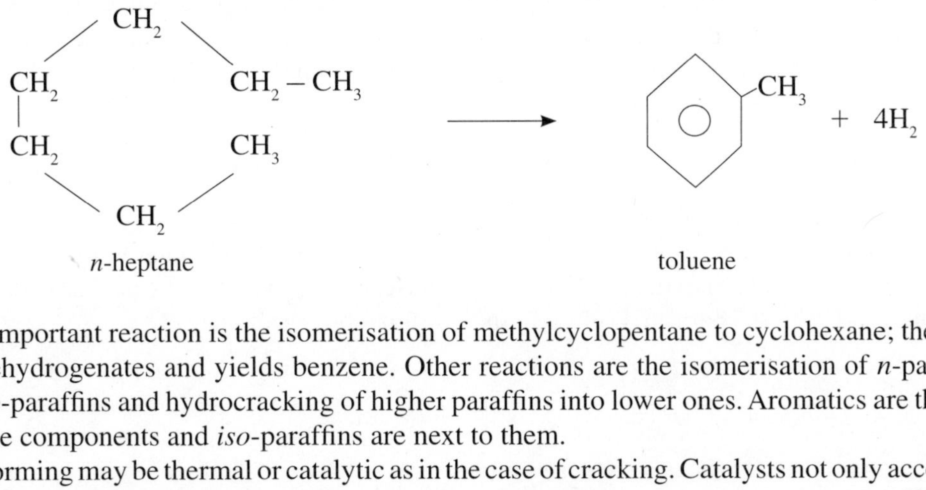

n-heptane toluene

An important reaction is the isomerisation of methylcyclopentane to cyclohexane; the latter then dehydrogenates and yields benzene. Other reactions are the isomerisation of *n*-paraffins into *iso*-paraffins and hydrocracking of higher paraffins into lower ones. Aromatics are the best gasoline components and *iso*-paraffins are next to them.

Reforming may be thermal or catalytic as in the case of cracking. Catalysts not only accelerate the process but also improve the yield and quality of the resultant gasoline (reformate). Thermal reforming has been almost completely replaced by catalytic reforming.

Two types of reforming catalysts are used commercially, namely non-precious metal oxide-type, such as, molybdena or chromia supported on alumina and the precious metal acid oxide-type, such as, platinum on a silica–alumina or alumina base. Platinum is more active and selective. It is, however, easily poisoned by compounds containing sulphur, nitrogen and arsenic and needs pretreatment of the feed for the removal of the impurities. The catalyst may be in the form of powder, beads or pellets. Fixed bed, moving bed and fluid bed processes are in use. Platforming, catforming, hydroforming, fluid hydroforming and thermofor reforming (TCR) are the various commercial reforming processes. The feed for catalytic reforming is drawn from naphthas obtained in the various refinery processes, for example straight-run distillation, catalytic cracking and coking. The best results are obtained with the straight-run naphthas of high naphthene content.

Besides the main product (reformate), reforming also produces lighter bydrocarbons, hydrogen and traces of very high boiling materials. The hydrogen-rich gaseous product is recycled with the naphtha feed. The reformate has a wider boiling range than the original naphtha. The increased volatiaity is an added advantage gained by the reforming process.

Reforming is normally a once-through process. The reactor conditions are: temperature 450 to 520 °C and pressure 15 to 45 kgf/cm^2. Fluid bed and moving bed processes are run at the high temperature ranges using 15 kgf/cm^2 pressures. High pressures upto 45 kgf/cm^2 are for the fixed bed process. For the first time in India, the catalytic reforming process was established at the Bharat Petroleum Refinery at Mumbai. Subsequently this has been adopted by other refineries at Koyali, Cochin, Chennai and Haldia.

Rebuilding processes

***Polymerisation, Alkylation and Isomerisation*:** These processes aim at the production of superior gasolines and are thus complementary to catalytic cracking and catalytic reforming. The cracker gases in the refinery are rich in olefins. These compounds undergo polymerisation in the presence of catalysts, for example, phosphoric acid or sulphuric acid at suitable temperatures and pressures, and ultimately give products rich in branched-chain hydrocarbons (polymer gasoline).

$$\underset{\textit{iso}\text{-butene}}{2CH_3C(CH_3){=}CH_2} \longrightarrow \underset{\textit{di}\text{-isobutene}}{CH_3C(CH_3)_2CH_2C(CH_3){=}CH_2}$$

In the alkylation process, a gaseous *iso*-paraffin, for example, *iso*-butane is reacted with an olefin (propene or butene) in the presence of sulphuric acid or hydrofluoric acid as catalyst and the result is a larger *iso*-paraffin boiling in the gasoline range (alkylate).

$$\underset{\textit{iso}\text{-butane}}{C_4H_{10}} + \underset{n\text{-butene}}{C_4H_8} \longrightarrow \underset{\textit{iso}\text{-octane}}{C_8H_{18}}$$

The isomerisation process converts *n*-paraffins into *iso*-paraffins in the presence of aluminium chloride catalysts. Commercially it is mostly practised to obtain *iso*-butane from the butane fractions. The *iso*-butane is then used in the production of good quality gasoline by the alkylation process. Pentanes and hexanes are also being increasingly isomerised to supply excellent high volatile gasoline components, such as, *iso*-pentances and *iso*-hexanes.

The polymerisation and isomerisation processes are also used in the production of pure petrochemicals.

***Hydrogenation*:** There are essentially three classes of hydrogenation processes used in petroleum refining, namely, hydrotreating, aromatic saturation and hydrocracking.

1. *Hydrotreating* is a mild operation aimed at removing impurities like sulphur, nitrogen, oxygen, halides and trace metals. It is also used to stabilise petroleum products by saturating olefinic compounds. Hydrofining, autofining and ultrafining, are the various processes falling into this class. Hydrotreating is applied to products of all boiling ranges from light naphthas to lubricating oils.
2. *Aromatics* are undesired constituents of kerosene, stove oil, diesel fuel and lubricating oil. Saturation of aromatics by hydrogenation results in the upgrading of these products. This process is still of limited use in industry.
3. *Hydrocracking* is also known as destructive hydrogenation and leads to the formation of saturated compounds of lower molecular weight by the cracking of heavy fractions and residues under a substantial pressure of hydrogen. Hydrogen suppresses the formation of tar and coke and also raises the crackability of polycyclic aromatic compounds. The process can achieve high conversions into gasolines and high quality diesel fuels. A mild hydrocracking of petroleum residues can supply feed for other refinery processes.

The important catalysts for the hydrogenation processes are cobaltmolybdena for hydrotreating, nickel for saturation of aromatics, tungsten sulphide for vapour-phase hydrocracking and an iron catalyst or bauxite for liquid-phase hydrocracking.

The normal range of conditions for hydrotreating are: temperature 300–400°C, pressure 15–17 kgf/cm^2, space velocity up to 20 v/v.h. More severe conditions are required for hydrocracking, namely temperature range of 400–550°C and pressure range of 70–700 kgf/cm^2. Hydrocracking of residues and heavy fractions boiling above 300°C is conducted in the liquid phase.

The development of the hydrogenation process in refineries has been possible largely owing to the availabilty of surplus hydrogen-rich gases from the catalytic reforming of naphthas. Only Chennai and Haldia refineries have adopted hydrotreating processes. It is planned to have a hydrocracker at the refinery at Mangalore.

Purification processes

Petroleum products have to reach the consumer in a certain degree of purity and quality. The impurities and undesired components may come from the original crude or may be developed during various stages of processing the crude and its fractions. Even some feedstocks for catalytic processes require purification so that the catalyst deactivation is minimised. Depending upon the nature of the crude and the ultimate use of the product, the required purification may involve a simple treatment or elaborate processing.

The specific aims of the purification processes are:

1. To minimise (i) acidity, (ii) carbon residue, (iii) catalyst deactivation, (iv) corrosivity, (v) gum formation, (vi) interaction with paints and (vii) noxious combustion products;
2. To improve (i) burning characteristics of kerosene and diesel fuels, (ii) gasoline properties, for example, lead susceptibility and octane rating, (iii) colour and colour stability, (iv) odour and (v) pour point of diesel fuel;
3. To purify the product and recover the following for sale or further processing: (i) alkyl phenols, (ii) hydrogen sulphide, (iii) mercaptans, (iv) naphthenic acids, (v) nitrogen bases and (vi) sulphonates.

The chief components responsible for colour include sulphur compounds, nitrogen bases and certain hydrocarbons such as high molecular weight aromatics. Objectionable odours are imparted by hydrogen sulphide, naphthenics and aliphatic fatty acids which may have originated from the crudes or appeared during refinery processing. Acidity gives rise to the problem of corrosion, gum formation and discolouration of products and interferes during the processing of various feedstocks. Corrosivity may also be due to non-acidic substances, for example, mercaptans and polysulphides. Gum formation is promoted by polysulphides, acidic materials, metallic contaminants, aryl mercaptans and olefinic bodies.

The burning quality of kerosene deteriorates owing to both the presence of aromatics which cause soot formation, and to the presence of sulphur compounds, unstable hydrocarbons, metallic naphthenates and sulphonates which enhance the tendency for the wick to char.

The *octane rating* indicates the quality of gasoline. Straight-chain paraffins have poor octane rating while aromatics and isoparaffins have a very high octane ratting. Premier gasoline may be made by removing the undesired straight-chain paraffins and increasing the content of aromatics. Gasoline also improves in octane rating by the addition of small quantities of tetraethyl lead. This property is known as lead susceptibility. Polysulphides, disulphides and mercaptans seriously reduce the lead susceptibility of gasoline.

The flow characteristics of diesel are reduced by the dissolved wax which crystallises out even at ordinary temperatures. This renders the pour point objectionably high. Methods of dewaxing can improve the quality of diesel fuel in this respect.

Treatment with acids and alkalis

Sulphuric acid: Owing to its chemical properties, sulphuric acid is the most versatile refining agent known, as it is also inexpensive and readily available. Nitrogen bases, asphaltic substances, di-olefins, olefins and acetylenes are readily removed by sulphuric acid. Aromatics are also removed at a high concentration. Although many sulphur compounds, for example, thiophenes, mercaptans and sulphides are also readily removed, there are some others like elemental sulphur, carbon disulphide and certain high molecular weight sulphur compounds which are not attacked. By removing the impurities, sulphuric acid improves colour, stability and odour and reduces corrosiveness and gum formation of petroleum products.

The use of sulphuric acid in petroleum refining has declined steeply owing to the development of newer processes. The two major drawbacks of the sulphuric acid treatment are

1. The large refining losses due to the formation of acid sludge,
2. The problem of disposal of acid sludge.

It is still the most common refining agent. Petroleum products of all boiling ranges—naphtha to lubricating oil—are given finishing treatment with it.

Acid of 93% strength (66° Be') finds the most general use. A more dilute acid is used when the aim is only to improve the colour of light distillates. Lubricating oils may need up to 98% strength. High temperatures enhance refining losses. Lower temperatures yield products of a better colour and are favourable for the removal of sulphur compounds. Gasolines are treated at 20–30°C. Higher temperatures (30–50°C) are used for treating kerosene because the removal of aromatics, unsaturated hydrocarbons and asphaltic materials is favoured at those temperatures. Owing to high viscosity of lubricating oils, still higher temperatures (40–80°C) are needed to ensure intimate contact with the acid. Contact time also varies with the nature of the product. The heavier products need a longer time of contact.

The general procedure for acid treatment is similar. Raw stock and acid are brought together in desired proportions at a suitable temperature. After thorough contacting the phases are separated by settling. The oily phase is neutralised and the spent acid is discarded. The product may need fractionation to remove high-boiling material produced during acid treatment.

Other acids: The colour, stability and odour of distillates are improved by treatment with hydrofluoric acid or acetic anhydride. These reagents however have limited use.

Caustic soda: A wash with caustic soda is a common treatment after processing with sulphuric acid. The alkali wash is also widely used to remove naturally-occuring acidic substances, such as hydrogen sulphide, carbonyl sulphide, lower aliphatic mercaptans, phenols, aryl mercaptans, fatty acids and naphthenic acids.

***Sweetening processes*:** Mercaptans and hudrogen sulphide have an objectionale odour and are said to make an oil *sour*. By definition, a stock is sweet if it passes the *doctor test*. In this test, the oil sample is shaken with an equal volume of sodium plumbite solution and a small amount of sulphur. The mercaptans first react with sodium plumbite to form oil soluble lead mercaptides. The mercaptides on further reaction with elemental sulphur form sweet smelling, oil soluble disulphides and give a black precipitate of lead sulphide. The hydrogen sulphide also gives this black precipitate with the plumbite solution. If the test does not produce even a dark discolouration at the interface, the sample is sweet. The processes by which the foul smelling stocks are rendered sweet smelling or odourless are known as the *sweetening processes*. In essence these convert mercaptans, hydrogen sulphide and also elemental sulphur into oil soluble sulphur compounds without any objectionable odour and thus differ from those processes where the aim is to materially remove the mercaptans and other sulphur-containing materials from the stock.

$$\underset{\text{mercaptan}}{2RSH} + \underset{\text{sodium plumbite}}{Na_2PbO_2} \longrightarrow \underset{\text{lead mercaptide}}{(RS)_2Pb} + 2NaOH$$

$$(RS)_2Pb + S \longrightarrow \underset{\text{disulphide}}{RSSR} \longrightarrow Pbs$$

The principle of the doctor test forms the basis of the doctor sweetening process. There are other processes also using copper chloride, hypochlorite and lead sulphide, respectively. All these processes are almost obsolete. Firstly, the disulphides harm lead susceptiblity of gasolines and secondly, the need for reduction of mercaptans to doctor sweet (0.0004%) is questioned. However, some cheap sweetening processes, such as air-inhibitor and air-solutiser processes are still in use. In these processes, air oxidises the mercaptans in the presence of additives like phenylene diamine (inhibitor), methanol (solutiser), caustic soda, caustic potash and doctor solution.

***Mercaptan dissolving processes*:** These remove the mercaptans from the stock by dissolving out and thus have an advantage over the sweetening processes where disulphides are left behind. Treatment with caustic soda removes the lower mercaptans. The high molecular weight mercaptans have only limited solubility in this reagent. However, some solutisers like fatty acids, cresols and naphthenic acids markedly enhance the solubility of the higher mercaptans, particularly in caustic potash. This forms the underlying principle of a number of commercial processes for the removal of mercaptans from distillates.

***Desulphurisation processes*:** There are two general methods of reducing the amount of sulphur compounds in petroleum stocks:

1. Treatment with chemicals and solvents,
2. Catalytic conversion into hydrogen sulphide with or without the supply of external hydrogen.

Treatment with acids and alkalis and catalytic hydrotreating were discussed in the earlier sections. The use of solvent extraction and clay treatment for desulphurisation will be taken up in the subsequent sub-sections.

***Solvent extraction*:** The difference in solubility of suitable solvents forms the basis of many refinery processes which either resolve a stock into two or more products or purify a product from harmful impurities. One of the oldest processes of petroleum refining—solvent extraction—has quite a few major uses in modern refineries:

1. Upgrading lubricating oils by removing aromatic materials and sulphur compounds,
2. Separating reduced crude and residue into lubricating oils and asphalts (de-asphalting),
3. Upgrading kerosene, heating oils and diesel fuels by removing aromatics, olefines and sulphur compounds,

4. Improving cracker feedstocks by removing asphaltic and other coke-producing materials (decarbonisation),
5. Separating wax-bearing oils into wax and wax-free oil (dewaxing),
6. Recovering benzene, toluene and xylenes from suitable naphthas and
7. Separating *iso*-butene from butane–butene streams.

The important solvents employed in one or more of the above uses are given below.

Liquid sulphur dioxide: It is a versatile solvent applied to a wide variety of light distillates, for example, heavy naphtha, kerosene and diesel fuel (Edeleanu process). Mixed with benzene, it is also applicable to lubricating oils. At its boiling point, –10ºC, liquid sulphur dioxide completely dissolves aromatics, unsaturates, mercaptans and many other sulphur compounds, but the paraffins and naphthenes remain insoluble in it. A disadvantage is that moisture cannot be tolerated in the system because the solvent then becomes highly corrosive.

Furfural: This solvent is being increasingly used in the upgrading of lubricating oils. It is also applied to light distillates. Aromatics are highly soluble in it. Low moisture content in the system does not pose a problem. In fact, a small amout of water is purposefully left in the solvent for increasing its selectivity.

Phenol: The solvency properties of phenol are similar to those of furfural, but phenol is toxic and solidifies at ordinary temperatures. However, solvent recovery is easier and cheaper. Phenol is extensively used in upgrading lubricating oils. Furthermore, a mixture of phenol and cresols, selectro, is used as a solvent in the double solvent, Duosol process of upgrading lubricating oils. Selectro dissolves napthenic hydrocarbons and the other solvent (liquid propane) dissolves paraffinic hydrocarbons.

Liquid propane: Propane, an abundant material in refineries is a useful solvent in many ways. It extracts paraffinic hydrocarbons from oils and precipitates asphaltic or resinous materials. It is widely used in de-asphalting, dewaxing and upgrading of lubricating oil upgrading (Duosol) processes. It is also used to facilitate treatment with sulphuric acid.

Methyl ethyl ketone (MEK): It is used in the dewaxing processes. Wax-bearing oils are dissoved in the solvent and then chilled. The wax crystallises out in an easily filterable form. The ketone is used as a mixture with benzene and toluene which increases the solubility of the oil in the solvent.

Glycol: Aqueous solutions of diethylene glycol (8%–12%) are used in the extraction of benzene, toluene and xylenes (light aromatics) from the reformate or cracked naphtha by the well-known Udex process. The product is used as a premium gasoline or further processed for the recovery of pure aromatic hydrocarbons.

Dilute sulphuric acid (65%): It is used in the extraction of *iso*-butene from butane–butene stream in the refinery. The hydrocarbon is recovered from the solution by dilution with water to 45% acid, followed by stripping.

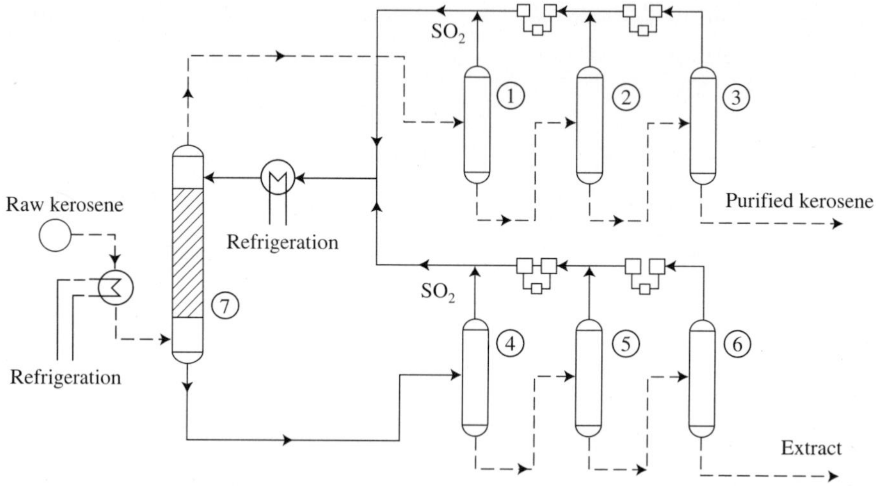

Figure 5.5 Edeleanu process of kerosene purification
1, 4: High pressure 2, 5: Low pressure 3, 6: Vacuum 7: Extraction tower

A typical sulphur dioxide plant for producing premium kerosene is shown in Fig. 5.5. The raw feed is first dried and cooled, and then charged to a packed tower near the bottom. The refrigerated solvent is fed at the top. Both the overhead raffinate and the bottom extract streams are passed through triple-effect evaporators for the complete recovery of sulphur dioxide. The first stage in each of these two evaporators is operated at a pressure of about 6 to 7 kgf/cm^2 which is sufficient to condense the solvent vapours without compression. The other two stages are operated at much lower pressures (25 mm Hg in the last stage) and need compression for liquefying the solvent vapours. There is a small drying tower in the SO_2-stream for the continuous removal of water. This is the only piece of equipment requiring special corrosion resistant material of construction. Solvent losses amount to about 0.4 kg per kilolitre of the charge. By using a solvent-to-feed ratio of 100 and extraction temperature of – 18°C with a feed of 14% aromatic content, it is possible to get a raffinate of only 2% aromatics and extract of 82% aromatics in a commercial plant.

Clay treatment

Clays adsorb asphaltic, resinous and acidic materials from heavy oil and waxes. This property is utilised to remove traces of asphalts and resins from low-asphalt content stocks and waxes, and therby improved their colours. Colour stability, oxidation resistance, and de-emulsibility are also improved. High asphalt oils are fast de-asphalted by acid treatment or solvent extraction and then given a clay treatment. Waxes are dissolved in a solvent like propane and the solution is acid treated. Conventional acid treatment may be followed by clay treatment for the neutralisation of the charge by the adsorption of sludge and acidic materials. Clays are also used to adsorb arsenic compounds from catalytic reformer feeds.

Another use of clays in refineries is for their catalytic properties. Fuller's earth and bauxite catalyse the polymerisation of colour forming and gum forming materials in cracked gasoline.

The treated stock is freed from the polymer by distilling (*rerunning*). The final product is resistant to decolourising and gum forming on storage.

Fuller's earth and bauxite also catalyse the conversion of organic sulphides and mercaptans to hydrogen sulphide and hydrocarbons at 300–350°C. They are used to desulphurise straight-run gasolines and kerosenes. The sulphur in cracked gasoline is largely thiophenic, and thiophenes are not attacked by clay treatment. Catalytic desulphurisation by hydrotreating is essentially a variation and improvement of the clay treatment. External hydrogen enhances the process of desulphurisation and minimises coke formation.

- The liquid phase clay treatment may be carried out either by percolation or by contact filtration. The former consists in percolating the liquid charge through a fixed or moving bed of the clay. In the filtration method, the liquid is mixed with the adsorbant and then filtered.
- The vapour-phase treatment usually involves passing vapours over a fixed bed of clay. Under desulphurisation conditions, coke deposits deactivate the calalyst. The latter can be regenerated by burning off the coke.

The spent clay from the processes other than desulphurisation are not usually regenerated. Their disposal is a problem. There is a limited demand for the spent clay in the manufacture of cinder bricks. Clays may be regenerated by controlled combustion.

Deasphalting

Crudes contain asphalt in concentrations ranging from 0% to 65%. They are the highest molecular weight compounds in petroleum. Some of them have boiling points in the range of lubricating oils and heavy gas oils. Asphalts readily oxidise and form carbonaceous sludge and hence must be removed from lubricating oils. These are also undesirable in heavy gas oils to be used as catalytic cracking feedstock because they form coke. The processes by which asphalts are removed from heavy stocks are known as deasphalting. They may be atmospheric distillation, vacuum distillation, acid treatment and solvent extraction.

Owing to heavy losses of hydrocarbons, acid treatment for deasphalting is almost obsolete. The bulk of the asphalts are removed by atmospheric and vacuum distillation. However, a good deal of desirable materials for lubricating oils is still left in the residue from vacuum distillation. Solvent deasphlating is the best process from this point.

Dewaxing

Petroleum waxes are solid hydrocarbons (melting point 35–95°C), which are soluble in crude oils and their fractions. They boil in the wide temperature range of distillate fuels and lubricating oils, and hence cannot be separated by distillation. On the other hand, the dissolved waxes precipitate out at low temperatures and therefore the wax-bearing diesel fuels and lubricating oils must be dewaxed. Wax is the valuable by-product of the dewaxing processes.

Dewaxing is done by chilling the oils when wax crystallises out. The wax crystals are separated by filtration, centrifuging or settling. The separation stage is difficult, tedious, time-consuming and uneconomicl. However, if the oil is first dissolved in a suitable solvent and the solution chilled, the crystallised wax can be more conveniently separated from the solution in vacuum filters. Methyl ethyl ketone (MEK) and propane are the two solvents that are widely used in dewaxing processes.

There is another method of dewaxing, which is based on the fact that urea forms solid, filterable complexes, called *adducts*, with large paraffins having little or no branching. Urea dewaxing is very effective on lighter stocks.

Dewaxing processes are particularly important in the refining of the highly waxy Indian crudes.

Deoiling

A large quantity of oil is left in the crude waxes produced in dewaxing processes. The oil content may be as high as 50% if the chilling is done without a solvent. Hence the importance of the deoiling process. A finished wax should not contain more than 0.5% oil. There are three types of deoiling processes, namely sweating, resettling and solvent extraction.

1. In the *sweating* process, the crude wax, *slack wax*, is melted and cast into thick sheets. These sheets are then subjected to slow heating on pans with a perforated false bottom. As the temperature rises, a mixture of oil and low-melting wax drains out. The heating is continued until the remaining wax has the desired melting point. The sweating process may be repeated for the production of good quality wax.
2. In the *resettling* process, the slack is diluted with cold naphtha and recrystalised by chilling. Final separation of the wax crystals is by centrifuging.
3. Propane or a mixture of methyl ethyl ketone and an aromatic solvent may be used in the *solvent extraction* process. These selective solvents give better separation than naphtha. The steps in solvent deoiling are similar to those in solvent dewaxing.

5.2.6 Important Petroleum Products

- The main products of a petroleum refinery are motor gasoline, aviation gasoline, kerosene, jet fuels, diesel fuels and fuel oils.
- A few other products are also obtained in large quantities, either directly from a refinery or by further processing feedstocks obtained from a refinery; these are lubricating oils, petroleum wax, bitumen, industrial naphthas, coke and liquefied petroleum gas (LPG).
- A third category of petroleum products includes medicinal oils, transformer oil, spray oils, insecticidal oils, and various types of machine oils.

Petrochemicals form an entirely different class of products.

Gasolines

Gasolines are the liquid fuels used in reciprocating spark-ignition internal combustion engines. These are broadly of two types: motor gasoline and aviation gasoline.

Table 5.9 Specifications for motor gasoline (IS: 2796-2000)

Characteristic	83 octane gasoline	93 octane gasoline
Colour, visual	orange	red
Octane number (research method), min	83	93
Ledd content (as lead, g/l, max)	0.56	0.80
Distillation		
(i) Initial boiling point	not limited but to be reported	
(ii) Recovey up to 70°C, per cent by volume, min	2	2
(iii) Recovery up to 125°C, per cent by volume, min	10	10
(iv) Recovery up to 180°C, per cent by volume, min	90	90
(v) Final boiling point, °C, max	215	215
(vi) Residue, per cent by volume, max	2	2
Reid vapour pressure at 38°C, kgf/cm^2, max	0.70	0.70
Sulphur, total per cent by weight, max	0.25	0.20
Residue on evaporation, mg/100 ml, max	4.0	4.0
Oxidation stability, minutes, min	360	360
Copper strip corrosion for 3 hours at 50°C	not worse than no.1	
Density at 15°C	not limited but to be reported	

Motor gasoline is a mixture of low boiling hydrocarbons with a final boiling point of about 215°C. It is made in a refinery by blending a number of products, such as

1. Staiglt-run (virgin) gasoline obtained directly by the distillation of crude oil,
2. Cracked gasoline produced by the cracking of gas oils,
3. Reformate obtained by the reforming of heavy naphtha,
4. Components produced by the alkylation and polymerisation of cracked gases,
5. Natural gasoline obtained as a condensate from natural gas.

Butane is added (up to about 10%) to facilitate easy start of cold engines. Natural gasoline and liquefied petroleum gas are the sources of butane.

Commercial motor gasoline contains a number of additives also. These are *anti-knock agents* such as tetraethyl lead (TEL), *combustion deposit modifiers*, such as tricresyl phosphate (TCP), and *antioxidants*, such as 2, 6-ditertiary butyl-4-methyl phenol (2, 6 B_4M) and *N*, *N^1*-disecondary butyl-*p*-phenyline diamine, to prevent gum formation. Sometimes a few other additives are also used. Metal deactivators are used to arrest gum formation due to trace metals like copper and iron. Cleansing agents are added to prevent deposits in the fuel supply system. Similarly, anti-rust agents and anti-icing agents are also added.

Table 5.9 gives the specifications for motor gasoline (IS:2796-2000). Gasolines are dyed to impart distinctive colours to different brands and grades.

Aviation gasolines are characterised mainly by higher volatility and higher octane rating than the motor gasolines. Although a higher proporation of TEL may be added, say, upto 4.6 ml per US gallon, the other additives are lesser in number and quantity. Aviation gasolines are composed mostly of *iso*-paraffin alkylates and aromatic reformates.

Table 5.10 Specifications for kerosene (IS:1459-1974)

Characteristic	Requirement
Acidity, inorganic	Nil
Burning quality	
(i) Char value, mg/kg of oil consumed, max	20
(ii) Bloom on glass chimney	not darker than grey
Colour (Saybolt), min	10
Colour-strip corrosion for 3 hours at 50°C	not worse than no.1
Distillation	
(i) Per cent recovered below 200°C, min	20
(ii) Final boiling point, °C, max	300
Flash point (Able),°C, min	35
Smoke point, mm, min	20
Sulphur, total, per cent by mass, max	0.25

Kerosene

Kerosene is used mainly in oil lamps, stoves and cookers. The desirable constituents are paraffins. Therefore, it is obtained as a straight-run distillate from selected crude oils. No additives are used to improve its quality. The purification processes are sufficient to yield kerosene with desired properties (Table 5.10).

Jet fuels

Kerosene-type fuel oils with low freezing point are mainly used in commercial jet airliners. Military jet aircraft use a 30:70 blend of a kerosene fraction and a low octane straight-run heavy naphtha fraction. Light gasoline or butane is added for adjusting the volatility. Antioxidants, metal deactivators and corrosion inhibitors are employed as additives to jet fuels.

Diesel fuels

Straight-run distillates from paraffin and mixed base crude oils are the most desired components of diesel fuels; but these are often utilised as cracing stock for gasoline production. Moreover, there is some overlap in the boiling ranges of kerosenes and diesel fuels. The latter are, therefore, prepared from heavy distillates obtained from catalytic cracking units. These distillates are rich in aromatics and iso-paraffins; the low cetane number is improved by the use of additives. The suitable volatility is obtained by blending with light fractions. Residual oils are also used in significant proportions in meeting the large demand of diesel fuels. The increased cost of maintenance—due to the use of poor fuels in diesel engines—is more than offset by the low cost of these fuels (Table 5.11).

Fuel oils

The wide range of liquid fuels used in boilers and furnaces is covered by the term fuel oil (Table 5.12). At one extremity we have a *light fuel oil* produced from a *cycle gas oil* of cracking

Table 5.11 Requirements for automotive diesel fuel

Sl. No.	Characteristics	Requirements	
		Bharat Stage II	Bharat Stage III
1.	Acidity, inorganic	nil	nil
2.	Acidity, total mg of KOH/g, max	to report	to report
3.	Ash per cent by mass, max	0.01	0.01
4.	Carbon residue (Ramsbottom) on 10% residue, per cent by mass, max	0.30	0.30
5.	Cetane number, min	48	51
6.	Cetane index, min	46	46
7.	Pour point, max		
	(a) Winter	3°C	3°C
	(b) Summer	15°C	15°C
8.	Copper-strip corrosion for 3 h at 100°C	not worse than No.1	not worse than No.1
9.	Distillation, per cent (v/v), recovered:		
	(a) at 350°C, min	85	–
	(b) at 360°C, min	–	95
	(c) at 370°C, min	95	–
10.	Flash point		
	(a) Abel, °C, min	35	35
	(b) Pensky-Martens closed cup, °C, min	66	66
11.	Kinematic viscosity, centistokes at 40°C	2.0–5.0	2.0–4.5
12.	Sediment, per cent by mass, max	0.05	–
13.	Total contamination, mg/kg	–	24
14.	Density at 15°C, kg/m^3	820–860	820–845
15.	Total sulphur, mg/kg, max	500	350
16.	Water content		
	(a) per cent (v/v)	0.05	–
	(b) mg/kg, max	–	200
17.	Cold filter plugging point (CFPP), max		
	(a) Winter	6°C	6°C
	(b) Summer	18°C	18°C
18.	Total sediments, mg/100 ml, max	1.5	–
19.	Oxiidation stability, g/m^3, max	–	25
20.	Polycyclic aromatic hydrocarbon (PAH), per cent by mass, max	–	11
21.	Lubricity corrected wear scar diameter (wsd 1.4) at 60°C, microns, max	460	460
22.	Oxygen content, per cent by mass, max	0.6	0.6

units, and at the other extremity we have *heavy fuel oils* produced from residual stocks of crude distillation units, catalytic cracking units, thermal cracking units and cokers. Several intermediate grades of fuel oils may be made by the blending of distillates and residual stocks. Visbreaking of residual stocks is a very important process in the production of fuel oils. Cycle gas oil is given such treatment as filtration through clay and hydrofining. Sludge dispersing agents and tricresyl phosphate are added to improve the combustion in burners. In the case of residual fuel oils, magnesium-bearing additives may be used largely to eliminate corrosion and fouling.

Table 5.12 Specifications for fuel oils derived from petroleum or shale (IS:1593-1982)

Characteristic	Grades			
	Low viscosity	Medium viscosity 1	Medium viscosity 2	High viscosity
Flash point (Pensky–Martens) (closed),°C, min	66	66	66	66
Kinematic viscosity, centistokes at 50°C, max	80	125	180	370
Water content, per cent by volume, max	1.0	1.0	1.0	1.0
Ash, per cent by mass, max	0.1	0.1	0.1	0.1
Sulphur, total per cent by mass, max	3.5	4.0	4.0	4.5
Sediment, per cent by mass, max	0.25	0.25	0.25	0.25
Acidity, inorganic	nil	nil	nil	nil

Lubricants

Lubricants include a wide range of products. Base oils are obtained as distillates or residual products in the refinery and are blended to meet specific requirements of individual operations. The conventional additives include fatty oils, lard and fatty acids. Some of the additives contain constituents like sulphur, lead, graphite, molybdenum disulphide or metal soaps. Anti-oxidants and anti-corrosion agents are also sometimes added to protect the metal surfaces from corrosion.

5.2.6 Properties and Testing of Petroleum and Petroleum Products

It has been seen in the previous section that the petroleum products include a wide range of substances. The properties that make them suitable for specific uses differ from case to case. No single set of test procedures is therefore available for evaluating them. Some of the common and special properties of petroleum products and the principle of the corresponding test procedures are discussed below. Indian specifications for some products are given in Tables 5.9 to 5.12

Specific gravity

Specific gravity is used in calculations involving weights and volumes. Because of the closeness of the inherent chemical nature of the main components of petroleum products, it has been possible to correlate their specific gravity with other important parameters. Higher specific gravity means higher concentration of carbon in relation to hydrogen. Hence, heavier oils have lower calorific value (gross) on a weight basis.

The reverse is true if the calorific value is expressed on a volume basis. The following equation is useful in the calculation of gross calorific values of petroleum oils:

$$C_G = 12400 - 2100d^2 \tag{5.1}$$

where, C_G is the gross CV, kcal/kg, and d is the specific gravity at 15.5/15.5°C.

In many empirical correlations, specific gravity is used as *Degree API* defined in the following way by the Americal Petroleum Institute:

$$°API = \frac{141.5}{Sp.gr.\ (15.5°/15.5°C)} - 131.5 \tag{5.2}$$

This is a slight variation of the Baume scale. Among hydrocarbons, the aromatics have higher specific gravity than the paraffins. Therefore a knowledge of specific gravity can appreciably predict the quality of a given product.

Viscosity

Viscocity is a very important property of fluids and a measure of their resistance to flow. In petroleum technology, this property is widely used in design calculations for transportation, combustion in burners and internal combustion engines and lubrication. It is an important specification parameter for many petroleum products.

Viscosity may be determined in a U-tube viscometer and other convenient apparatus. The conventional method for oils is to measure kinematic viscosity and express this in centistokes or some arbitrary units. Standard equipment and standard procedure are used to measure the time required for a fixed volume of oil to flow through an orifice of fixed dimension at a certain temperature. Different standards vary in the design of the equipment and the procedural details. The different standard viscometers used for oils are: Engler, Redwood No. I, Redwood No II, Saybolt Universal and Saybolt Furol. The result is usually expressed as the number of seconds required for the flow. In the Engler viscometer, the viscosity is expressed as a ratio of the time of flow of equal volumes of oil and water and is called degree Engler. It is possible to convert the data of one standard into those of another and also into stokes (Table 5.13).

Viscosity is one of the most important criteria of fuel oils. It significantly influences the ease of handling and transport and the nature of storage. The viscostiy of the oil supplied to atomising oil burners has a direct effect on a burner's performance, and hence on its combustion efficiency. Residual oils have high viscosity at ordinary temperatures and need preheating before atomisation. The viscosity of fuel oils falls as the temperature rises but becomes nearly constant above 120°C. For atomisation, there is little gain in heating the oil beyond 120°C. The maximum viscosity for easy atomisation in conventional burners is 25 centistokes or 100 sec Redwood No. I. For easy pumping in usual pipes, the maximum is 1,200 centistockes or 486 sec Redwood No. II.

There are two opposing effects of oil preheating on the capacity of the burner. While decreased viscosity facilitates flow of oil through the service line and atomisation in the burner, decreased density results in the delivery of lesser weight of oil to the burner. There is therefore an optimum temperature beyond which additional preheat actually lowers the burner capactiy.

The performance of diesel fuels also greatly depends upon their viscosity. Too low a viscosity causes excessive leakage at the injection pistons while too high a viscosity produces coarse oil droplets which result in the formation of engine deposits owing to incomplete combustion

Table 5.13 Viscometer conversion factors

Given viscosity	Kinematic viscosity, centistokes	Engler, degrees	Redwood I, seconds	Redwood II, seconds	Saybolt universal, seconds	Saybolt fural, seconds	Standard tar viscometer, 10 mm, seconds	Standard tar viscometer, 4 mm, seconds
Kinematic viscosity, centistokes	1.0	0.131	4.08	4.408	4.57	0.457	0.0025	0.074
Engler, degrees (from 6° upwards)	7.64	1.0	31.2	3.12	34.9	3.49	0.0191	0.565
Redwood I, seconds (from 90 s upwards)	0.245	0.0321	1.0	0.100	1.12	0.112	0.000612	0.0181
Redwood II, seconds (from 30 s upwards)	2.45	0.0321	10.0	1.0	11.2	1.12	0.00612	0.181
Saybolt universal, seconds (from 120 s upwards)	0.219	0.0287	0.93	0.089	1.0	0.10	0.000547	0.0162
Saybolt furol, seconds (from 40 s upwards)	2.19	0.287	0.93	0.893	10.0	1.0	0.00547	0.162
Standard tar viscometer 10 mm cup, seconds (from 10 s upwards)	400	52.4	1630	163	1830	183	1.0	29.6
Standard tar viscometer 4 mm cup, seconds (from 100 s upwards)	13.5	1.77	55.1	5.51	61.6	6.16	0.034	1.0

The action of lubricating oils in a particular system depends upon their particular viscosity. Owing to oxidation and other reasons, viscosity changes appreciably which necessitates the periodical replacement of the lubricant. The effect of temperature on viscosity of oils is well known. If a system is subjected to a variation of temperature, and if the viscosity of the lubricant varies beyond the tolerable limit, the purpose of lubrication is defeated. It has been found that the extent of variation in viscosity for a given temperature fluctuation is not the same for all lubricating oils. Paraffinic stocks tend to resist alteration in viscosity with temperature change while naphthenic stocks show large dependence of viscosity on temperature. This viscosity–temperature relationship of lubricating oils is denoted as the viscosity index (VI) and is expressed as a number on an arbitrary scale. A highly paraffinic Pennsylvania oil has a VI of 100 and a naphthenic Gulf Coast oil has a VI of 0. If L, H and U are the viscosities in CS of the 0 VI, 100 VI and the given oil, respectively, at 100°F.

$$VI = \frac{L - U}{L - H} \times 100$$

The viscosity of the two standard oils and the unknown sample must be the same at 210°F. The kinematic viscosity, U, of the sample is first determined at 210°F and the values of L and H are then read from the standard ASTM chart corresponding to that viscosity of the sample.

Oils with a high VI are preferred where a wide variation in temperature is encountered as in the outdoor use of crank case oils and hydraulic oils. Certain additives, for example, polybutenes improve the VI of oils. For high VI oils (VI>100), another parameter called viscosity index extension (VI_E) is used.

$$VI_E = \frac{(\text{antilog } N) - 1}{0.0075} + 100$$

$$\text{where, } N = \frac{\log H - \log U}{KVT_{210}}$$

H = kinematic viscosity at 100°F of an oil having a VI of 100 using the standard method and having the same viscosity as the unknown oil at 210°F.

U = kinemative viscosity at 100°F of the unknown oil.

KVI_{210} = kinematic viscosity of the sample at 210°F

$H > U$

At very high pressures the viscosity of oils increases considerably. At a pressure of 70 kgf/cm^2, the increase is by 15%. Most oils begin to solidify at 3,500 kgf/cm^2.

Viscosity is not an additive property. Therefore the viscosity of oil blends has to be experimentally determined or calculated with the help of empirical blending charts.

Distillation range

Distillation range is the difference in temperature between the temperature at the initial boiling point and at the end point of a distillation test. It pertains to distillate products such as gasoline and kerosene. Residual fuel oils are not normally subjected to a distillation test. The singlemost important criterion for crudes is their distillation characteristics.

In the distillation test for petroleum products, a measured volume of the oil is distilled at the specified rate which is suitably increased as the distillation proceeds. The *initial boiling point* (IBP) is the temperature when the first drop falls from the condenser and the final *boiling point* (FBP) is the maximum temperature recorded at the end of distillation. *Mid-boiling point* is the temperature at which 50% of the oil distills off. Distillation is stopped at 360°C because further heating causes cracking of the residue. If necessary the distillation may be contined under vacuum. The distillation range is of value in the evaluation of petroleum products.

Distillation parameters like mid-boiling point often find application in correlations for other important parameters.

Distillation data are conveniently used to construct very useful curves, for example, true boiling point (TBP) curves, and equilibrium or flash vaporisation curves. The TBP curves give the variation of boiling point with the yields of distillates. Each point in the flash vaporisation curve gives the total amount of vapour in equilibrium with the liquid at the particular temeprature. These curves are used in the design of continuous distillation equipment in refineries.

Flash point and fire point

The *flash point* is defined as the minimum temperature at which the given oil evolves just sufficient vapour to form an inflammable mixture with air, as shown by the formation of a momentary flame (flash) when an external source of fire is brought in contact with the vapour. The *fire point* is the minimum temperature at which the oil vapours will continue to burn instead of just flashing. For the same product the fire point is higher than the flash point. It is customary to determine and report both the temperatures, although the lower temperature (flash point) is itself a measure of the fire risk of an oil or any inflammable liquid in bulk.

The permissible conditions of storage and use are different for products with flash points below 23°C (gasoline), between 23°C and 66°C (kerosene) and above 66°C (fuel oils and gas oils). The oils of the first category are regarded as *dangerous and highly inflammable* for transport and storage, and are subject to strict government regulations. Fuel oils with flash points above 66°C are regarded as safe.

Besides depending on such inherent properties of the oil as the volatility and the inflammability limits of the vapour–air mixture, the flash point also depends upon the design of the apparatus, the test procedure and the barometric pressure. Two types of apparatus are commonly used:

1. The Pensky–Martens apparatus for oils with flash point above 50 °C,
2. The Abel apparatus for oils with flash points below this temperature.

Kerosene is tested in the Abel, and fuel oil, diesel fuels and gas oils in the Pensky–Martens apparatus. Both these apparatuses are of the closed-cup type. The oil is heated in a closed cup under specified conditions and a pilot flame is introduced into the vapour space by opening a shutter in the lid. The American Tag apparatus is similar to Abel.

Flash points of lubricating oils, some crude oils and residues are often determined by the open-cup tests. Either the Pensky–Martens cup or the Clevelend open-cup is used.The open-cup test is liable to error and gives only approximate values. The open-cup flash point is higher than the closed-cup value for the same sample.

The flash point and the fire point give indications of the range and nature of the boiling point curve. The flash point of lubricants helps to indicate the relative amount of low-boiling mateial present in it.

These temperatures must not be confused with the completely unrelated spontaneous ignition temperature (SIT) (see Chapter 8).

Pour point and cloud point

These are used as criteria for fuel oils lubricating oils and diesel fuels used in cold surroundings. The *pour point* is defined as the temperature 5°F (2.8°C) higher than that at which the oil ceases to flow when cooled and tested according to prescribed conditions. The cessation of flow results from an increase in viscosity or from the crystallisation of wax from the oil. Heavy fuel oils have quite a high pour point and heating facilities are usually necessary in order to ensure the satisfactory flow of the oil. Fuel oils of wax-bearing crudes have much higher pour point than those derived from crudes of low wax content. A low pour point is the desired property of an oil in respect of handling in cold atmosphere.

The *cloud point* is defined as the temperature at which an oil becomes cloudy when it is cooled in a specified manner. This temperature is higher than the pour point (usually by 5°C to 6°C). The cloud is due to wax crystals. If the oil is moist then also it may become cloudy due to the immisibility of oil and water at the low temperature. The cloud point is more significant than the pour point for diesel fuels where the wax crystals can plug the filters in the fuel injection system and stop the flow even when the oil is above its pour point.

Asphaltic substances act as pour point depressants, that is, they reduce the pour point of an oil by inhibiting the growth of wax crystals. Pour points may also be reduced by increasing the proportion of lighter hydrocarbons in an oil. In general, naphthenes and aromatics have a lower pour point than paraffins. Unfortunately the former are the undesired components of fuel oils and diesel fuels.

Smoke point and char value

These empirical criteria are determined to assess the burning quality of kerosene as an illuminant or as a fuel. A good quality kerosene should be clean in respect of smoke in the flame and the deposits, char or toadstool on the wick. Smoking is due primarily to aromatic hydrocarbons; char deposits are related to both aromatics and disulphides. The smoking tendencey of kerosene is determined by the smoke point test, and the deposit forming tendency by the char value test.

The *smoke point* is the maximum flame height in millimetres at which kerosene will burn without smoking in a standard apparatus. Good quality kerosene shows a smoke point of 20 to 25 mm. Careful acid treatment raises the smoke point to above 30 mm.

The *char value* is the amount of char in milligrams, for every kilogram of kerosene burned, which is formed on the wick of a standard lamp burning under prescribed conditions. The char value of a high grade kerosene should not exceed 20 mg/kg.

The formation of smoke and char depends upon the type of burning appliance also. Some appliances need kerosene of high smoke point and low char value. Yellow flame, wick-fed appliances, for example, railway signal lamps which are required to burn for prolonged periods without attention, fall into this category. Blue flame appliances, for example, space heaters, can tolerate kerosenes of lower smoke point and higher char value.

The burning quality of kerosene is also judged by its tendency to form a white or grey deposit or *bloom* on the glass chimney of lamps, which leads to a reduction in the illumination power. Disulphides are believed to be responsible for the bloom.

Carbon residue

This property is important for oils used in gas production, in diesel engines, in burners and for lubricating oils used in motors. When an oil is heated both evaporation and cracking set in. The heavier, complex compounds form, on decompostion, some carbonaceous deposits known as carbon residue. There are two methods for the determination of this residue, Conradson's and Ramsbottom's. The Conradson test is used for fuel oils and the Ramsbottom test for lubricating oils. The distillate fuels have a low carbon residue and the test is applied on the 10% residue after distilling off 90% of the fuels. Kerosenes have so low a residue that this test is meaningless and the char value is determined in lieu of carbon residue.

Fuels for high speed diesel engines should have a Conradson carbon residue of not more than 0.1% on a 10% distillation residue. Low and medium speed engines can tolerate much higher values of Conradson carbon. The nature of the carbon deposits in diesel engines is very important. If the deposit is fluffy it is easily removed with the exhaust gases. On the other hand compact deposits get firmly attached to the engine parts.

Atomising burners are practically insensitive to the carbon residue of the fuel. However the gas oils used in the vaporising burners should have Conradson carbon value not exceeding 0.05% on the 10% distillation residue.

Sulphur content

All crudes contain sulphur. Even after passing through various purification processes the petroleum products contain sulphur. In general the sulphur content increases with the increase in the boiling range of the product. Some typical values are: kerosene 0.05%–0.2%, gas oil and diesel oil 0.3%–1.5% and fuel oil 2%–4%. The high value, 4% is for the heavy furnace oils of high sulphur crudes.

The sluphur content of oils is best determined by using a bomb calorimeter. Approximate values are obtained by the lamp method in which a known weight of the oil is burned in an air current and the combustion products passed through an alkaline solution, the sulphur being finally estimated in the solution as sulphuric acid.

The sulphur content should be as low as possible. Sulphur is largely present in oils in the form of corrosive organic compounds. On combustion it gives off foul gases.

Moisture

The solubility of water in oils is quite low—0.005% to 0.05% under ordinary conditions. It may however be present in the dispersed state in residual fuel oils up to 1%. Water in oils is undesirable. Its separation from oils is facilitated by preheating. The Dean and Stark method is used for the determination of moisture in oils. The sample is heated with xylene which forms an azeotrope with water. The distilled vapours on condensation and settling give water as a separate layer which can be measured. The Karl Fischer test can be used to accurately determine water content of 50 to 1000 ppm in oils. The method consists of electrometric titration with a special reagent.

Ash

The amount of ash in petroleum products is normally very low. The distillates should have zero ash but certain additives to gasolines or diesel fuels raise its value. Even a residual fuel should not have ash exceeding 0.1%. Some constituents of petroleum ash have very harmful effects. The presence of vanadium in fuel oil causes discolouration of the molten glass tanks, while sodium causes bonding of ash constituents on boiler superheaters and also attacks refractories.

Calorific value

The ultimate analysis of petroleum fuels does not vary widely. Therefore their calorific value also varies within a comparatively narrow range, 10,000–11,300 kcal/kg gross, the higher figure being for gasolines and the lower for heavy furnace oils. Petroleum products are high in hydrogen content, 11.8% to 14.5%, and hence their net calorific value is less than the gross by as much as 620–760 kcal/kg. The calorific value of oils is determined in a bomb calorimeter. Empirical formulae are available for calculating the calorific value of petroleum fuels from their specific gravity (Eq. 5.1).

Octane number

Maximum power is derived from gasoline when it burns silently and relatively slowly. Under certain engine conditions the combustion may begin smoothly and then the whole of the unburnt fuel may burn rapidly with the formation of pressure waves (detonation). This leads to knocking of the gasoline engine. Much of the generated power is wasted and the engine life is shortened when knocking occurs. A good gasoline should resist knocking. The anti-knock quality of fuel is measured in a standard engine in terms of the relative performance of two standard fuels and then expressed as its *octane number*. By definition, the octane number of a gasoline is equal to the percentage by volume of *iso*-octane (2,2,4-trimethyl pentane) in a mixture of *n*-heptane and *iso*-octane having the same knocking tendency as the sample being tested. *n*-Heptane is of poor anti-knock quality and is assigned zero octane number. On the other hand, *iso*-octane has excellent anti-knock quality and is assigned an octane number 100. If a gasoline matches with a 80/20 blend of *iso*-octane and *n*-heptane, its octane number is 80. Secondary reference fuels are used in practice for experimentally determining the octane number of gasoline.

The anit-knock quality of a gasoline primarily depends upon the types of hydrocarbons present. Aromatics have the best anti-knock quality and *iso*-paraffins, naphthenes, olefins and *n*-paraffins follow in succession. All the refinery processes for the upgrading of gasolines aim at the conversion of other hydrocarbons into aromatics and *iso*-paraffins.

A certain minimum anti-knock quality is required for knock-free operation of a gasoline engine. This requirement primarily depends on the engine compression ratio. In the USA and Canada, the motor gasoline is marketed in two grades: regular with octane number 94 and premium with octane number 100. The motor gasolines in India have 83 and 93 as the specified octane numbers. Aviation gasolines have octane numbers exceeding 100.

The octane rating of commercial gasolines is not entirely due to the hydrocarbons only. Certain additives are put into the gasoline to raise its octane number. The most effective and common additive is tetraethyl lead (TEL) which is used in admixture with organic halides. The halides are added to prevent the deposition of lead by forming volatile lead halides. Paraffins and naphthenes have higher lead susceptiability than aromatics.

Other anti-knock agents are tetramethyl lead (TML) and tertiary butyl acetate. Methyl-cyclo-pentadienyl manganese tricarbinol may be used in conjuction with TEL for boosting the octane number of motor gasoline.

Cetane number

There is a time lag (delay period) between the injection of diesel fuel into hot compressed air and its ignition. If the delay period is unduly large there will be accumulation of too much fuel in the cylinder, which will eventually burn with an undue rapidity. A rapid rise in the cylinder pressure will follow and cause a 'diesel knock'. The delay period is inherently connected with the type of hydrocarbon. The preferred hydrocarbons in the order of decreasing ignition quality (increasing delay period) are: *n*-paraffins, olefins, naphthenes, *iso*-paraffins and aromatics. This order is the reverse of the order for gasoline anti-knock quality.

The ignition quality of diesel fuel is measured in a standard engine by matching against blends of two reference fuels and expresssed in terms of cetane number. A *n*-paraffin (*n*-hexadecane or cetane, $C_{16}H_{34}$) is given 100 cetane number and an aromatic (d-methyl naphthalene) is given zero cetane number. If the given fuel matches with a 40/60 blend of cetane and α-methyl naphthalene, it is assigned a cetane number of 40. In practice, secondary reference fuels are used in the test because the primary references are costly and not easily available.

High speed engines, above 1,500 rpm, need high cetane number (45–50) fuels. A low speed engine is not so demanding since there is more time available for the combustion. A cetane number of 25 to 30 may suffice.

Diesel fuels are composed largely of distillates. Cracked fractions are rich in *iso*-paraffins and aromatics, and hence the amount of cracked fractions in good diesel fuels should be low.

The cetane number of diesel fuels rich in aromatics and *iso*-paraffins may be improved by the addition of alkyl nitrates (amyl and butyl nitrate), carbamates (isopropyl-*n*-methyl-*n*-nitrocarbamate), peroxides (ditertiary butyl peroxide) and metal-organic compounds. These additives are less effective towards catalytically cracked stocks than towards straight-run stocks.

Aniline point

This is defined as the temperature at which a liquid just becomes completely miscible with an equal volume of aniline. Aromatics are easily miscible with aniline and have a low aniline point. On the other hand, paraffins have high aniline point. This test is therefore an indication of the type of hydrocarbon present in a petroleum product. Hexyl benzene has an aniline point of -12°C while cetane has a value of 95.1°C. A high aniline point is an indication of high suitability as a diesel fuel.

The aniline point of a diesel fuel is usually converted into an empirical diesel index with the help of the following formula:

$$\text{Diesel index} = \frac{\text{Aniline point, }^{\circ}\text{F} \times \text{Degree API}}{100}$$

Numerically the diesel index is usually three units higher than the cetane number. It is not a very reliable parameter, however its ease of determination makes it popular in the evaluation of the ignition quality of diesel fuels.

5.2.7 Petroleum Refining in India

Petroleum refining in India started in 1893 with the setting up of a small refinery at Margherita in Assam, which is now non-existent. Set up in 1899, the Digboi refinery represents the first major refinery in the country. With an initial refining capacity of 0.25 million tonnes of crude, this was the only refinery up to 1954 and could meet 7% of the demand of petroleum products in the country. The crude was available from the Digboi fields.

In the mid 1950s, three coastal refineries were set up by foreign-controlled oil distributing companies, two on the west coast and one on the east coast. The crude rrequirements of these refineries were met by imports. Following the discovery of oil in new areas in Assam and Gujarat, three government-controlled refineries were set up between 1962 and 1965 at Guwahati, Barauni and Koyali in succession. These were all based on indigenous crudes. In the absence of substantial new oil finds, and with the growth in the demand of petroleum products in the country, three more coastal refineries were set up on the basis of imported crudes, two in the south and one in the east.

After the discovery of the large oil reserves in the Bombay High offshore areas, the import of crude oil was sharply reduced and many refineries started operation with indigenous crude. One more refinery was set up at Bongaigaon in Assam in 1981 and, for the first time, a large refinery was built at Mathura in Uttar Pradesh in 1984 to process indigenous crude from

Bombay High and Gujarat. The government has approved two more large refineries, one on the west coast and the other in the deep north. The list of all the refineries is given in Table 5.3.

The growth of the petroleum industry in India may be judged from Tables 5.2 and 5.4. Both the production of crude oil and the throughput of refineries have increased remarkably. The country has to import more than 10 million tonnes of crude from the Middle East while the indigenous production is hovering around 30 million tonnes. The production of all the petroleum products taken together is now slightly in excess of their total consumption of over 40 million tonnes.

There is some imbalance in the patterns of production and consumption of petroleum products in India (Table 5.14).

The existing refineries can meet the present requirement except in certain products like kerosene and lubricants. Compared with other countries, the per capita production of petroleum products is still very low.

The flow-sheet of a refinery in India is shown in Fig. 5.6

Lubricating oils are made in the refineries at Digboi, Barauni, HPCL (Mumbai), Chennai and Haldia. There are two separate lube oil blending plants, one each at Kolkata and Mumbai.

5.3 LIQUID FUELS FROM SOURCES OTHER THAN PETROLEUM

5.3.1 Coal Tar and Coal Tar Fuels

Where distillation facilities are not available, crude coal tar is used as a furnace fuel. Like other crude fuels, it possesses a number of disadvantages:

- the presence of volatile components which are lost on storage and which increase the fire hazard,
- the presence of varying amounts of water which reduce the calorifc value, leading to corrosion and irregular combustion,
- the presence of solid impurities such as coke dust which may cause choking of the burners and pipe lines.

Because of the above considerations and because of the fact that coal tar is a storehouse of chemicals, its direct use as a fuel is discouraged. It is preferable to distil the tar into various fractions and then prepare liquid fuels from certain distillate fractions and their blends with the residual pitch. These fuels are known as coal tar fuels (CTF).

The coal tar fuels are designated by numbers which are the temperatures in degrees Fahrenheit to which they must be heated to reach a condition suitable for atomisation, that is having a maximum viscosity of 25 centistokes. The range of coal tar fuels covers CTF 50, CTF 100, CTF 200, CTF 250, CTF 300 and CTF 400. For example, CTF 200 is the fuel which is to be preheated to 200° F for ensuring a decreased viscosity for smooth atomisation. CTF 50 does not need any preheating.

CTF 50 and CTF 100 are the blends of the creosote oil, 230 to 270°C, and anthracene oil, 270° to 350°C, fractions. CTF 400 is the distillation residue with a softening point of 75–80°C,

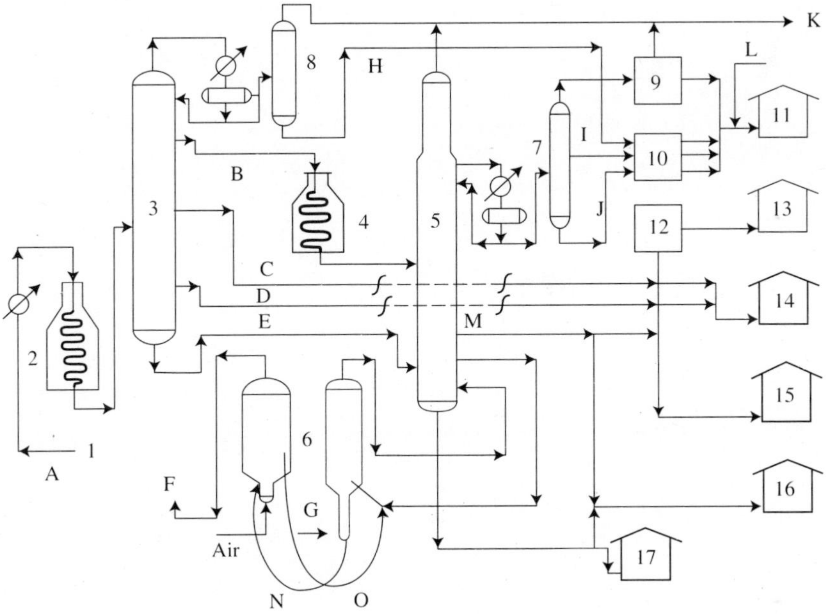

Figure 5.6 Flow-sheet of a refinery in India

A. Crude
B. Heavy virgin naptha
C. Kerosene or light gas oil
D. Heavy gas oil
E. Atmospheric bottoms
F. To stack
G. Stripping steam
H. Light virgin naphtha
I. Light cracked
J. Heavy cracked
K. Fuel gas
L. Gasoline additives
M. Light cracked gas oil
N. Regenerator
O. Reactor

1. Starting point
2. Furnace
3. Crude tower
4. Reformer furnace
5. Fractionator–absorber
6. Catalytic cracking unit
7. Debutaniser
8. Stabiliser
9. Catalytic polymerisation
10. Naphtha treating
11. Gasoline
12. Kerosene treating
13. Kerosene
14. Auto diesel fuel
15. Industrial diesel fuel
16. Fuel oil
17. Hot heavy stock

that is ‘medium soft’ pitch. CTF 200, 250 and 300 are the blends of the creosote oil, anthracene oil and residual pitch. While CTF 400 is used as a liquid fuel in installations close to the site of production, a hard pitch is also manufactured for use as a fuel in pulverised form.

The most popular coal tar fuel is CTF 200 which is normally a 50:50 pitch-creosote blend. A typical ultimate analysis is: C 89.1, H 6.4, N 1.1, S 0.5, O 2.9 and ash 0.07. Other CTFs

Table 5.14 Growth of the Indian petroleum industry

Item	Unit	2000–01	2002–03	2003–04	2004–05	2005–06	2006–07*
1. Reserves (Balance recoverable)							
(i) Crude oil	Mt	703	741	733	739	786	756
(ii) Natural gas	10^9 ml^3	760	751	854	923	1101	1090
2. Consumption							
(i) Crude oil (in terms of refinery crude throughput)	Mt	103.44	112.56	121.84	127.42	130.11	146.55
(ii) Petroleum products (excl. RBF)	"	100.07	104.13	107.75	111.63	113.21	119.55
3. Production							
(i) Crude oil	"	32.43	33.04	33.37	33.98	32.19	33.99
(ii) Petroleum products	"	95.61	104.14	113.46	118.58	119.75	135.26

*: Provisional: As on 1 April of initial year.
Source: Public Sector Undertakings, Ministry of Finance, Govt. of India.

show slight variation from these figures. In general, carbon content increases and hydrogen and oxygen content decrease in the series CTF 50 to 400. The C:H ratio increases from 11.0 to 16.5 in this series. These high values are due to the aromatic nature of the coal tar fuels.

Table 5.15 gives the important properties of coal tar fuels. The contents of sulphur and ash are very low. Besides, the harmful element, vanadium is negligible in them. The alkali content is also very low.

Coal tar fuels have two outstanding features—low sulphur content and high flame emissivity. The latter property is associated with the aromatic nature of these fuels. Superiority of coal tar fuels to other liquid fuels is best displayed in open hearth furnaces in many metallurgical processes where low sulphur content and high flame emissivity are fully exploited. Other applications of CTFs include the productioin of aluminium and cement in rotary furnaces, the making of glass and steam raising.

Mention should be made of the use of the light oil fraction, upto 180°C of coal tar in the production of motor benzol. Motor fuel and diesel fuel are also obtainable from the creosate oil fraction by suitable treatment of hydrogenation.

5.3.2 Liquid Fuels from Coal and Coal Tar Fractions by Hydrogenation

The conversion of coal into oil is essentially a process of raising the H:C ratio. The destructive hydrogenation of coal in the presence of catalysts yields oils and is known as the Bergius–Pier

Table 5.15 Properties of coal tar fuels

Property	CTF 50	CTF 100	CTF 200	CTF 250	CTF 300	CTF 400	Pulverised grade
Gross calorific value, kcal/kg	9,150–9,750	9,150–9,750	9,000–9,450	9,000–9,300	8,900–9,200	8,750–9,050	8,750–9,050
Viscosity • Redwood No.I at 37.8 °C, seconds	60 max	100 max	1,000–1,500	–	–	–	–
• Standard tar viscometer, 100 mm cup, seconds	–	–	–	–	70–120 at 30°C	40–80 at 55°C	–
Softening point, °C, ring and ball	–	–	–	–	33–38	78–85	110
Density at 37.8°C kg/l	0.95–1.05	0.95–1.05	1.09–1.21	1.09–1.21	1.14–1.26	1.19–1.31	–
Insoluble matter (in toluene), per cent by weight	nil	nil	5–10	5–15	10–20	10–20	–
Ash, per cent weight	nil	nil	0.1	0.1	0.2	0.2	0.3
Flash point(Pensky–Martens closed) °C	–	–	greater than 65°C	–	–	–	–
Sulphur, per cent by weight	–	–	less than 1.0	–	–	–	–

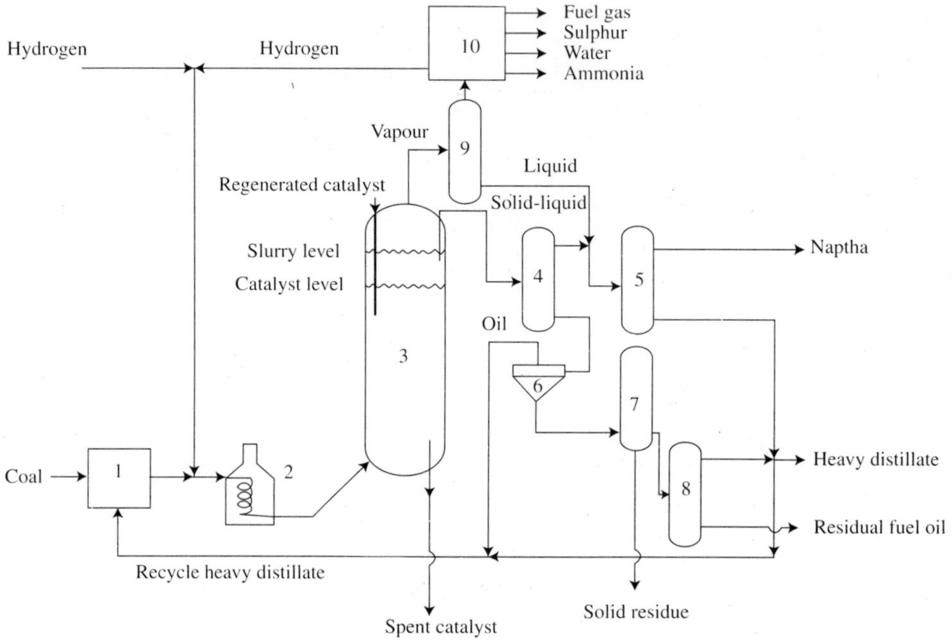

1. Slurry preparation 2. Pre-heater 3. Ebullated bed reactor 4. Flash separator 5. Still 6. Hydroclone 7. Liquid–solid separator 8. Vacuum still 9. Condenser 10. Gas treatment and separation

Figure 5.7 H-Coal process

process. The first commercial plant based on this process was installed in Germany in 1927 to convert brown coals into gasoline. During the Second World War Germany had a number of such plants which supplied the majority of her liquid fuel requirements.

The production of liquid fuels by the hydrogenation of coal is not yet competitive with petroleum refining. Currently there is no commercial plant, although efforts are being made to improve the technical and economic aspects of coal hydrogenation.

The conventional coal hydrogenation process operates in two stages:

1. Liquid-phase hydrogenation to produce a *middle oil,*
2. Vapour-phase hydrogenation of the middle oil to gasoline.

The coal is first ground and made into a paste with a heavy recycle oil and catalyst which is a compound of iron, tin or molybdenum. The paste is preheated and then treated with hydrogen at high presure, 250–350 atm or more at a temperature of 450°C to 500°C. The unreacted coal is filtered off and the liquid product distilled. The middle oil is further hydrogenated. The light oil gives gasoline, while the heavy oil is recycled. The main product of the second stage—vapour-phase catalytic hydrogenation of the middle oil—is gasoline. The heavier products are recycled. Diesel fuels and fuel oils may also be produced if desired, but the cost of these products are much higher than the corresponding petroleum products under existing techno-economic conditions in the world today.

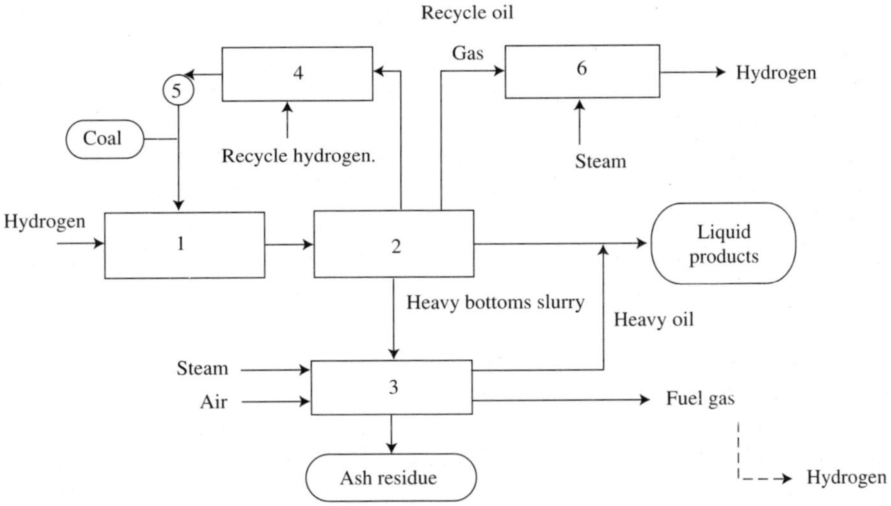

1. Non-catalytic liquefaction 2. Atmospheric and vacuum distilliation
3. Flexicoking 4. Catalytic solvent hydrogenation 5. Preheating 6. Steam reforming

Figure 5.8 EDS process

In the new processes of coal hydrogenation under development, the main objective is to accomplish the process at lower pressures by selecting a suitable catalyst and by varying the process flow-sheet.

The H-Coal process is an advanced process of coal liquefaction by catalytic hydrogenation. It has been developed in the USA. A simplified flow-sheet is given in Fig. 5.7.

The H-Coal process is a catalytic hydroliquefaction process using an ebullated bed reactor. Crushed coal is mixed with recycle oil into a paste, compressed to 200 bar and added, with hydrogen, first to a preheater and then to the reactor. The hydrodynamics of the reactor are so maintained that the catalyst level is well below the slurry level. The hydrogenation takes place at 450°C and 200 bar. The catalyst is in the form of pellets of cobalt molybdate on alumina. The products are taken out of the reactor and separated into gases, liquids and solid residue. The liquids on further processing give gasoline, diesel, fuel oil and other products. The gases are treated for hydrogen recovery by steam reforming. The solid residue is gasified and treated for hydrogen production.

The advantages of the H-Coal process relate mainly to the ebullated bed reactor which is based on the commercial H-Oil process. The main criticism is due to the likely contamination and degradation of the catalysts coming in contact with coal and its mineral content. A variation of the coal hydrogenation process is the coversion of coal into a solution in tetralin or a similar solvant (Pott–Broche process) and then treating the solution with hydrogen in the presence of catalyst under suitable conditions of temperature and prerssure. One advanced process under this category is the Exxon Donar Solvent (EDS) process developed in the USA.

The EDS process is based on petroleum industry technology and is stated to be simple and flexible enough to accept a range of coals and produce a spectrum of products to meet market demands. Figure 5.8 shows the simplified block diagram. The coal is ground, mixed with recycle solvent oil, preheated and added with hydrogen to a plug-flow reactor operating up to 465°C and 130 bar; conditions may be altered to suit coals and product requirements. The unique feature of the process is that the recycle solvent is hydrogenated in a separate catalytic reactor before being used for the slurry preparation. There is no extrernal catalyst in the hydrogenation of coal. Its mineral matter, however, has a catalytic effect. All the products are recovered by distillation and there is no filtration stage. The primary products can be hydrogenated further. The distillation residues are subjected to a patented process, *flexicoking*. The residues are coked to produce further liquid hydrocarbon products and the resultant coke is gasified for the generation of fuel gas and hence, hydrogen. The conversion of coal may be deliberately limited to 70% to provide enough residue to balance the hydrogen requirement.

Of the coals of different ranks, the bituminous rank (82%–87%) carbon are the easiest to hydrogenate. Brown coals need more severe conditions. Coal tar—both high and low temperature—are easier to hydrogenate owing to higher H:C ratio than coal. There are a number of commercial hydrogenation plants in West Germany, East Germany, Czechoslovakia and Poland for the conversion of coal tar and petroleum residual oils into gasoline, diesel fuel and other products.

The coal tar is usually hydrogenated in a two-stage process. The first stage consists of a liquid phase hydrogenation at 250 atm and 460–470°C in the presence of iron or molybdenum catalyst. The second stage is vapour phase hydrogenation using fixed bed catalysts, molybdenum or tungsten compound under the conditions of high temperature (400°C) and pressure (300 atm). Many variations of the tar hydrogenation process have been developed. While gasoline is the main target product, it is also intended to produce diesel fuels and fuel oils by reducing the severity of reactor conditions.

A novel development in the processing of low temperature tar is its conversion into a material similar to crude petroleum oil by performing mild hydrogenation at about 50 atm. This *synthetic crude* can then be processed in a petroleum refinery, either alone or in blends, with natural crudes for the production of high octane gasoline, diesel fuel and other products.

It has also been proposed to convert suitable coals into synthetic crude in the above line and then process it in a petroleum refinery.

The concept of a coal refinery appears tangible. By using a chosen route coal may be converted into liquid fuels, chemicals and other products in such a coal refinery.

5.3.3 Liquid Fuels from Coal by the Fisher–Tropsch Process

The process is based upon the catalytic hydrogenation of carbon monoxide when the following reactions take place:

$$n\text{CO} + (2n + 1)\text{H}_2 \rightarrow \underset{\text{paraffin}}{\text{C}_n\text{H}_{2n+2}} + \text{H}_2\text{O}$$

$$n\text{CO} + 2n\text{H}_2 \rightarrow \underset{\text{olefin}}{\text{C}_n\text{H}_{2n}} + n\text{H}_2\text{O}$$

By the selection of a suitable temperature, pressure and catalyst, the yields of hydrocarbons in the boiling range of gasoline and diesel fuel are maximised. Owing to the straight chains, the gasoline is of a low octane number and needs catalytic reforming before use. For the same reasons, the diesel fuel is of excellent quality (cetane number exceeding 100). The other products range from gaseous hydrocarbons to solid wax.

There are many variations of the original Fischer–Tropsch process. By changing the process variables, the mixtures of carbon monoxide and hydrogen are made to yield many chemicals, for example, alcohols, ketones and fatty acids.

While the catalytic reactions between CO and H_2 remain the main basis of the Fischer–Tropsch process, its commercial success largely depends upon the methods of preparation of the mixture of CO and H_2 (synthesis gas) and its purification from impurities like sulphur compounds, carbon dioxide, and other compounds which cause serious catalyst poisoning. About 80% of the operational costs of the process is taken up by the manufacture and purification of synthesis gas.

Synthesis gas is produced by the gasification of solid fuels like coal and coke by adjusting the H_2:CO ratio to 2:1. The gasification processes will be discussed in Chapter 6.

The original catalyst for the Fischer–Tropsch synthesis was cobalt. Now cheaper iron oxide catalysts are used. The process is carried out at 25–30 atm pressure, the temperature being up to 270°C for the Arge fixed-bed reactor. All the reactions are highly exothermic. Therefore the control of the reactor temperature is critical.

The Fischer–Tropsch process was developed on a commerecial scale in Germany before the Second World War and supplied a sizeable proportion of her liquid fuel requirements during the war. At present there are three large commercial Fischer–Tropsch plants in the world for the manufacture of liquid fuels from coal and are located in South Africa (SASOL plants, I, II and III). Many other countries have investigated the Fischer–Tropsch process on pilot and semi-commercial scales. The process is in a highly developed stage and can be widely adopted whenever the economics are favourable. The SASOL I plant has been on stream since 1955. The three plants together supply about 70% of South Africa's motor fuel requirements. The success of the plant is largely due to the plentiful supply of cheap coal and the lack of crude oil resources in South Africa.

In the SASOL plants, the raw material is sub-bituminous coal and the Lurgi process is used to gasify it. The purification of the synthesis gas from sulphur compounds and CO_2 is achieved by washing with methanol at – 60°C at an elevated pressure (Rectisol process). The purified gas contains 56% H_2, 27% CO, 14% CH_4 and small quantities of carbon dioxide and nitrogen. It is either passed through an Arge fixed-bed reactor or reformed with oxygen and steam with the help of nickel catalyst at 20 atm and 1,100°C to convert methane into additional quantities of hydrogen and carbon monoxide before passing into a Synthol entrained-bed reactor. The higher temperatures in the Synthol unit give increased yields of gasoline and comparatively low yields of heavier oils and waxes. The Arge unit uses lower temperatures and produces a smaller yield of gasoline, but larger yields of heavier products. The gasoline from the Synthol unit is of much higher octane rating.

Besides fuels (LPG, gasoline, diesel oil and wax oil), the SASOL plant also produces chemicals (methanol, ethanol, methy ethyl ketone and acetone). A simplified flow diagram of this plant is given in Fig. 5.9.

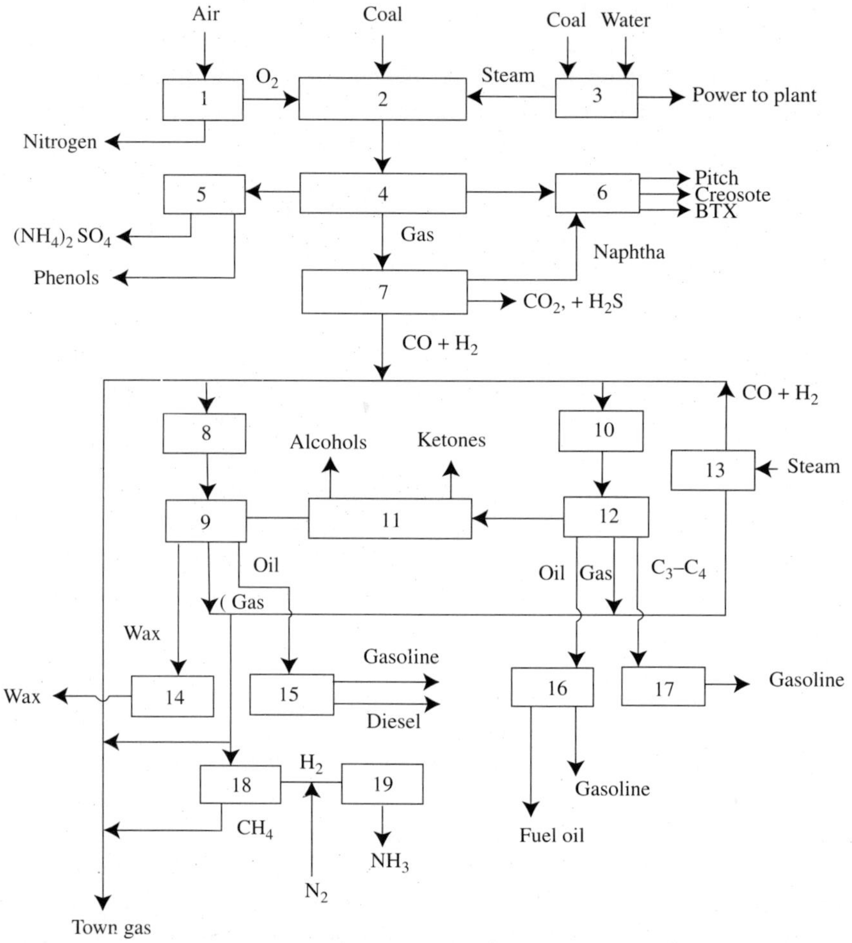

1. Oxygen plant 2. Lurgi gasifier 3. Power plant 4. Separation 5. Phenasolvan unit 6. Tar and oil work-up 7. Rectisol unit 8. Fixed-bed Arge reactor 9. Separation 10. Entrained bed synthol reactor 11. Oxygenate work-up 12. Separation 13. Steam reformer 14. Hydrogenation 15. Oil work-up 16. Oil work-up 17. Oligomerisation 18. Cryogenic separation 19. Ammonia synthesis

Figure 5.9 Simplified flow-diagram of Fisher–Tropsch process

A recent development of great potential is the productioin of gasoline from methanol on a commercial scale by the Mobil process. A new route has thus been opened up for conversion of coal into synthetic petroleum products via methanol (Fig. 5.10). Coal is gasified and the raw gas is treated for generating a mixture of carbon monoxide and hydrogen called *syngas*. Methanol is synthesised from the syngas and then converted over a special zeolite catalyst at about 400°C and 20 atm into gasoline and by-products. The catalyst may be in the form of

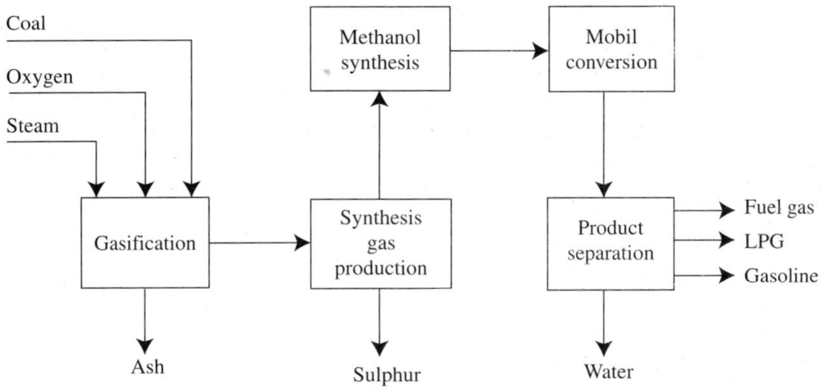

Figure 5.10 Mobil process for the production of gasoline from methanol

fixed bed or fluidised bed. Hydrocarbons are produced from methanol at a yield of over 99% of the theoretical, and the share of gasoline in these hydrocarbons is about 80%, the by-products being liquefied petroleum gas and light gas.

There is a 6,00,000 tonnes per year (tpy) gasoline commercial plant, based on natural gas in New Zealand using the Mobil process. West Germany has a demonstration scale Mobil plant based on bituminous coal.

5.3.4 Motor Benzol

Crude benzol is a valuable by-product of the coal carbonisation industries. It contains mainly hydrocarbons together with some acids and bases. A wash with alkali removes the acids and another with 40% sulphuric acid removes the bases. The neutral material on distillation gives the first fraction containing carbon disulphide and some low boiling unsaturated hydrocarbons. The next fraction, 80–150°C, represents refined benzol and the heaviest fraction, >150°C, consists mainly of high boiling unsaturated hydrocarbons.

The refined benzol is essentially a mixture of benzene, toluene and xylenes. While it can be rectified to get pure aromatic hydrocarbons, there is a demand for its use as a component of motor gasolines owing to its high anti-knock quality (about 115 octane number). By blending 50% of motor benzol, the octane number of a gasoline may be raised from 70 to 80. Motor benzol is not used as a motor fuel by itself owing to high initial boiling point, 80°C, high freezing point, −50°C and large carbon deposits. Benzol can be used also as a ternary blend with gasolines and alcohols as motor fuel.

The light oil of coal tar distillation is another source of motor benzol. In fact crude benzol and the light oil are mixed together before making the recovery of motor benzol. Crude benzol is water white in colour and is classified as 90s benzol, 75s benzol, and so on, depending upon the percentage distilling below 100°C. The gasolines in India do not contain motor benzol.

5.3.5 Alcohols

Methyl and ethyl alcohols can be used in gasoline engines. Cost considerations prohibit the use of any alcohol in burners. The octane numbers of methyl and ethyl alcohols are quite high, 114 and 99, respectively, but their calorific values are low, 5,335 and 7,120 kcal/kg, respectively. The flash point is −1.1°C for methyl alcohol and +12.8°C for ethyl alcohol. The lead susceptibility is moderate.

The importance of ethyl alcohol is due to the fact that it is produced from carbohydrates which are perennially synthesised in nature. Modern technology produces almost anhydrous alcohol at a reasonable cost. The consumption of alcohols does not lead to any void, unlike the utilisation of fossil fuels which exhausts the 'store' of the earth.

Alcohols are used in motor gasoline blends up to 10%–20% called *gasohol*. They serve to raise the octane rating of straight-run gasolines. There is a strong case for their large uses in countries where petroleum is scarce and alcohols are cheap and abundant. The estimated consumption of ethyl alcohol as fuel in Brazil in 1982 was about 3.7 billion litres out of the total production of 4.2 billion litres. Ethanol production is predicted to reach 26.4 billion litres in 2008 in Brazil.

5.3.6 Shale Oil

Shale oil is obtained by the destructive distillation of oil shale. The latter is widely distributed over the surface of the earth. It is a fragile, agrillaceous sedimentary rock resembling slate. As mined it is dark brown or grey-to-black in colour.

The reserves of oil shales in the world amount to well over 20,000 million tonnes. Considerable deposits are in the USA, the old USSR, France, Sweden, Tasmania, Australia, Manchuria, the UK, and South Africa. Commercial activities are mostly confined to Estonia (old USSR), Scotland (UK), Australia, Sweden and Manchuria. The present shale oil production is negligible owing to the availability of petroleum. Oil shales are very important standby resources for liquid flels.

Oil shales are essentially a mixture of an organic mass and mineral matter. The mineral content varies from 33% to above 70%. The organic mass is called *kerogen*. It is composed of the remains of a diversity of vegetable life. Oil shales differ widely from each other in respect of the nature and content of kerogen. The elementary analysis of the kerogen of an Estonian shale is: C 76.5–77.5, H 9.1–9.3, S 1.7–2.2, Cl 0.5–0.7, N 0.2–0.5, and O 10.5–12.0, all in per cent daf.

No appreciable quantity of oil is obtainable by simple solvent extraction of the shales. A process of thermal decompostion breaks down the complex structure of kerogen and produces oil. The oil yield depends upon the nature of the shale and the method of processing and may be as high as 50% of the shale in some cases. More common yields are 10% to 20%.

Shale mining is similar to coal mining. Most of the world's oil shale is extracted by the underground method. Opencast mining is also practised. The run-of-mine shale is prepared through crushing and screening and then charged into retorts, of which there are many designs.

Vertical externally heated retorts are preferred. The retort temperature varies from 480°C at the top to 750°C at the base.

The gases and vapours are condensed, and the recovered products are: crude oil, ammonia liquor, crude naphtha (adsorbed from the gas after the recovery of oil) and gas. The crude oil is dark green in colour and has a specific gravity of 0.860 to 0.890. Compared to crude petroleum, it contains larger amounts of unsaturates, nitrogen bases and phenolic compounds. Like petroleum, shale oils are also of three types such as paraffin-base, asphalt-base and mixed-base. The oils from Scottish, Australian and Manchurian shales are of paraffin base. Those from Estonian, Swedish and Tasmanian shales are asphaltic. Oils from Colorado shale (USA) may be classified as of mixed-base.

The refining of crude shale oil is basically similar to petroleum refining. Motor gasoline and diesel fuel are the main products. These are comparable to the corresponding petroleum products. Wax is recovered from the wax-bearing oils.

The disposal of spent shale is a big problem. Brick making is one use. Very large idle land areas are required for dumping the spent shale.

5.3.7 Oil from Tar Sands

Tar sands are also known as oil sands and bituminous sands. These are sand deposits impregnated with dense, viscous petroleum-like material or bitumen. Tar sands are found throughout the world, the largest deposits being in Athabasca, Canada and in Ornico, Venezuela. The former contains more than 1,86,000 million kilolitres and the latter more than 1,60,000 million kilolitres of bitumen. The world's total deposits are about 4,50,000 million kilolitres of bitumen in tar sands.

The production of synthetic crude oil from tar sands is now a commercial reality. The first venture of 7,200 kl/day design capacity came on stream in 1967 and the second began in 1978 and nearly approached its rated capacity of 20,500 kl/day in 1981. Both are in Canada.

The properties of the bitumen recovered from Athabasca tar sand are: specific gravity (°API) 1.00–1.014 (8–10), viscosity at 60°C, 37 p, pour point 10°C. Elemental analysis: C 83.1, H 10.6, S 4.8 N 0.4, and O 1.1. Hydrocarbon type: saturated 22%, aromatic 21%, resins 39% and asphaltenes 18%, respectively; vanadium content 250 ppm and nickel content 100 ppm. Distillation temperatures: initial boiling point 260°C and 50% boiling point 530°C.

The sands are obtained by surface mining. Bitumen may be recovered from them by several methods such as (i) direct coking, (ii) anhydrous solvent extraction, (iii) cold-water separation process, and (iv) hot-water process. In the hot-water process, the tar sand is heated and mixed with water to form a pulp of 60%–85% solids at 80–90° C. In a settler the sand sinks to the bottom and the bitumen floats up in the form of a froth. The middlings are subjected to froth flotation in air for scavenging another quantity of bitumen-laden froth. The froth is from the bitumen which contains 1%–2% mineral (dry basis) and 5%–15% water (wet basis).

Bitumen is a hydrogen-deficient oil. The primary upgrading process improves the H:C ratio by either carbon removal (coking) or hydrogen addition (hydrocracking) and produces lighter products which are more easily processed downstream. The secondary upgrading involves further hydrogenation which causes further cracking and also removal of sulphur and nitrogen. The upgraded or synthetic crude is finally refined to marketable products such as gasoline, jet fuel and fuel oil by conventional means.

Gasification of liquid fuels

Gas oils and residual oils from petroleum refineries are conveniently converted into gaseous fuels either in separate oil gasification units or in the carburetted water gas generator units. Naphtha and gas oils are also converted into synthesis gas for the manufacture of chemicals and fertilisers.

The different processes of liquid fuel gasification will be taken up in Chapter 6.

5.4 STORAGE AND HANDLING OF LIQUID FUELS

Liquid fuels of low flash point, such as gasoline, are stored in tanks fitted with roofs which float on the surface of the liquids. The elimination of any air space above the fuels reduces fire hazards and evaporation losses. These fuels are transported from the main installations to the distribution depots and to consumers either by pipeline or by tankers and other containers. The storage tanks in the filling stations are sited below ground. All installations handling liquid fuels with low flash point have to meet various prescribed requirements.

Liquid fuels with flash point well above room temperature, such as kerosene, diesel fuel and fuel oils, are stored in tanks provided with fixed conical shaped roofs. Heating coils are fitted to the storage tanks of the heavier grades of fuel oils and CTFs to maintain them in an easily pumpable condition.

All fuel oils contain water which settles to the bottom of the tank and is removed through a drain cock. In the case of CTF, the water floats on the fuel and is removed by a draw-off arrangement.

The lighter oils are usually pumped to burners without preheating. Heavier oils require heating in two stages, first in the storage and then at the time of delivering to the burners. Tank heating is achieved by low pressure steam coils situated as low as possible in the tank. No lagging of these tanks is usually necessary. Hot oil pipe line should be lagged and there should be electric or steam tracer lines within the lagging to balance the heat loss from the pipes through the lagging. The preheating of the heavier oils before pumping and before atomisation should be only up to the temperature required by viscosity considerations and fixed by plant trials. Excessive heating should be avoided for attaining heat economy. Steam heaters are preferred to oil preheaters.

Liquid fuels are supplied to furnaces by gravity feed system or ring main system. The former comprises either an overhead feed tank or a ground level feed tank with a pump in the delivery line. The burners get the fuel through a single feed line. This system is used only when one or two burners are involved.

The ring main system is preferred for medium and heavy fuel oils burnt in a number of burners fitted to one or more furnaces. This ensures the minimum pressure and temperature differences between the burners. Figure 5.11 shows a typical flow diagram of a ring main

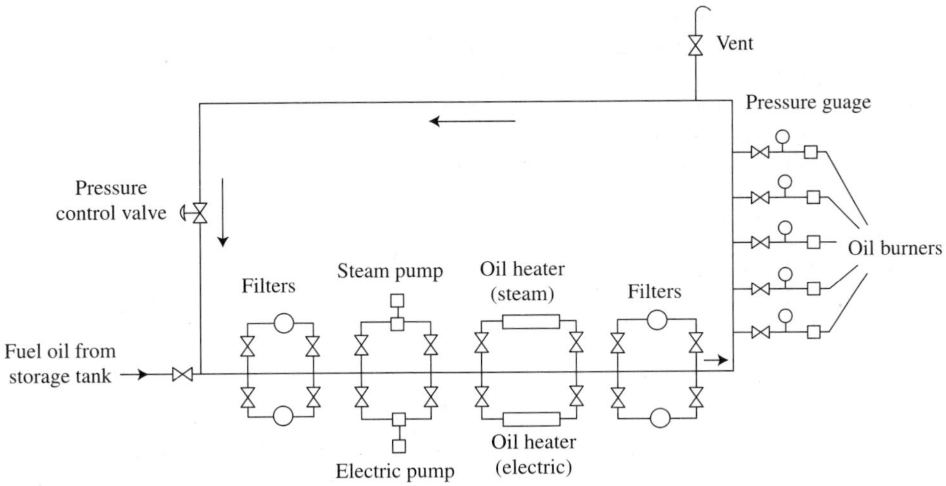

Figure 5.11 Ring main system

system. The oil is circulated through a ring main round the factory at a rate of two to three times the actual consumption of the burners. Branch lines are used to feed the individual burners. There are oil filters both at the suction and the discharge sides of the pump filter for protecting the pump as well as the burner nozzles from dirt. There is an oil heater in the ring main and no separate heating is necessary in the branch lines. The return leg of the ring main is necessary in the branch lines. It is connected to the suction end of the pump and not to the storage. This increases the heat economy. Duplicate pumping, heating and filtering units are provided to ensure continuous plant operation.

The oil in the ring main system described above is at the atomisation temperature. In a variation of this circulation system there is no heating in the ring main and the oil is kept at a pumpable temperature only. The branch lines are provided with small heaters to feed the burners with the oils at the correct atomisation temperature.

In the case of very highly viscous oils, the fuel handling system consists of concentrically steam-jacketed pipelines, heated pumps and heated filters.

Bibliography

1. Bureau of Indian Standards, New Delhi
   IS: 1447 Pt 1 (2000) Methods of sampling of petroleum and its products
   IS: 1447 Pt. 2 (1992) Sampling of LPG
   IS: 1447 Pt. 3 (1992) Sampling of Semi-solid and solid petroleum products
   IS: 1447 Pt. 4 (1989) Sampling of Petroleum Coke for laboratory analysis
   IS: 1448 Methods of test for petroleum and its products (Pts 1–148)
      Pt 3 (1984) Aniline point
      Pt 4 (1984) Ash, sulphated ash and water soluble ash

Pt 5 (1970) Burning quality
Pt 6 (1984) Calorific value by bomb calorimeter
Pt 7 (2004) Calorific value by calculation
Pt 8 (1967) Carbon residue by Ramsbottom method
Pt 9 (1960) Cetane number
Pt 10 (1970) Cloud point and pour point
Pt 17 (1960) Disel index
Pt 18 (1991) Distillation
Pt 20 (1998) Flash point by Abel
Pt 21 (1992) Flash point by Pensky–Martens
Pt 23 (1991) Hydrocarbon types by FI adsorption
Pt 25 (1976) Viscosity
Pt 31 (1968) Smoke point
Pt 32(1992) Density
Pt 33 (1991) Sulphur by bomb
Pt 34 (1979) Sulphur by Lamp method
IS: 2796 (2000) Specification for motor gasoline

2. Eqbal J and Sarkar S. Coal Refinery–A Tangible Concept, *Petroleum and Hydrocarbons*, Vol. 7. 1 (1972).
3. Hobson GD and Pohl W, (eds.). *Modern Petroleum Technology*, Fourth edition, London: Institute of Petroleum, 1975.
4. Huxtable WH, (ed.). *Coal Tar Fuels*. Second edition. London: Association of Tar Distillers, 1961
5 Petroleum, *The Wealth of India, Raw Materials*, Vol. 7, 318. New Delhi: Council for Scientific and Industrial Research, 1966.
6. Petroleum Refining, *The Wealth of India, Industrial Products*, Pt 6, 233, New Delhi: Council for Scientific and Industrial Research, 1965.
7. *Standard Methods for Testing Petroleum and its Products*. London: Institute of Petroleum, 1960.
8. Williams DA and James G. *Liquid Fuels,* London: Pergamon, 1963.

6 Gaseous Fuels

6.1 INTRODUCTION

Just as coal powered the industrial revolution of the nineteenth century and oil fuelled the growth of the twentieth century, gas is emerging as the preferred source of energy to power the economics of the twenty-first century, and India is no exception. High growth in population, increasing urbanisation, rising incomes and standards of living of the people of India are expected to result in high and steady consumption growth in natural gas along with petroleum in the country.

The remarkable features of gaseous fuels are the absence of mineral impurities, the consistency in quality and the convenience and efficiency in use. On an industrial scale, a gas handling system is the least expensive of all the fuels. It comprises essentially of a network of piping along with safety and control equipment. Except in the case of liquefied petroleum gas, no storage space is usually required as the suppliers are responsible for delivery. However, large consumers have gas holders and many of them produce their own gas. Whether purchased from a supplier or made by a large consumer, the gaseous fuel is available on tap at the actual point of use by means of distribution pipes. It is an ideal fuel in a region where there is a concentration of population. The popularity of gaseous fuels may be judged by the adoption of these fuels in domestic and commercial undertakings on an ever increasing scale in many countries.

6.2 TYPES OF GASEOUS FUELS

The following classification of fuel gases is based on the mode of occurrence and the method of production.

Fuel gases occurring in nature
- Natural gas and many other variations of it
- Methane from coal mines

Fuel gases made from solid fuels
- Producer gas
- Water gas
- Gases from other gasification processes, including substitute natural gas (SNG)
- Coal gas
- Blast furnace gas
- Wood gas

Fuel gases made from petroleum

- Refinery gases
- Liquefied petroleum gas (LPG)
- Gases from oil gasification processes

Fuel gases made by fermentation of organic wastes

Two important industrial gaseous fuels are not included in the above list. These are hydrogen and acetylene which find wide use in metal cutting, welding, and many other processes. Hydrogen will be taken up later in detail.

From the utility point of view, the gaseous fuels are often referred to as a righ gas or lean gas, depending upon whether their calorific value (CV) is above 4,000 kcal/Nm^3 or below about 1,500 kcal/Nm^3, respectively. Hydrocarbon gases and coal gas are examples of a rich gas, while blast furnace gas and producer gas are examples of a lean gas. Water gas is a fuel of medium calorific value, about 2,800 kcal/Nm^3.

The use of the terms rich gas and lean gas has undergone change recently with the greater utilisation of very high calorific value gases like natural gas and LPG. There are now five categories of gases on the basis of calorific value.

Name of the gas	Calorific value
1. Liquefied petroleum gases	~30,000 kcal/Nm^3
2. Natural gas and SNG	8,400–9,100 kcal/Nm^3
3. Town gas	4,500–5,300 kcal/Nm^3
4. Industrial fuel gas (IFG)	2,200–3,100 kcal/Nm^3
5. Low calorific value gas (LCVG)	700–1,700 kcal/Nm^3

LPG, NG and SNG may be bracketted as high Btu gases, and town gas and IFG as medium Btu gases, while LCVG may be termed a low Btu gas.

Calorific value alone is not a satisfactory parameter for the technological classification of gaseous fuels. The thermal output of a heating appliance depends on the calorific value and gravity of the gas. Both these properties are combined into one fairly good classification parameter known as the *Wobbe index* or *Wobbe number* which is a dimensional group and thus CV/$\sqrt{}$specific gravity depends on the units chosen. The speed of combustion of a given fuel gas largely depends upon its flame speed, and should be used as an additional parameter for a rational classification of gaseous fuels.

The International Gas Union (IGU) has classified different gases into families on the basis of the Wobbe index (Table 6.1.). Each family covers a range of gas qualities which can be interchanged for a given heating applicance with the minimum deviation from the designed thermal rating of the appliance.

The rich gases—CV above 4,000 kcal/Nm^3—may be classified into three groups on the basis of the flame speed. For this purpose, a flame speed factor called the *Weaver flame speed factor* is expressed on an arbitrary scale in which hydrogen is assumed to have a factor of 100. The three groups are:

High flame speed gases

- CV = 4,000 to 5,000 kcal/Nm^3
- Weaver flame speed factor = 32 to 45

Table 6.1 IGU gas families

Family sub-division	Wobbe index	
	kcal/Nm^3	MJ/Nm^3
Gas family 1	4,250–8,550	17.8–35.8
• Coke oven gas		
• Manufactured gas		
• Hydrocarbon air mixtures		
Gas family 2	8,550–12,824	35.8–53.7
• Sub-divsion I	8,550–12,322	35.8–51.6
• Sub-division II	12,322–12,824	51.6–53.7
Gas family 3	17,074–20,823	71.5–87.2
LPG		

Low flame speed gases

- CV = 7,500 to 10,000 kcal/Nm3
- Weaver flame speed factor = 13 to 25

Intermediate type gases

- CV = 5,000 to 7,500 kcal/Nm^3
- Weaver flame speed factor = 25 to 32

Of the various components of a fuel gas, the inert gases (nitrogen and carbon dioxide) and carbon monoxide reduce the flame speed and the Wobbe index, whereas, hydrogen does the reverse. Hydrocarbons reduce the flame speed but raise the Wobbe index.

A proper appraisal of the classification of gaseous fuels is necessary to derive the best performance of heating appliances. Either the composition of the fuel or the design of the appliance may be altered if the situation demands.

6.2.1 Natural Gas

Natural gas is a mixture of paraffinic hydrocarbons, in which methane is the principal component. It occurs in gas fields and also in association with crude petroleum in oilfields. Natural gas also contains ethane, propane, butane and pentane in varying proportions. In addition, small quantities of iso-paraffins and naphthenes may also be present. The vapours of the hydrocarbons, other than methane and ethane, can be separated out as a condensate. When the natural gas contains very small quantities of recoverable condensate, say, less than 15 g/m^3, it is known as *dry natural gas*. If the content of condensate is appreciable, say, above 50 g/m^3, the gas is known as *wet natural gas*. The latter is commercially treated for the recovery of the condensate which is a valuable product known as *natural gasoline* and finds wide use as a blending agent for motor fuels. The recovery of natural gasoline is accomplished by compressing and cooling, or absorption in an oil, or adsorption on a porous solid. During the process of recovery, pentane is completely recovered from the gas while small quantities of butane and propane are left behind in it. The residual gas is in effect a dry natural gas and known as stripped natural gas to distinguish it from the naturally occurring dry gas. Similarly,

Table 6.2 Production of natural gas (million cubic metres)

ITEM	1990–91	2000–01	2003–04	2004–05	2005–06	2006–07	2007–08*
1	2	3	4	5	6	7	8
Onshore							
Gujarat	1696	3149	3517	3710	3831	3294	2931
Assam/Nagaland	2010	2204	2204	2249	2408	2526	2598
Arunachal Pradesh	29	32	44	40	48	35	30
Tripura	70	376	508	497	480	520	534
Tamil Nadu	64	200	605	678	906	1130	1169
Andhra Pradesh	47	1604	1927	1707	1663	1525	1567
Rajasthan	Nil	160	168	213	242	242	255
Total	**3916**	**7725**	**8973**	**9094**	**9578**	**9272**	**9084**
OIL**	1518	1861	1887	2010	2270	2265	2341
ONGC**	2398	5555	5779	5658	5751	5876	5877
JVC/Private**	0	309	1307	1426	1557	1131	366
Offshore							
ONGC (Mumbai High)	14082	18465	17805	17313	16823	16567	16329
JVC/Private	0	3287	5184	5356	5801	5908	6861
Total	**14082**	**21752**	**22989**	**22669**	**22624**	**22475**	**23190**
Grand Total	**17998**	**29477**	**31962**	**31763**	**32202**	**31747**	**32274**

*: Provisional; **: Included under Onshore
Source: Ministry of Petroleum and Natural Gas, Govt. of India, http://petroleum.nic.in /petstal. pdf

Table 6.3 Reserves of natural gas (billion cubic metres)

Area	2003	2004	2005	2006	2007
Onshore	327	339	340	330	270
Offshore	527	584	761	745	785
Total	854	923	1101	1075	1055

Note: The data relate to 1 January 1990 and thereafter 1 April of each year.
Source: Ministry of Petroleum and Natural Gas, Govt. of India, http://petroleum.nic.in /petstal. pdf

when natural gas comes out along with crude oil, it is known as associated gas, as in the case of Bombay High.

The proved reserves of natural gas in the world are about 100 trillion m^3 with the following break-up: the old USSR 43.2%, Iran 13.5%, the USA 5.7%, Qatar 4.3%, Saudi Arabia 3.5%, Algeria 3.1% and Norway 3.0%. The old USSR and the Middle East account for two-thirds of the total reserves.

In the last two decades, natural gas has gained importance in India. The production of natural gas has increased from almost nil at the time of Independence to 32 billion cubic

Table 6.4 Sectoral consumption of natural gas (million cubic metres)

Industry	2003–04	2004–05	2005–06	2006–07	2007–08*
Energy purposes					
1. Power Generation	11478	12099	11878	11963	12037
2. Industry Fuel	3099	3569	3780	3205	3324
3. Tea Plantation	142	142	151	170	160
4. Domestic Fuel	93	343	75	443	
5. Captive use/LPG Shrinkage	4865	4944	5048	5034	5618
6. Others	1263	231	1120	40	1258
Total	**20940**	**21328**	**22052**	**20855**	**22436**
Non-energy purposes					
1. Fertilizer Industry	7889	8173	7762	8497	9822
2. Petro-Chemicals	1128	1236	1175	1377	1432
3. CNG	1	0	0	0	
4. Others	948	38	36	639	638
Total	**9966**	**9447**	**8973**	**10513**	**11892**
Grand total	**30906**	**30775**	**31025**	**31368**	**34328**
Percentage of grand total					
Energy purposes	67.8	69.3	71.1	66.5	65.4
Non-energy purposes	32.2	30.7	28.9	33.5	34.6

*: Provisional.
Source: Ministry of Petroleum and Natural Gas, Govt. of India, http://petroleum.nic.in/petstal. pdf

metres. The bulk of India's natural gas production comes from the western offshore region, particularly from Bombay High. Assam, Andhra Pradesh and Gujarat are also major producers of natural gas (see Tables 6.2 and 6.3).

The sectoral consumption of natural gas is given in Table 6.4. Power generation consumed about 11.9 billion cubic metres in 2006–07. The captive use/LPG shrinkage was about 5.6 billion cubic metres.

The actual composition of natural gas varies widely from field to field, the methane content varying from below 50% to above 95%. The composition of some gases is given in Table 6.5.

Natural gas in certain localities contains large amounts of nitrogen and carbon dioxide. In some places small but recoverable amounts of helium are found. Similarly, appreciable amounts of hydrogen sulphide are also present in some gases which may be treated for the produc.ion of elemental sulphur. Since hydrogen sulphide is corrosive and has a bad odour, gas containing this compound is known as a foul natural gas. The hydrogen sulphide-free gas is called a *sweet gas*.

After delivery from wells, natural gas is processed to remove fine solids. It is then treated for the recovery of gasoline and liquefied petroleum gas (LPG). Finally, hydrogen sulphide is removed to obtain gas for distribution. In cold climates it may be necessary to dehydrate the gas to avoid choking of lines owing to the formation of crystalline hydrates of methane, ethane, propane and *iso*-butane. These hydrates are stable solids below 19°C.

Natural gas is delivered to the consumers by a network of pipelines of diameter 51–56 cm from the production centre, at pressures usually varying between 14 and 28 kg/cm^2. However,

Table 6.5 Composition and properties of natural gas

	Type		
	dry and sweet	wet and sweet	dry and sour
Composition, per cent by volume			
CH_4	96.91	77.28	45.64
C_2H_6	1.33	11.18	0.21
C_3H_8	0.19	5.83	–
Butanes	0.05	2.34	–
Pentanes	0.02	1.18	–
CO_2	0.82	0.80	53.93
H_2S	–	–	0.01
N_2	0.68	1.39	0.21
Total	**100**	**100**	**100**
Gross calorific value, kcal/Nm^3, dry	9,000	11,200	3,260
Specific gravity (air = 1)	0.574	0.741	1.077

considerable interest has been shown in the ocean transportation of liquefied natural gas (LNG) from the countries in which it is a surplus fuel. Successful shipments have been made of Sahara gas to the UK. Japan also imports LNG in commercial quantities.

The bulk users are thermal power stations as fuel, and fertiliser production plants as a source of hydrogen. Natural gas is a preferred pipeline fuel for domestic and industrial heating in ovens, furnaces and boilers. Special burners are required to use methane owing to its high calorific value, high air requirement, moderate specific gravity and low flame speed. It is widely used as a source of hydrogen and carbon in chemical industries, in the production of ammonia, methanol, acetylene and carbon black. Table 6.5 gives the composition and properties of natural gas.

Due to the low energy density, for use as a vehicular fuel, it is compressed to a pressure of 200–250 bars to facilitate storage in cylinders mounted in vehicles and so it is called *compressed natural gas* (CNG).

India's recoverable resources of more than 690 billion cubic metres make natural gas a long-term substituted fuel for use in petrol and diesel engines. Low exhaust emissions, low noise, less maintenance, not prone to adulteration and driver's comfort are some of the attractive features of CNG as an automotive fuel.

Liquefied natural gas is first made from natural gas. It essentially facilitates storage and transportation of the fuel. This LNG is again reheated to CNG and filled on board a vehicle for use as fuel.

Given the availability and the infrastructure, CNG qualifies to be one of the most prominent alternative fuels. It is better than conventional fuels both in life cycle emissions and vehicle exhaust emissions. Higher octane value of CNG reduces knocking problems of a vehicle.

Natural gas is neither corrosive nor toxic, its ignition temperature is high, it is lighter than air, and has a narrow flammability range, making it an inherently safe fuel compared to other fuel sources. Natural gas cannot contaminate soil or water. It will always rise into the atmosphere, unlike other fuels, which are heavier than air and can pool, either as a liquid or a vapour, upon the ground. Natural gas contains a distinctive odorant (mercaptan), which allows natural gas to be detected at 0.5% concentration in air—well below levels that can cause drowsiness due to inhalation and well below the weakest concentration that can support combustion. Due to high ignition temperature of natural gas (540°C), simple exposure to a hot surface (such as exhaust manifold) is unlikely to lead to a fire.

After several trials and field studies by various institutions, and learning from the experience of other countries, India initiated the use of CNG as an automotive fuel. Recently large numbers of buses, autos and taxis in Delhi have switched over to CNG. CNG vehicles are also present in Mumbai and Vadodora. Other cities are also planning to introduce CNG in a phased manner. The cost of CNG is about a third of the cost of petrol in terms of CV. Comparing the emissions from vehicles that use conventional fuel, the CO emission from vehicles running on CNG is less by 70%–90%, hydrocarbon emission by 40%–60% and CO_2 emission by 10%.

It is also possible to use a bifuel. Vehicles can be operated in the dual mode like petrol–CNG and diesel–CNG. The major disadvantage is the loss of luggage space. CNG cylinders take up a lot of storage space and generally have to be placed in the boot of the car. The body of the cylinders too have to be made of good grade steel. Finally, based on its characteristics CNG may slightly hamper vehicle performance.

The sale of CNG in some states of India is shown in Table 6.6. Delhi accounts for the most (about Rs. 3 lakhs per year). Next comes Maharashtra (Rs.1.8 to Rs.1.3 lakhs per year). Gujarat sold the least—Rs 4000 in 2004–05 and only Rs. 1000 the previous year.

6.2.2 Hydrogen

Hydrogen is the most abundant element in the universe. It is an energy carrier—it contains the energy that powers the sun which sustains all life on earth. It is found in all living organisms. Hydrogen is a fuel which can be used to power internal combustion engines and fuel cells. When hydrogen is combined with oxygen in a fuel cell, electricity is produced. The water is usually emitted as a waste product.

As hydrogen is only found combined with other elements, it needs to be separated from them for use as a fuel for fuel cells.

Methods of hydrogen production

1. Electrolysis of water,
2. Reforming gaseous and liquid hydrocarbon fuels,
3. Gasification of organic and plastic materials into synthetic hydrogen-rich gases,
4. Anaerobic digestion, or biological conversion of hydrocarbons in the absence of oxygen—usually by bacteria,
5. Fermentation or biological conversion of hydrocarbons, usually by bacteria or yeasts,
6. Certain algae emit hydrogen as a waste product from their metabolism.

If the electricity used to drive the electrolysis process is derived from renewable/alternative energy sources, the resulting hydrogen can be called 'renewable hydrogen', and therefore

Table 6.6 State-wise consumption of CNG

Item	Delhi	Maharashtra	Gujarat			Utter Pradesh			Andhra Pradesh		Tripura	Total
	Delhi IGL	Maharashtra MGL	Vadodara GAIL	Surat GGCL	Ankleshwar GGCL	Lucknow GAIL GGL	Agra GAIL GGL	Kanpur GAIL CUGL	Vijaywada GAIL BGL	Hyderabad Gail BGL	Agartala GAIL TNGCL	
No. of vehicles												
Cars	130149*	53706	850	22065	2724	130	135	95	324	244	0	210422
Autos	80276	128596	3000	36212	4427	6695	5377	2730	1188	1505	314	270320
LCV/RTVs	5507	1695	0	0	0	116	25	72	0	0	21	7436
Buses	11665	945	50	462	35	136	201	344	103	8	0	13949
Others/Phat-phat	360	976	0	1177	80	0	0	4494	8	16	0	7111
Total	**227957**	**185918**	**3900**	**59916**	**7266**	**7077**	**5738**	**7735**	**1623**	**1773**	**335**	**509238**
Average Consumption												
TPD	1055.25	736.32	22.63	124.5	16.8	30.64	23.9	34.1	10.04	5.76	1.075	
MMSCMD	1.336	1.05	0.029	0.18	0.02	0.044	0.032	0.001	0.013	0.008	0.001	

Note:

* Includes taxies.

- IGL: Indraprastha Gas Ltd.
- MGL: Mahanagar Gas Ltd.
- GGCL: Gujarat Gas Company Ltd.
- GAIL: GAIL (India) Ltd., Vadodara
- BGL: Bhagyanagar Gas Limited, Vijaywada, Hyderabad
- GGL: Green Gas Ltd., Lucknow, Agra
- CUGL: Central UP Gas Ltd., Kanpur
- TNGCL: Tripura Natural Gas Company Ltd., Agartala
- MMSCMD: Million standard cubic metres per day
- TPD: Tonnes per day

Source: Ministry of Petroleum and Natural Gas, Govt. of India, http://petroleum.nic.in /petstal.pdf

carbon-neutral, thus not contributing to the greenhouse effect and resulting global warming. Renewable hydrogen can also be produced by gasifying biomass and other materials that do not contribute to increasing overall carbon levels in the atmosphere.

The key challenges for commercialistion of fuel cells and hydrogen infrastructure technologies are the following:

1. Cost and durability of fuel cells,
2. Storage of hydrogen,
3. Production and delivery of hydrogen,
4. Public acceptance of hydrogen as a fuel through certified procedures and training manuals for fuel cells.

Fuel cells are an important enabling technology for the hydrogen economy and have the potential to revolutionise the way we power our nation, offering cleaner, more-efficient alternatives to the combustion of gasoline and other fossil fuels.

Fuel cells have the potential to replace the internal combustion engine in vehicles and provide power in stationary and portable power applications because they are energy efficient, clean and fuel-flexible. Hydrogen or any hydrogen-rich fuel can be used by this emerging technology.

6.2.3 Methane from Coal Mines

Methane is associated with coal in situ and its quantity varies from mine to mine. It is a source of dangerous explosions in gassy mines. This gas is often known as *firedamp*. Owing to its porous nature, coal adsorbs methane formed during coalification. At the time of mining, the gas is gradually desorbed. Unless this is removed by proper ventilation, explosive mixtures accumulate inside the mine. This methane-rich gas may be recovered through a system of boreholes and used as a fuel. In the UK in 1961, about 220 million cubic metres of mine gas were produced. Belgium, Germany and France have developed processes for the recovery of mine gas. However, the advent of natural gas from the North Sea has completely ruined this industry in Europe.

The methane content of coal mine gas is 93%–94%, the other notable constituents being ethane, carbon dioxide and nitrogen.

6.2.4 Producer Gas

The combustible gaseous mixture obtained by blowing air, or a mixture of air and steam through an incandescent bed of solid carbonaceous fuel is known as producer gas and the plant in which it is made is called a gas producer, or simply a producer. It is a fuel of low calorific value, the principal components being carbon monoxide and nitrogen. Hydrogen is also present in significant amounts when air and steam are used in place of air alone. The advantages of this lean fuel are: (i) advantages of firing gaseous fuel, (ii) low price and (iii) ease of production. It is the most widely used industrial fuel gas where cheap solid fuels are abundant while natural gas is not.

Bituminous coal, anthracite and coke are used as the raw material in gas producers. With suitable variations in the plant, other fuels, namely peat, lignite and even wood or organic solid waste material may also be used.

The exact composition of producer gas depends on the type of fuel, the composition of the blast and other operation variables. The data on some typical producer gases are given in Table 6.7.

Principal reactions in an air-blown producer

When air comes in contact with solid amorphous carbon above its ignition temperature, say, above 500°C, the following two main reactions occur:

$$C + O_2 + 3.76N_2 \rightarrow CO_2 + 3.76N_2 \quad \Delta H = -97{,}000 \text{ kcal*} \tag{6.1}$$

$$CO_2 + C \rightleftarrows 2CO \quad \Delta H = +38{,}270 \text{ kcal} \tag{6.2}$$

The net overall reaction is

$$2C + O_2 + 3.76N_2 \rightleftarrows 2CO + 3.76N_2 \quad \Delta H = -58{,}730 \text{ kcal} \tag{6.3}$$

As long as oxygen is in excess, carbon dioxide is formed by the reaction (6.1), which then undergoes reduction into carbon monoxide by reaction (6.2). Reaction (6.2) is known as the *Boudouard reaction* and is actually the single most important reaction controlling the process.

The Boudouard reaction is an endothermic reaction. Its equilibrium constant greatly increases with the rise in temperature. The content of carbon monoxide in the equilibrium mixture of carbon monoxide and carbon dioxide at 1 atm pressure sharply increases beyond 500°C as seen below

Temperature, °C	400	500	600	700	800	900	1000
CO, per cent	0.9	6.5	26.4	63.0	89.5	97.5	99.6

Reaction (6.2) is a heterogeneous reaction. The kinetics are such that the equilibrium is reached at 1,300°C in a *few seconds* and at 1,100°C in about *one minute*. Only two-thirds of the equilibrium are attained in one minute at 1,000°C.

Increased pressure inhibits the formation of carbon monoxide. But the effect of pressure is not great and the air blast in normal practice is slightly above atmospheric pressure. Therefore, the inhibitory effect of pressure is negligible.

The type of carbon does not practically disturb the equilibrium content of carbon monoxide but it greatly influences the kinetics of the reaction. In order of reactivity, charcoal ranks very high, as shown by the following data on the percentage attainment of equilibrium after one second contact time at 1,100°C in a given test unit: Charcoal 91.3, coke 13.5 and anthracite 10.5. The lower reactivity of coke and anthracite is overcome in practice by increasing the contact time by carrying out the reaction at temperatures above 1,300°C.

The above discussion shows that the producer gas process is favoured by a high temperature of the fuel bed, adequate time of contact (thick fuel bed) and reactive fuel. All factors which hinder good contact between carbon and gaseous reactants must be avoided. If the ash of the fuel

---

* using β-graphite, $\Delta H = -94{,}050$ kcal

Table 6.7 Composition and properties of some manufactured gaseous fuels

	Air blast with coke	Producer gas			Water gas		Lurgi gas from bituminous coal		Coke oven gas	Blast furnace gas	Segas/Onia-Gegi Gas using		
		mixed coal	blast coke	with anthracite	blue	carburetted	crude	purified			fuel oil	gas oil	naphtha
Composition, per cent by volume													
CO_2	1.0	4.0	5.0	6.0	4.7	5.6	28.0	0.8	2.0	11.0	9.2	5.8	3.0
C_nH_m (unsaturated)	–	0.4	–	–	–	7.0	0.2	0.2	2.6	–	6.6	4.2	5.1
O_2	–	–	–	–	–	0.4	–	–	0.4	–	0.5	0.4	0.8
CO	33.5	29.0	29.0	26.0	41.0	30.5	22.4	31.5	7.4	27.0	14.8	8.3	15.1
H_2	1.5	12.0	11.0	17.0	49.0	37.0	38.0	53.3	54.0	2.0	49.1	50.3	57.6
CH_4	–	2.6	0.5	1.2	0.8	14.0	10.9	13.4	28.0	–	15.6	16.2	14.7
N_2	64.0	52.0	54.5	49.8	4.5	5.5	0.5	0.8	5.6	60.0	4.2	4.8	3.7
Total	100.0	100.0	100.0	100.0	100.0	100.0	100.0	100.0	100.0	100.0	100.0	100.0	100.0
Gross CV, kcal/Nm^3, dry	1,060	1,550	1,260	1,420	2,800	4,770	–	3,900	5,020	870	4,780	4,700	4,680
Specific gravity (air = 1)	0.98	0.87	0.90	000.85	0.55	0.63	–	0.48	0.38	1.01	0.56	0.52	0.40
Theoretical air, Nm^3/Nm^3	–	1.28	1.00	–	002.22	004.27	–	–	4.57	0.69	–	–	–

is of low fusion point, say 1,100°C, it melts in the high temperature zone and then resolidifies into *clinker*, in the cooler parts. The clinker disturbs the air distribution, engulfs unburnt carbon and lowers the efficiency of the process. Similarly, coking coal produces massive lumps in the bed when it is used as the producer fuel. The process, however, can be carried out by poking the bed and breaking the masses into small pieces.

After the gases leave the fuel bed, the carbon monoxide partly dissociates into carbon dioxide and carbon.

$$2CO \rightleftarrows CO_2 + C \quad \Delta H = -38{,}270 \text{ kcal} \tag{6.2 A}$$

This is actually the reversal of the Boudouard reaction and is know as the *Neumann reversal reaction*. The result of this reaction is a drop in the calorific value of the product gas, owing to the conversion of the combustible carbon monoxide into the non-combustible carbon dioxide in the gaseous space above the fuel bed.

Demerits of air-blown producer gas

Reaction (6.3) shows that air-blown producer gas is a mixture of about one-third carbon monoxide and two-thirds nitrogen. The maximum carbon monoxide content is calculated as only 34.7%, the rest being the inert nitrogen. By rearranging reactions (6.1) and (6.3) we get

$$\underset{\text{producer gas}}{(2CO + 3.76N_2)} + \underset{\text{air}}{(O_2 + 3.76N_2)} \rightarrow \underset{\Delta H = -1{,}35{,}270 \text{ kcal}}{2CO_2 + 7.52N_2} \tag{6.4}$$

The maximum calorific value of air-blown producer gas is calculated from this reaction:

$$\frac{135270}{5.76 \times 22.4} = 1{,}050 \text{ kcal/Nm}^3 \text{ of the gas}$$

This is a very low grade fuel. The yield of this gas is computed from reaction (6.3).

$$\frac{5.76 \times 22.4}{2 \times 12} = 5.37 \text{ Nm}^3\text{/kg carbon gasified}$$

The maximum cold gas efficiency of the air-blown producer is calculated from reactions (6.1) and (6.4).

$$\frac{135270}{2 \times 97000} \times 100 = 70\%^*$$

This low cold gas efficiency is a serious demerit of the process.

As much as 30% of the heating value of the carbon is liberated during the formation of the producer gas. This causes intense local heating which fuses the ash and leads to serious clinkering problems.

---

* Based on β-graphite, the maximum cold gas efficiency is 72%.

Use of steam in air blast

All the three demerits of the air-blown producer are removed by using a wet blast, that is, adding steam to the blast of air. The two reactions between amorphous carbon and steam are

$$C + H_2O \rightleftarrows CO + H_2 \qquad \Delta H = +28{,}440 \text{ kcal} \qquad (6.5)$$

$$C + 2H_2O \rightleftarrows CO_2 + 2H_2 \qquad \Delta H = +18{,}600 \text{ kcal} \qquad (6.6)$$

Reaction (6.5)—the water gas reaction—is active at and above 1,000°C while reaction (6.6) predominates between 500°C and 600°C. Both the reactions are endothermic and convert a part of the sensible heat liberated by carbon in the bed into the potential heat in the combustibles formed, namely CO and H_2 . Since carbon dioxide is an undesired component of the product, the reaction (6.6) should be suppressed by maintaining a high bed temperature.

In the space above the fuel bed the following gaseous reaction—known as the *water gas shift reaction* or *shift conversion*—takes place between undecomposed steam and carbon monoxide.

$$CO + H_2O \rightleftarrows CO_2 + H_2 \qquad \Delta H = -9{,}840 \text{ kcal} \qquad (6.7)$$

This is an undesired reaction which raises the concentration of the inert CO_2 in the product and reduces its calorific value.

Reaction (6.5) produces carbon monoxide and hydrogen in equal proportions. The equlibrium composition, in per cent of either of these two components, in the reaction product at different temperatures at atmospheric pressure is give below:

Temperature, °C	400	500	600	700	800	900	1000
CO or H_2, per cent	2.99	12.67	30.61	44.14	48.52	49.58	49.85

Higher pressure inhibits the reaction. But the inhibitory effect of a blast at slightly above atmospheric pressure is negligible.

In addition to the above reactions, methane formation or a methanation reaction between carbon and hydrogen also takes place.

$$C + 2H_2 \rightleftarrows CH_4 \qquad \Delta H = -20{,}840 \text{ kcal} \qquad (6.8)$$

This is an exothermic reaction and is therefore largely suppressed by the high temperature conditions of the bed. However, high pressure favours the reaction. The equilibrium concentration of methane at 1,000°C is 0.9% at 1 atm, 3.5% at 4 atm and 11.8% at 16 atm. In the usual producer practice, the methane content of the product is small. High pressure processes have been developed where methane content is significantly high (10%–13% in crude gas).*

The advantages derived by the use of air–steam mixture used as the blast are:

1. The total contents of the combustibles ($CO + H_2 + CH_4$) is raised and the inerts ($N_2 + CO_2$) content is lowered. The calorific value of the gas is significantly increased upto about 1,250 kcal/Nm^3.

---

* Based on β-graphite, the ΔH values for reactions (6.5), (6.6) and (6.8) are: +31,390 kcal, +21,550 kcal and –17,890 kcal, respectively.

2. A part of the sensible heat liberated by the combustion of carbon is converted into the potential heat of hydrogen and cabon monoxide. The cold gas efficiency of the mixed blast producer is much higher at about 82% than that of the air blown producer.
3. The endothermic reactions between steam and carbon prevent the sharp increase in fuel bed temperature even in localised spots, and hence the tendency of the ash to form clinker is greatly diminished.

The amount of steam blown through the bed is an important factor affecting the performance of a producer (Fig. 6.1). Initially the increase in the amount of steam raises the total combustibles content and calorific value of the gas and also the thermal efficiency of the producer. However, too much absorption of the heat of the fuel bed by the endothermic reactions of steam reduces the bed temperatures to below 1,100°C and hence, lowers the degrees of carbon dioxide reduction and steam decomposition. Practice has shown that beyond an optimum supply, any increase in the amount of steam reduces the thermal efficiency of the producer, the total combustibles content and the calorific value of the gas. However, if the ash of the fuel has a tendency to clinker, the steam supply should be increased beyond the optimum value. This step ensures the smooth continuous running of the producer at the cost of its efficiency and gas quality.

The wet blast may be made either by injecting steam to the air flow or by blowing air through water in the base of the producer. The first method is more convenient and is the normal practice. A controlled quantity of steam is blown into the air blast, raising its temperature to a predetermined level. This is technically called the *blast saturation temperature* (BST), on the assumption that the air is saturated with water vapour at this temperature. This assumption is valid for all practical purposes.

The amount of steam supply is best controlled by the BST. The optimum steam supply depends on the nature of the fuel but is usually at 55–62°C BST which corresponds to 15–22 kg of steam per 100 Nm^3 of dry air, respectively. In terms of the quantity of the solid fuel, the optimum steam supply is at 0.4–0.5 kg per kilogram of carbon gasified.

Reaction zones in a gas producer

The fuel bed in a normal producer rests on a metallic grate. The former may be divided into a number of reaction zones: ash zone, oxidation zone, primary reduction zone, secondary reduction zone, and drying-cum-carbonisation zone, depending upon the sequences of reactions that take place in the counter-current movement of the solid fuel and the blast (Fig. 6.2). The air–steam blast is preheated by the ash zone which also serves to make the blast distribution uniform and protect the grate from intense heat. The oxygen is consumed within 75–100 mm of the bed, which constitutes the oxidation zone. Once the reactions start, the compositions of the gaseous stream go on changing from point to point along the bed depth (Fig. 6.3). Carbon dioxide forms at the expense of oxygen. Its concentration reaches a maximum at the top of the oxidation zone when carbon monoxide begins to appear. The monoxide continues to form by the reactions (6.2) and (6.5) initially, then by reaction (6.2) alone, and approaches an equilibrium with the dioxide at the top of the bed. Soon after the appearance of carbon

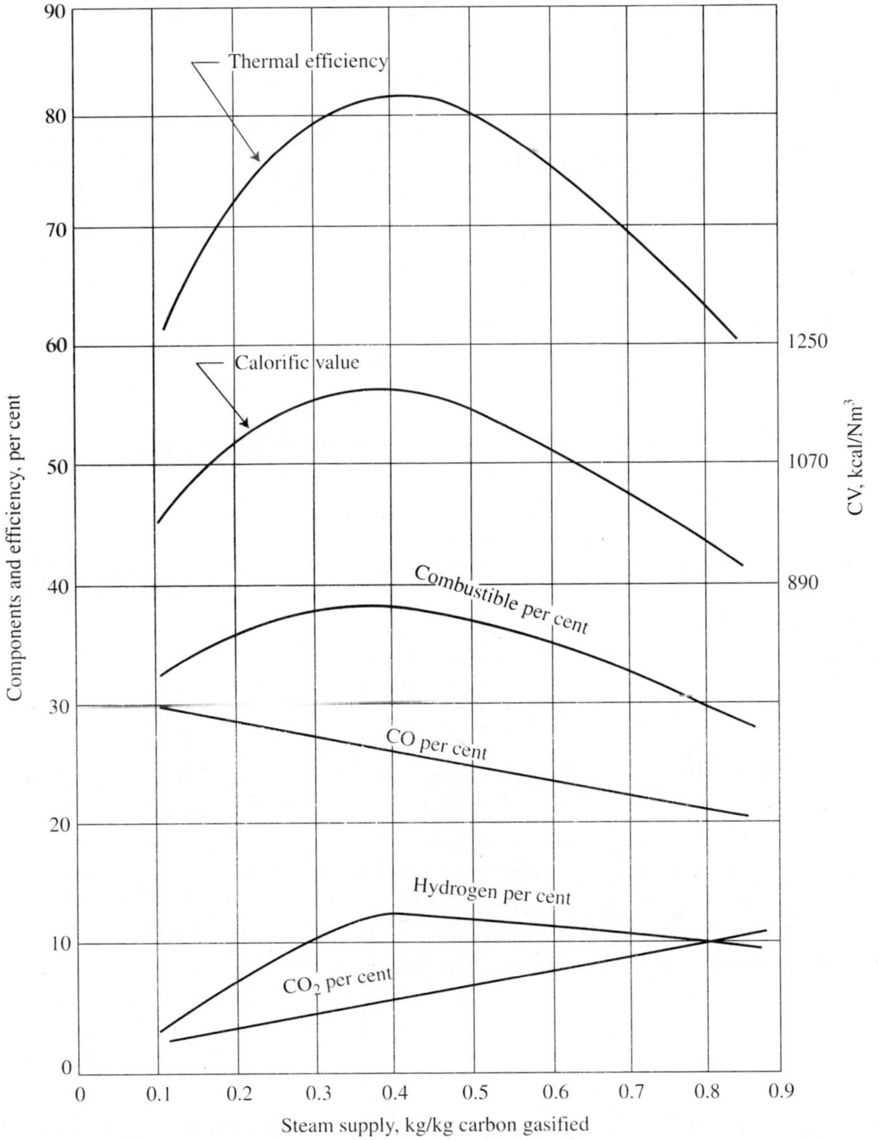

Figure 6.1 Effect of quantity of steam on gas producer performance

monoxide, the steam decomposition begins and then vigorously continues for about 25 mm of the bed depth, which is called the *primary reduction zone*. About 40 mm layer of the bed above the zone constitutes the *secondary reduction zone* where the steam is decomposed by carbon monoxide. The uppermost layer of the bed is the *drying and carbonisation zone*. The water vapour and the volatile matter of the fuel are added to the escaping gas. Therefore, fuels with significant yield of volatile matter result in the production of enriched gases. The top-most zone essentially serves as the preheating zone for precarbonised fuels like coke.

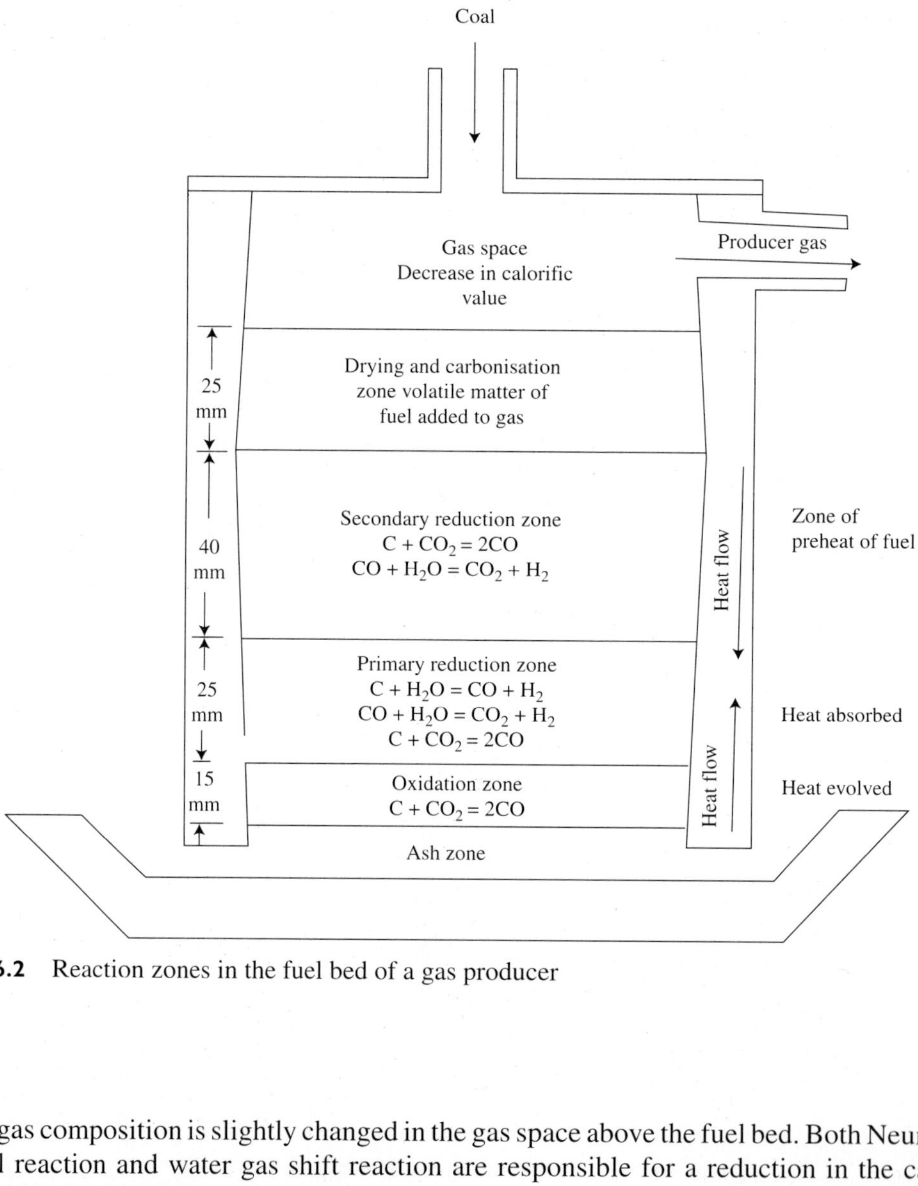

Figure 6.2 Reaction zones in the fuel bed of a gas producer

The gas composition is slightly changed in the gas space above the fuel bed. Both Neumann's reversal reaction and water gas shift reaction are responsible for a reduction in the calorific value of the final product.

Figure 6.2 is a schematic representation of the zones. In practice there is no well-defined boundary between two neighbouring zones. Moreover, the gas flow is not uniform across the bed cross section. It is maximum at the periphery and minimum at the centre. Therefore, temperature profiles are concave-shaped—a horizontal section of the bed having the temperature falling from the periphery to the centre. The reaction zones are similarly concave-shaped and not horizontal.

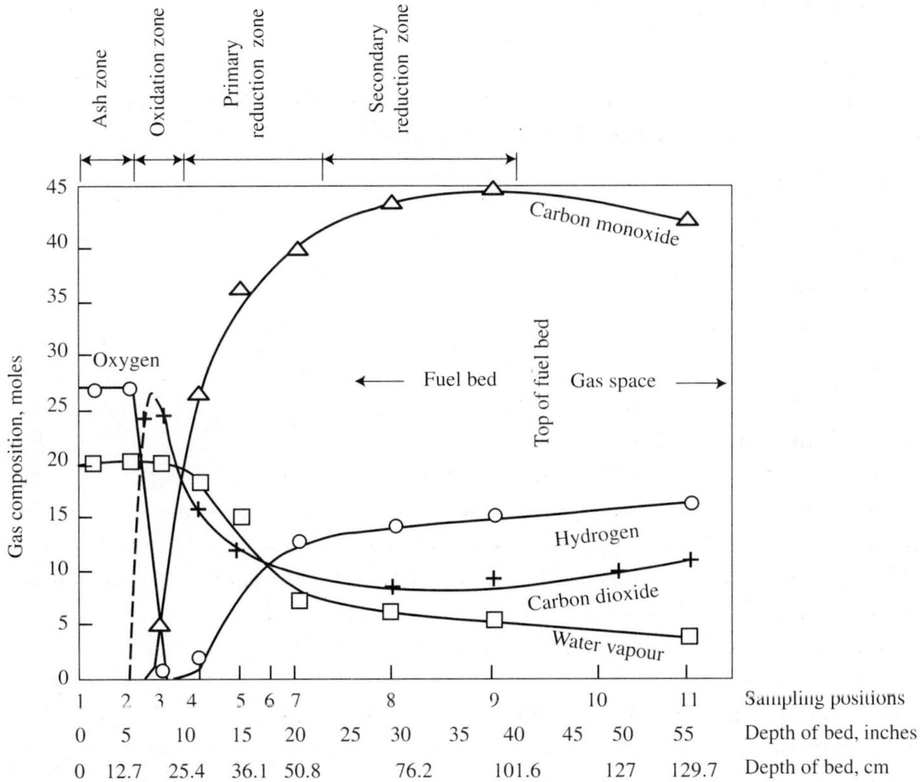

Figure 6.3 Composition of gases in the fuel bed of a gas producer

Impurities in the raw gas

Where the producer is an internal part of the furnace setting, the hot raw gas is directly fired in the furnace. In other plants, the gas is cooled and freed of some impurities and then stored in gas holders.

The usual impurities are undecomposed steam, tar dust, hydrogen sulphide and ammonia. Most of the steam and tar is condensed on cooling. Ammonia may be removed and recovered as ammonium sulphate. But recovery is not economical. No special treatment is normally given for the removal of hydrogen sulphide and other impurities. Since these are corrosive and harmful, their content is kept low by using low-sulphur coals in gas producers. Dust is partly removed with the condensates and should be completely removed with cyclones or other types of dust catchers.

When raw gas is directly fired in a furnace, the tar contributes to the calorific value of the gas and raises the temperature and emissivity of the flame. The dust particles also increase the flame emissivity. But dust is a major cause for fouling of gas mains carrying hot raw producer gas. Therefore, some form of a dust catcher is desirable even in producers supplying hot raw gas to furnaces.

Gasification rate in a producer

The specific gasification rate is expressed as kilograms of fuel gasified per hour per square metre of the bed cross sectional-area. It is desirable to achieve a high gasification rate with good gas quality. This is limited by the following factors:

- the form in which ash is discharged,
- the use of air with 79% inert nitrogen,
- the height of the fuel bed,
- the rate of reduction of carbon dioxide and steam.

In normal producers, the ash is removed in the solid form. The ash fusion point seriously limits the bed temperature and hence the gasification rate. In high-speed slagging producers, the ash is discharged in the molten state and therefore the gasification can be carried out at high temperatures of about 1,500–1,800°C.

The presence of large amounts of nitrogen in the blast leads to high wastage of heat in the form of sensible heat of the gases leaving the reaction zone, and hence automatically limits the bed temperature. The use of oxygen or oxygen-enriched air helps to maintain high bed temperatures and hence, high gasification rates. The gas quality is also improved. The operations without inert nitrogen have an unfavourable factor of the increased partial pressure of the carbon dioxide gas entering the reduction zone, which displaces the equilibrium of reaction (6.2) towards carbon dioxide. This effect, however, is more than compensated by the other merits of a high-oxygen blast.

A conventional producer needs a certain minimum bed height of 0.6–3 m for achieving the desired reduction of carbon dioxide and steam. Deep beds, however, offer resistance to the blast and hence limit the gasification rate.

Customary fuels such as coal, coke or anthracite are characterised by the relatively slow rate of reduction of carbon dioxide and steam. The situation may be improved by the addition of catalysts (0.01%–0.1% of Na_2CO_3, K_2CO_3 or Ni) to the fuel and the use of highly combustible fuels, finely divided fuels, or underground fuel beds in situ. The last two methods involve entirely different types of gasification processes. A preheated blast increases the gasification rate. However, the benefits of a preheated blast can be realised only in the slagging producers. Temperature control is the problem in other processes.

Besides the reactivity, two other properties of the fuel greatly affect the gasification rate. These are size and size distribution, and hardness. Uniform size, say 20–40 mm, promotes uniform gas flow across the bed cross sectional area. The presence of excessive amounts of fines leads to high carry-over losses and prevents the use of a high blast rate. If the fuel is soft, it is degraded during handling and operation and thus produces non-uniform fuel with excessive fines. Modern high capacity producers prefer closely sized and hard coke as fuel.

Some typical gasification rates for different types of modern producers are:

slagging ash producer	2.5 t/m^2 h
atmospheric and pressure producer	0.2–0.4 t/m^2 h
suction producer	0.6–0.9 t/m^2 h

The clean gas yield in a large conventional producer may be 4–4.5 Nm^3/kg of dry fuel.

Types of producers

The design of a gas producer may vary in respect of the following features: (i) direction of travel of blast and fuel, (ii) effective pressure in the producer, (iii) method of ash removal, (iv) method of blast introduction, (v) fuel feeding system, (vi) bed stirring devices and (vii) shell construction.

In *up-draft* producers, the blast travels upwards and the fuel downwards. Ash is removed from the bottom of the bed and the gas from the top. This is the most conventional type of gas producer. The earlier discussion on reaction zones pertains to this. Both the fuel and the blast travel downwards in the *down-draft* producers. Hence, the volatile matter of the fuel undergoes decompostion during its travel through the bed. This is a rare type of producer. The *double-draft* producer is also rare—the blast enters both from the top and the bottom. Another type of producer has found use in gas-fired vehicles: it has a *cross-draft*, where the fuel travels downwards but the blast is introduced on the side of the bed and allowed to travel practically horizontally before the gas is taken out from the other side.

Depending upon the pressure conditions, there may be atmospheric (balanced-draft), suction or pressure producers. The atmospheric-type is by far the most common owing to the elements of safely and control. The suction-type is of limited capacity and finds use in the manufacture of power gas and in heating built-in type gas retorts. Pressure producers are used to supply methane-rich gas to furnaces. The gas pressure at the top of an atmospheric producer bed is usually slightly positive, 25–125 mm w.g.

In conventional producers, ash is removed in the solid state by a mechanical ash removing system. The producer shell is placed on a water pan (seal) and either the shell or the pan is rotated. The ash falls on the pan and is continuously removed by placing an obstruction in the path of the relative movement of the ash. When both shell and pan are rotated, the ash is ploughed from the pan intermittently. Older types of producers have stationary grates where ash is manually removed at definite intervals. In slagging producers, ash is allowed to melt and then removed by tapping at set intervals as in a cupola or blast furnace. Some high-speed producers allow the ash to be carried away with the gas and collected in dust pockets in the gas line.

The most common position for feeding of the blast in up-draft producers is below the grate. The old producers had a flat, inclined or stepped bar grate. The newer units have a conical-shaped or mushroom-like grate. The blast inlet is just below the hollow centre of such a grate and enter into the fuel bed through holes or slots in the grate. The slagging producers receive a high velocity blast through cooled tuyeres in the side of the producer. There is no distribution problem in the downward blast.

The producers usually receive intermittent fuel-feeds through hoppers. A more complicated arrangement such as a worm feed may be necessary for special fuels. There are variations in the method of uniform distribution of the fuel and the levelling of the bed. The producer bed needs occasional stirring to maintain its uniform porosity. The rotation of the shell or water pan provides agitation. Coking coals need extra stirring from the top by rotary or oscillating, water-cooled mechanical poking rods.

Producers are usually built of cylindrical steel shells. The shell may be lined with firebrick and may have a water jacket to protect it from intense heat. Low pressure and sometimes high-pressure steam is raised in the water jacket. Larger capacity producers have water-jacketted shells. The walls of the brick-lined producers should be kept relatively cool to avoid the formation of wall clinker by the slagging of bricks with coal ash.

An up-draft modern producer may have a height of 2–3 m and a diameter 1–3.75 m, while the fuel bed may be of 1.75–2.75 m depth. The brick-lining may be 0.25–0.5 m thick.

Efficiency of a producer

The thermal efficiency of a gas producer may be expressed in one of three ways.

1. Cold gas efficiency, defined as

$$\frac{\text{potential heat of gas}}{\text{total heat of gasified fuel}} \times 100$$

$$= \frac{Q_g V_g}{Q_f + \bar{C}_{p,f}\,\Delta t_f} \times 100$$

2. Hot gas efficiency, defined as

$$\frac{\text{total heat of gas}}{\text{total heat of gasified fuel}} \times 100$$

$$= \frac{(Q_g - \bar{C}_{p,g}\,\Delta t_g)V_g}{Q + \bar{C}_{p,g}\,\Delta t_g} \times 100$$

3. Comprehensive efficiency, defined as

$$\frac{\text{total heat of gas} + \text{heat in surplus steam}}{(\text{total heat of fuel} + \text{total heat of make-up steam} + \text{thermal equivalent of energy consumed})} \times 100$$

where,

Q_g = calorific value of gas, gross or net, kcal/Nm^3
Q_f = calorific value of fuel gasified, gross or net, but same as in Q_g, kcal/kg
V_g = gas yield, Nm^3/kg fuel
$\bar{C}_{p,g}$ = average specific heat of gas, kcal/Nm^3, °C
Δt_g = gas temperature above ambient temperature, °C
$\bar{C}_{p,f}$ = average specific heat of fuel, kcal/kg,°C
Δt_f = fuel temperature above ambient temperature, °C

In computing hot gas efficiency and comprehensive efficiency it is necessary to know the particular point in the gas main at which the reference is made because the gas composition (particularly with respect to the contents of tar, undecomposed steam and dust) and temperature vary accordingly.

The gasification efficiency is often expressed as the carbon efficiency and is defined as

$$\frac{\text{carbon in gas}}{\text{carbon in fuel}} \times 100$$

$$= \frac{C_g V_g}{C_f} \times 100$$

where,

C_g = carbon content of gas, kg/Nm^3

C_f = carbon content of fuel, kg/kg

V_g = gas yield, Nm^3/kg

This has only a limited significance since the carbon gasified to carbon dioxide does not contribute to the heating value of the gas.

The above expressions for the efficiency of gas producers are also useful in the calculation of efficiency of other gasifiers using solid and liquiid fuels as the raw materials.

6.2.5 Water Gas

Water gas is a gaseous fuel of medium calorific value, 2,800 kcal/Nm^3, generated by gasifying a solid incandescent source of carbon in superheated steam. The equipment is known as a water gas generator. The main reaction of the process is the reaction (6.5) resulting in the production of hydrogen and carbon monoxide in equal proportions of 50%. In practice it is impossible to completely exclude the reaction (6.6) which produces carbon dioxide and hydrogen. Other reactions which occur during the process are the methanation reaction (6.8), the water gas shift reaction (6.7) and the Boudouard reaction (6.2). A small amount of nitrogen is derived from the carbonaceous fuel and also from the air. The composition and properties of typical water gas are given in Table 6.7 along with other gaseous fuels.

This gas gives a blue flame owing to the high carbon monoxide content and is also known as *blue water gas*, or simply as *blue gas*.

The two important uses of water gas are in the manufacture of *town gas*, and chemicals and fertilisers. In the former case, it is first enriched with hydrocarbons to give a gas of higher calorific value, called *carburetted water gas*. In the production of fertilisers it is essentially used as a source of hydrogen for ammonia synthesis. Blue water gas may also be used as a furnace fuel. However, the installation of blue gas generators for the purpose of supplying fuels only to furnaces is not economical and cannot be justified.

The decomposition of steam by carbon is a strongly endothermic process. Attempts to supply the large heat requirement by external heating has proved uneconomical. The generation of blue gas is therefore an intermittent process consisting of two main stages:

1. The *blow* or blasting period during which the fuel bed is heated up by blowing air through it,
2. The *run* or gas-making period during which the blue gas is produced by passing steam through the incandescent fuel.

Reactions (6.1), (6.2) and (6.3) take place during the blow, and reactions (6.5), (6.6), (6.8), (6.7) and (6.2) take place during the run. The objective, however, is to accomplish only reaction (6.1) during the blow so that the maximum possible heat can be evolved by burning a given quantity of carbon into carbon dioxide. This requires a thin bed. On the other hand, the objectives during the run are to gasify the greatest possible quantity of carbon into carbon monoxide and also to decompose the greatest possible quantity of steam. These require a moderately thick fuel bed ensuring sufficient time of contact for the reaction (6.5). These requirements of the blow and the run are to some extent incompatible.

In modern generators, a deep fuel bed is used to achieve a good quality of blue gas and high productivity of the plant. The formation of carbon monoxide during the blow is minimised by increasing the air blowing rate. Moreover, the sensible and potential heat content of the blow gas is recovered in waste heat boilers. The blow gas of a modern generator may contain as high as 10% carbon monoxide, the dioxide constituting about 14%.

The fuel bed temperature undergoes fluctuations during the process cycle. It rises during the blow and falls during the run. The highest temperature during the blow is limited by the clinkering tendency of the ash, and the lowest temperature during the run is limited by the requirement of above 1,100ºC of the reaction (6.5). In order to obtain a gas of uniform composition, the duration of individual stages in the cycle should not be very large. The working cycle of a six-stage process is shown below:

1. Blow	1.00 min	
2. Steam purging		2 s
3. Up-run (steaming upwards)	1.25 min	
4. Down-run (steaming downwards)	1.25 min	
5. Up-run	0.50 min	
6. Air purging		2 s
Total	4 min	4 s

The change of direction of steaming helps to maintain a uniform bed temperature. Steam purging prevents the mixing of the blue gas with the blow gas and air purging ensures complete recovery of the blue gas remaining in the bed. The control of the cycle by valves is fully automatic. The normal fluctuation of the fuel bed temperature is 1,400°C at the end of the blow, to 1,150°C at the end of the run. As much as 45% of the carbon is consumed during the blow and only 55% during the run. Hence the importance of heat recovery systems

Modern blue gas generators may produce as much as 3,00,000 Nm^3/day of gas. The shell is water-jacketted as in the case of modern producers. The plant is operated in conjunction with a waste-heat boiler. The steam generated in the jacket and waste-heat boiler meets the plant requirement and also provides some surplus. The gas yield is about 1,800 Nm^3/t of daf coke. The steam consumption is about 0.55 kg/Nm^3 gas or about 1.0 kg/kg daf coke. The air requirement is about 1.8 Nm^3/Nm^3 gas or 3.2 Nm^3/kg daf coke.

The cold gas efficiency of a blue gas generator is about 63%. The overall efficiency, including the recovery of the potential heat of the blow gas, and the sensible heats of both blow gas and blue gas is about 78%. The gasification efficiency of a blue gas generator is the lowest of all fixed bed gasification processes and is below that of the carbonisation process used in coal gas production.

The blow gas is burnt on the outlet of the generator and the waste gases are passed through a waste heat boiler to the atmosphere. During the run the waste heat boiler receives the sensible heat of the blue gas. The dust in the blue gas is removed by passing it through a washer. Depending upon the use of the gas, further purification is resorted to for removing tar, hydrogen, carbon dioxide and other impurities.

The blue gas generation can be made continuous by blowing a mixture of steam and oxygen through the hot fuel bed. A separate blow period is thus eliminated. The resulting gas also passes through the hot fuel bed. A separate blow period is thus eliminated. The resulting gas has a higher proportion of carbon dioxide than normal blue gas.

When water gas is produced for ammonia synthesis, a proportion of air is added to the steam so as to get a mixture of hydrogen, carbon monoxide, carbon dioxide and nitrogen. This is known as *semi-water gas*. With suitable treatment, a mixture of nitrogen and hydrogen in a desired proportion is ultimately made from this gas.

The most common fuel for a water gas generator is hard coke with uniform lumps and without fines. The ash should be refractory in nature and its amount should be as small as possible. Where available, anthracite is also a good raw material for the process. Bituminous coal has the following disadvantages:

1. Lower capacity of the plant owing to difficulty in kccping thc bcd in good condition,
2. Formation of fines preventing the use of high blast pressure,
3. Formation of smoke from the volatiles.

Some developments have been made in overcoming these disadvantages in the total coal gasification processes.

In some processes using bituminous coal, lignite and peat, the blue gas generation is preceded by precarbonisation in an upper section of the unit. The gases are separately taken off. The blue gas is mixed with the carbonisation gas to produce an enriched gaseous fuel called *double water gas* or *double gas*.

The performance data of the semi-water gas generator at the Sindri fertiliser plant are: Internal hearth diameter: 3.05 m, raw gas produced 6,500 Nm^3/h, gasification rate 874 Nm^3/m^2h raw material consumption per Nm^3 of raw gas, coke 0.60 kg, steam 0.82 kg and air 1.3 Nm^3, cold gas efficiency 60%, carbon efficiency 58%, gas composition (in per cent): hydrogen 37.1, carbon monoxide 31.3, methane 0.6, nitrogen 23.0 and carbon dioxide 8.0.

6.2.6 Carburetted Water Gas

The calorific value of blue gas is too low for it to be an effective gaseous fuel in a town`s distribution system. At best it can act as a diluent of coal gas. However, it can be enriched with hydrocarbons to make carburetted water gas. The process is essentially a combination of solid fuel gasification and liquid fuel gasification. Modern oil gasification processes have benefitted from the technology of carburetted water gas.

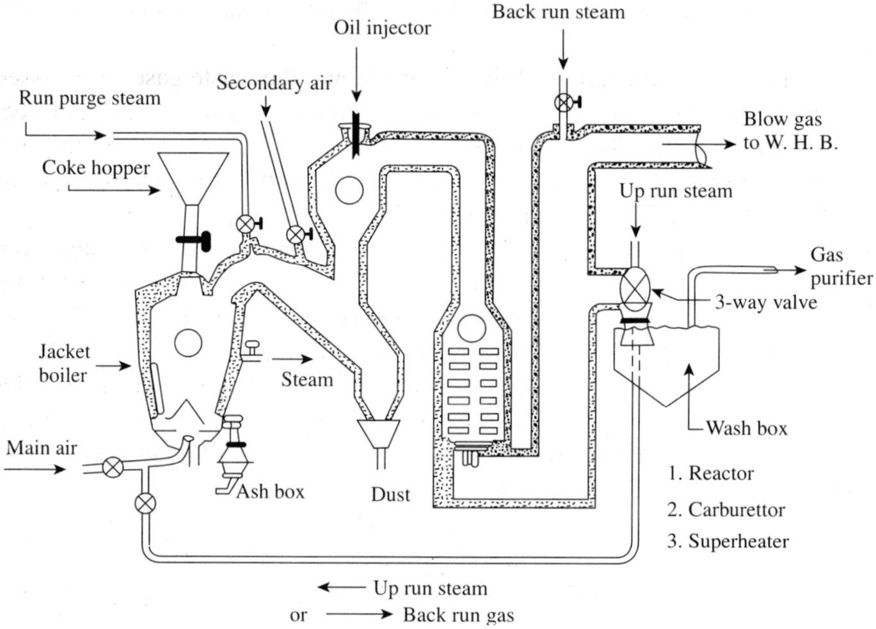

Figure 6.4 Carburetted water gas unit

Many petroleum oils undergo cracking at 700–750°C and produce methane, ethylene, propane and other unsaturated hydrocarbons, all of which have high calorific values. Carburetted water gas is made by passing blue gas through a hot chamber, or carburettor into which an oil is sprayed. The composition of the resultant mixture is obviously dependent upon many variables. The data for a typical commercial gas is given in Table 6.6. The carburetted water gas is used as a substitute for coal gas to meet the seasonal and peak loads of gas supply. Although the calorific value is similar, its higher specific gravity gives a poorer Wobbe index compared to coal gas.

A carburetted water gas plant consists of two additional units, namely, a carburettor and a superheater, besides the blue gas generator and other accessories. These two units are filled with chequered bricks (Fig. 6.4). The blow gas is burnt in the carburettor while the combustion is completed in the superheater. The combustion gases then enter the waste-heat boiler and finally escape through the chimney. During the run period oil is sprayed into the brickwork of the carburettor and the blue gas is passed through the chamber. Cracking is completed in the superheater. Since cracking is an endothermic process the brickwork temperature falls and this is again heated up in the blow period. The cyclic process of carburetting blue gas is well synchronised with the parent gas generation process. The resultant gas is passed out.

Some of the heat generated by the combustion of blow gas in a carburettor and superheater is fed back to the fuel bed for carrying out a back run. Steam is blown through superheater and carburettor into the top of the fuel bed. Oil is sprayed into the steam. It undergoes severe cracking on passage through the fuel bed. The back run increases the thermal efficiency of the plant.

The oil used in carburetting is usually a light petroleum fraction, 200° to 300°C. consisting principally of paraffins. Heavier oils are also used to some extent. These crack with the formation of coke. The gasification efficiency is lower but the heavy oils offer cheaper substitutes than the lighter fraction. The carburettor for heavy oils is often without any chequered bricks.

The requirements for producing 1 Nm^3 of a typical carburetted water gas are: coke 0.45 kg, oil 0.3 kg, steam 0.65 kg and air 1.5 Nm^3. The cold gas efficiency (with light oils) is about 65% which is slightly higher than that of a blue gas generator.

Complete gasification of coal

Gas-making processes using coke are only partial gasification processes involving two stages. First the coal is carbonised into coke in coke ovens and then the coke is gasified in the second stage. The methods of making producer gas and double water gas from coal are outwardly single-stage gasification pocesses. However, it is to be noted that carbonisation is allowed to precede gasification of coke in a separate zone of the same unit. As a result, large quantities of tar are produced and hence the gasification efficiency remains below what would have been otherwise. The processes of complete or total gasification of coal in a single stage maximises the conversion of coal into gaseous fuel and minimises the formation of tar. Obviously the gasification efficiency of such processes is higher than that of those mentioned earlier. Some of these processes are equally suitable for a wide range of feed material. The classification of various gasification processes is given in Chapter 4. A brief description of some of the more important of these follows.

The Lurgi process: This is a single-stage, fixed bed process gasifying a range of coarse or granular fuels (Fig. 6.5). A Lurgi generator consists of three main parts: charging box (top), shaft (middle) and ash receiver (bottom). The shaft is water jacketted. Solid ash is removed by a rotating grate. With the exception of medium and strongly coking coals, practically the whole range of solid fuels above 2 mm size are suitable. A mixture of superheated steam, 450–500°C and oxygen is used as a gasifying medium. The process is continuous and auto-thermic: the fuel–oxygen reaction supplying the heat. High pressure at 30 atm favours the methanation reaction (6.8) which raises the methane content of the gas. The hydrogenation of the volatile matter of coal under pressure also forms methane. The methane formation reactions are exothermic. Hence, the oxygen requirement of the gasification process is reduced.

A high steam–oxygen ratio, say, 10:1, is normally maintained to ensure that clinkering troubles are avoided. The steam–oxygen ratio is also used to control the hydrogen–carbon monoxide ratio. The large amount of steam is responsible for an increased content of carbon dioxide in the gas.

The Lurgi plant produces gas at a high pressure which facilitates its purification and subsequent distribution through pipes, and its use in the synthesis of ammonia, methanol and liquid fuel. High pressure also lowers the gas velocity through the bed and hence, reduces the dust *carry-over* from the bed surface and permits a reduction in the grain size of the fuel. The composition, in per cent of a crude gas and a purified gas from a commercial Lurgi plant using bituminous coal is given in Table 6.7. The calorific value of the purified gas is 3,900 kcal/Nm^3 and its specific gravity 0.48.

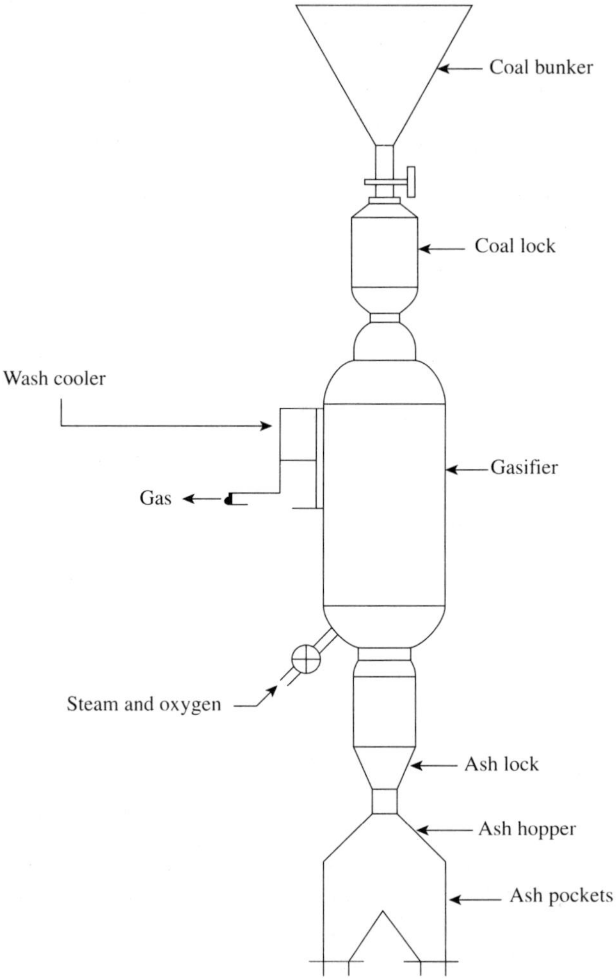

Figure 6.5 Lurgi gasification unit

Lurgi plants have been installed in many countries to manufacture fuel gas for public utility purposes and synthesis gas for the chemical industry. There is a pilot plant with a capacity of 0.8 tonnes of coal per hour (tph) at the Central Fuel Research Institute, Dhanbad. A slightly larger pilot plant of 1 tph has recently been installed at the Regional Research Laboratory, Hyderabad. Of the various complete gasification processes, the Lurgi pressure gasifier has the largest total installed capacity in the world. An individual unit has the capacity of 40 tph of coal.

The cold gas efficiency of a commercial Lurgi plant may be about 75%–80% of the heat in fuel. It may be further raised by recycling the tar in the gasifier. The rate of gasification may vary from 0.2–1.1 tonnes/m^2 per hour. The consumption of raw materials per Nm^3 of refined gas is: fuel (22% ash, 2.5% moisture) 0.85 kg, steam 1.4 kg and oxygen 0.21 Nm^3 or 0.3 kg.

The Lurgi gas produced as above is essentially an industrial fuel gas. It may be further treated for making substitute natural gas (SNG) near natural gas distribution networks to supplement existing supplies of the latter. Large demonstration plants have been installed at the individual capacity of, say, 7 tph coal and a pressure of 100 atm in the gasifier. The bed temperature is allowed to rise to 1,500°C and the ash is disposed off as molten slag. The specific productivity of such a gasifier is twice that of a standard Lurgi gasifier. At 100 atm pressure the methane content of the gas may be increased up to 18%.

The crude gas is treated for the removal of dust and tar, subjected to shift conversion (6.7) and purified from carbon dioxide and hydrogen sulphide. The gas is now ready for methanation over a nickel catalyst:

$$CO + 3H_2 \rightleftarrows CH_4 + H_2O; \qquad \Delta H = -49{,}271 \text{ kcal} \tag{6.9}$$

$$CO_2 + 4H_2 \rightleftarrows CH_4 + 2H_2O; \qquad \Delta H = 39{,}433 \text{ kcal} \tag{6.10}$$

After the condensation of steam the gas is essentially methane (SNG).

***The Winkler process*:** The Winkler gasification process is of the fluidised bed type. The generator is a brick-lined cylindrical shaft. The solid fuel below 8 mm size and 15% moisture is charged by feed worms. The gasification medium—mixture of oxygen and steam—is blown into the shaft from both the bottom and above the bed, the latter arrangement being provided for gasifying some of the entrained fuel particles. The normal bed temperature is 800–950°C and the space above the fluidised bed attains a temperature up to 1,000°C. Under these conditions the tar and the gaseous hydrocarbons are reformed to carbon monoxide and hydrogen. The product gas contains traces of methane, while phenol is absent. A slightly positive pressure is maintained inside the system. The hot gas is passed through a waste heat boiler having a steam superheater and a feed water preheater and then introduced into cyclones for dust removal. The gas is then sent to a water scrubber followed by a water seal for cooling and further dust removal. The final removal of dust is achieved in a Thiessen disintegrator. Further purification and processing are necessary for preparing synthesis gas.

The process is suitable for gasifying a wide range of solid fuels having medium to high reactivity, namely lignite, low temperature coke and char from coal and lignite and high volatile bituminous coal. High moisture fuels need predrying. Fuels of low reactivity like anthractie and coke, and fuels of low ash fusion temperature are unsuitable. The greatest demerit of the process is high dust carry-over which is as high as 20% even with reactive fuels. This naturally reduces the gasification efficiency. However, an elaborate dust recovery system permits the use of the dust in steam raising.

The performance data of a large commercial plant of 60,000 Nm^3/h capacity are given below:

Product gas

Composition, per cent					CV kcal/Nm^3 dry	Specific gravity (air = 1)
CO	H_2	CH_4	CO_2	N_2		
44.4	36.0	1.6	15.7	2.3	2,600	0.70

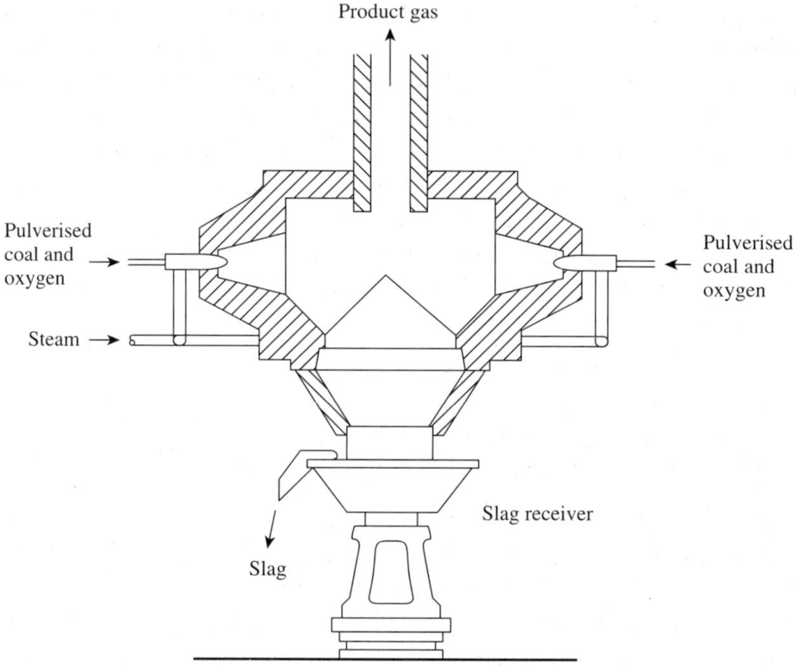

Figure 6.6 Koppers–Totzek gasifier

Gasification efficiency (carbon conversion) 80%, cold gas efficiency 58%, consumption per Nm^3 gas: lignite 1.1 kg, steam 0.31 kg and oxygen 0.36 Nm^3 or 0.51 kg.

Winkler gasifiers of commercial size have been installed in various countries to produce synthesis gas, water gas and also heating gas. Owing to its low calorific value and high specific gravity, the Winkler gas is not a good gaseous fuel and it is more suitable for the manufacture of chemicals, fertilisers and synthetic liquid fuels.

A Winkler gasification plant with three generators based on lignite was in operation at Neyveli since 1965. It had a total capacity of 8,60,000 Nm^3 gas per day. It supplied synthesis gas to the fertiliser plant. It was shutdown after the feedstock was changed to naptha.

Low reactivity fuels, and fuels with low ash fusion temperatures can be successfully gasified in two modified versions of the Winkler process. These are the Flesch–Winkler and the BASF–Flesh–Demag processes.

***The Koppers–Totzek process*:** This process gasifies pulverised fuel carried in suspension in oxygen at approximately atmospheric pressure in the presence of steam. The gasifier is a refractory-lined, horizontal, cylindrical vessel with conical ends (Fig. 6.6). Two streams of coal dust, dried to 2% moisture, and size below 0.1 mm in oxygen are fed through coaxial burners from the two conical ends. A considerable turbulence is created. Gasification proceeds rapidly at the very high temperature of 1,600°C in the chamber. Superheated steam is introduced around the burners to shroud the high temperature reaction zone and protect the burners and gasifier lining from the intense heat, and then to take part in the reactions.

The product gas leaves the gasifier at 1,000–1,300°C and is passed through a radiation boiler and a waste-heat boiler for the recovery of the sensible heat. About half of the ash and one-tenth of the fuel dust is carried with the gas. This is removed in dust catchers and the fuel is used again. Half of the ash is deposited on the gasifier wall as a fluid slag. It flows down and is finally removed from the bottom through a rotating quenching bowl in a granulated form.

At the high temperature of the reaction zone the volatile matter of the coal is converted into carbon monoxide, carbon dioxide and hydrogen. The methane content of the gas is only in traces, below 0.1%. Higher hydrocarbons are absent. The data about a gas obtained from a commercial Koppers–Totzek plant using a high ash coal are given below:

Composition, per cent					CV kcal/Nm^3 dry	Specific gravity (air = 1)
CO	H_2	CH_4	$CO_2 + H_2S$	N_2		
55.3	31.1	0.1	12.2	1.3	2,670	0.76

Koppers–Totzek plants have been installed in many countries for making synthesis gas for ammonia, methanol and others. Owing to high specific gravity and low calorific value, the gas obtained is not suitable as a fuel. The cold gas efficiency of the gasifier is 67%, carbon conversion 97% and the overall efficiency, including waste-heat recovery, 75.5%. The main points of attraction of the Koppers–Totzek process for the production of synthesis gas are:

1. Total gasification in a single stage,
2. Very low methane content and absence of tar in the gas,
3. Versatillity with regard to the quality of the solid fuel which includes dust and slacks produced during mining and handling a coal or lignite,
4. Operation at atmospheric pressure,
5. Simple design of the reactor,
6. Adaptability to the gasification of heavy oils.

There was a pilot plant with a capacity of 115 kg per hour of raw coal at the Central Fuel Research Institute, Dhanbad, which went into operation in 1959. Two commercial plants have been installed at Talchar in Orissa and Ramagundem in Andhra Pradesh in 1978 to supply synthesis gas for ammonia. Each plant has three gasifiers, each with a capacity of 40,000 Nm^3 per hour of raw synthesis gas. These are the world's largest coal gasifiers.

Some performance data of a commercial Koppers–Totzek plant are: gas yield 1,530 Nm^3/t coal (ash 35%), consumption per Nm^3 gas—coal 0.65 kg, steam 0.35 kg and oxygen 0.36 Nm^3 or 0.5 kg.

***The Texaco process*:** The Texaco coal gasifier, which is an entrainment system, is based on the well-proven Texaco synthesis gas generation process for the partial oxidation of heavy fuel oil. In the coal version, finely ground coal <0.074 mm in size is slurried with water and pumped into the gasifier where it is gasified with oxygen or air at a temperature of about 1,500°C and at pressures which can be up to 150 bar. The jet is directed downwards from the top of a cylindrical refractory-lined pressure vessel. At the bottom of the vessel the gas and the molten slag are quenched by a water spray and a slag quench bath. The gas leaving the reactor is cleaned and the fines containing carbon recovered for recycling with the coal slurry. Great care is taken to maintain a constant high solids content. The reactor temperature which is sensitive to the inputs, is closely controlled in the range of 1,100–1,370°C to suit the ash fusion temperature. Too high a temperature results in excessive refractory wear, whereas low temperatures give rise to slag lumps which would block the exit to the lock hopper. Both the slurry and oxygen are preheated.

The gas contains about 15% CO_2 and very little CH_4, and the rest is mainly H_2 and CO in a ratio of about 3:4. The calorific value is about 2,700 kcal/Nm^3 when blown with oxygen and about half of that if air is used. The gasification efficiency is 75%–80% and the oxygen requirement about 0.6–1.1 kg/kg of coal, depending on the type of coal.

***Underground gasification of coal*:** Coal can be gasified in situ without raising it from the mine. A number of small commercial plants had been installed in the old USSR which produced cheap gases of low calorific value mainly for electricity generation. The production in 1963 was 1.4 billion cubic metres with an average calorific value of 810 kcal/Nm^3 dry. Air and oxygen were used as the gasifying medium. Bituminous, sub-bituminous and mature brown coal deposits are suitable for underground gasification.

There are two systems of underground gasification: the shaftless system and the shaft system. Vertical channels are made in both systems for supplying the gasifying medium and taking out the product gas. In the shaftless system, no underground work is required while the shaft system needs extensive preparatory underground work. The major problem in the shaftless system is the making of satisfactory links underground between the inlet and the outlet holes. The natural permeability of the coal bed is not sufficient. Links can be achieved by establishing electro-links between two electrodes and carrying out electrical carbonisation of the coal in the path. A channel of hot coke is formed through which air is passed for gasifying the fuel between the two vertical boreholes. Another method involves the development of fractures in the coal mass by hydraulic or pneumatic means and then creating a path by *backward burning*.

There are a number of variants of the shaft system. In the more popular *stream method*, horizontal galleries are made at the bottom of the coal seam. Access galleries connect the horizontal galleries with the mine shafts. A fire is ignited at the seam bottom along the horizontal gallery and air is introduced through one access gallery. The product gas is taken out through the other access gallery. With the progress of gasification, the reaction zone advances upward along the coal bed. The stream method is usually adopted with the steeply pitched coal beds.

Another variant of the shaft system is the borehole producer method. Parallel galleries are constructed within the coal bed and then connected by making horizontal boreholes through the confined coal mass. Boreholes are ignited in succession or in small groups. Air is passed through one gallery and the product gas taken off through the other.

The chief points of attraction of underground gasification of coal are the elimination of mining and the utilisation of unworkable coal seams and left-over coal in mines. The product gas may be used not only as a cheap fuel but also in the manufacture of synthesis gas for ammonia, methanol and liquid fuels. Large-scale worldwide application of the process will be possible by the improvement of its technology and also by the ever-increasing cost of mining and gasification of coal by other routes.

The composition, in per cent, of a lean gas produced by underground gasification is given below

CO	H_2	CH_4	N_2	CO_2	C_nH_m	O_2
10.7	8.4	1.8	67.4	10.5	0.3	0.9

This gas has a calorific value of 820 kcal/Nm^3 dry, and specific gravity of 0.95 (air = 1).

6.2.7 Coal Gas

This is produced by the high temperature carbonisation of coal in gas retorts and coke ovens. Chapter 4 describes the manufacturing process and gives the average yield, compostion, calorific value and uses of the by-product coke oven gas; also see Table 6.3. The use of coal carbonisation pocesses as the main method of producting town gas or synthesis gas is rapidly decreasing. What is still attractive is the utilisation of the by-product gas obtained from coke plants. With a calorific value of 5,000 kcal/Nm^3 dry and specific gravity of 0.4, coal gas is a good fuel for distribution to homes and industries. It is also suitable for the production of synthetisis gas. Coke oven gas may be regarded as the cheapest raw material for ammonia production if the fertiliser plant is located near an integrated iron and steel plant.

There are two methods for obtaining hydrogen from coke over gas: low temperature separation and steam reforming. The fertiliser plant at Rourkela draws the coke oven gas from the steel plant and separates hydrogen by the low temperature method. The coke oven gas is compresed to 12–30 atm, Carbon dioxide is removed by water scrubbing and the gas is gradually cooled to the liquefying temperature. The unsaturates, methane and carbon monoxide are liquefied. The gaseous hydrogen is purified by washing with liquid nitrogen and used in the synthesis of ammonia. The unsaturates may be used in the manufacture of chemicals and the gas containing methane and carbon monoxide may be used as a rich fuel.

An additional quantity of hydrogen may be obtained from methane by steam reforming and from carbon monoxide by water gas shift reaction.

6.2.8 Blast Furnace Gas

The combustion zone of the blast furnace is near the bottom of the tall structure. The combustion gases ascend through the descending burden of coke, ore and flux. Carbon reduces carbon dioxide to carbon monoxide and decomposes steam into hydrogen and carbon monoxide. The reduction of iron oxide is achieved mostly by carbon monoxide and partly by hydrogen. The gases remain in contact with the charge in the upper part of the furnace, give away part of the sensible heat and leave the furnace top at a sufficiently high temperature say, 180°C to 200°C. The resultant gas is known as blast furnace gas.

Blast furnaces may be regarded as gas producers in which a few other reactions take place in addition to the usual producer reactions. The resultant gas has higher carbon dioxide content and lower hydrogen content than the producer gas. A typical gas analysis is shown in Table 6.7.

The gas is produced at a high yield, say, 2,400 Nm^3 per tonne of pig iron. The utilisation of the potential heat of the enormous quantity of the gas produced in an iron and steel plant is extremely important for the efficiency of the plant.

The high dust content, upto 20–30 g/m^3, is a deterrrent to its straight supply to combustion systems. The heavier particles are removed by dust catchers or cyclones. If necessary, further cleaning is done in wet cleaning units like stationary scrubbers, revolving cleaners or centrifugal machines like the Thiessen disintegrator, bag filters or electrostatic precipitators. The final degree of cleanliness is about 0.5 g/m^3 for industrial burners and 0.01 g/m^3 for gas engines. Owing to its low calorific value it is necessary to preheat the gas and air for obtaining high flame temperatures. It is also mixed with coke oven gas and used. In a modern iron and steel plant about 20% of the total volume of blast furnace gas is used to preheat the blast, the surplus gas is frequently used for steam raising purpose and for firing the open hearth, reheating furnaces and coke ovens.

6.2.9 Gases from Biomass

Biomass can be coverted to fuel gas as well as synthesis gas for chemicals, fertilisers and liquid fuels. The present emphasis is on the production of gasesous fuels from biomass. There are two routes to get gas from biomass: a thermochemical process and a biochemical process. The term biogas is reserved for the product obtained by the biochemical conversion of biomass.

Like coal, biomass can be partially gasified by carbonisation, which is essentially pyrolysis. This process can be controlled in such a way that either only charcoal, or charcoal and gas are produced as major products. In the former variant the medium calorific gas is used up in the charcoal plant itself. In the other variant, auxiliary fuel is used in firing the charcoal retort and the gas is released for other consumption. Wood is the main feedstock used in the carbonisation and total gasification of biomass. If agriwastes are used in carbonisation, the charcoal emerges in the form of dust and requires briquetting before it can be put to use. Typical compostion and properties of gas of wood carbonisation, *wood gas*, are given in Table 6.8.

In order to achieve total gasification it is necessary to use a gasifying medium like air, oxygen and hydrogen. Air is the most common of the three and wood logs and chips are the main feedstock for the process which is widely practised in many countries for producing a cheap fuel gas. The principle of wood gasification is similar to the principle of obtaining producer gas from coal. However, the wood producers have longer zones for drying and carbonisation than the coal producers. The wood producers are also suitable to gasify peat. A unit with a capacity of 70–80 tpd of dry wood chips is shown in Fig. 6.7. The performance of a commercial wood gas producer is given in Table 6.8.

Crude methyl alcohol may also be recovered in a yield of about 0.5% dry wood. A typical gas composition, in per cent, is: CO_2 6.0, CO 29.0, CH_4 2.5, C_nH_m 0.7, H_2 14.5, O_2 0.3 and N_2 47.0. Its calorific value is 1,660 kcal/Nm^3 dry and specific gravity 0.85 (air = 1). About

Table 6.8 Composition and properties of wood gas

Temperature of carbonisation,°C	~400	~1000
Yield of gas, Nm^3 per tonne of dry wood	125	550
per cent w/w dry wood	14.5	50.8
CV of gas, kcal/Nm^3 dry	3,320	3,280
Specific gravity of gas (air = 1)	0.68	0.50
Composition, per cent dry		
CO_2	30	20
C_3H_{10}	4	2
CO	25	25
CH_4	14	12
H_2	20	35
N_2	7	6

40% to 50% of the gas is consumed as a fuel for the producer. The rest is available for, say, thermal power stations. The liquid products yield chemicals on further treatment.

Wood gasification based on cheap wood and wood wastes is a process with sound economics.

Biogas

Anaerobic decomposition of organic wastes by suitable bacteria produces methan-rich gases known as biogas. Carbonaceous agricultural wastes of plant or animal origin such as animal dung, straw, hay and fruit peelings, and municipality sewage constitute organic wastes. Cattle dung, however, is the largest feedstock for biogas generation.

The anaerobic fermentation of biomass takes place in two stages. First, the complex organic compounds are decomposed into simple organic acids and these acids are later converted into methane and carbon dioxide. The acid-forming and methane-forming bacteria must remain in a state of dynamic equilibrium.

Fresh dung has a solid content of 20%. It is diluted to 7% to 9% with water and fed into a pit-like chamber. The temperature is kept at 30ºC to 40°C. In the initial acid forming stage, the pH is 6 when much carbon dioxide is given off. In two to three weeks time, as the volatile acids and nitrogen compounds are digested and methane is formed, the pH increases upto 8. Slight mixing improves the gas yield but violent agitation retards the process.

In a biogas plant the whole system is a continuous operation, that is, the matter to be fermented is fed in a semifluid form at one end, and the fermented spent liquor is extracted at the other end periodically, without disturbing the system. The plant is a composite unit of a digester and gas holder. The gas holder floats on the top of the digester wherein the gas is collected and delivered at constant pressure to the gas appliances through a pipeline.

The composition and yield of biogas depend upon many factors, including the material used as feed for fermentation. On an average from cattle dung one gets a gas consisting of 55% to 60% methane and 40% to 45% carbon dioxide with negligible amounts of hydrogen sulphide, hydrogen and nitrogen. The yield of gas is about 0.092 m^3/kg fresh dung in summer and 0.036 m^3/kg in winter. It is necessary to stimulate gas production in winter by warming up the water used for diluting and mixing the cattle dung. Some nutrients like urine, oil cakes or molasses may also be added.

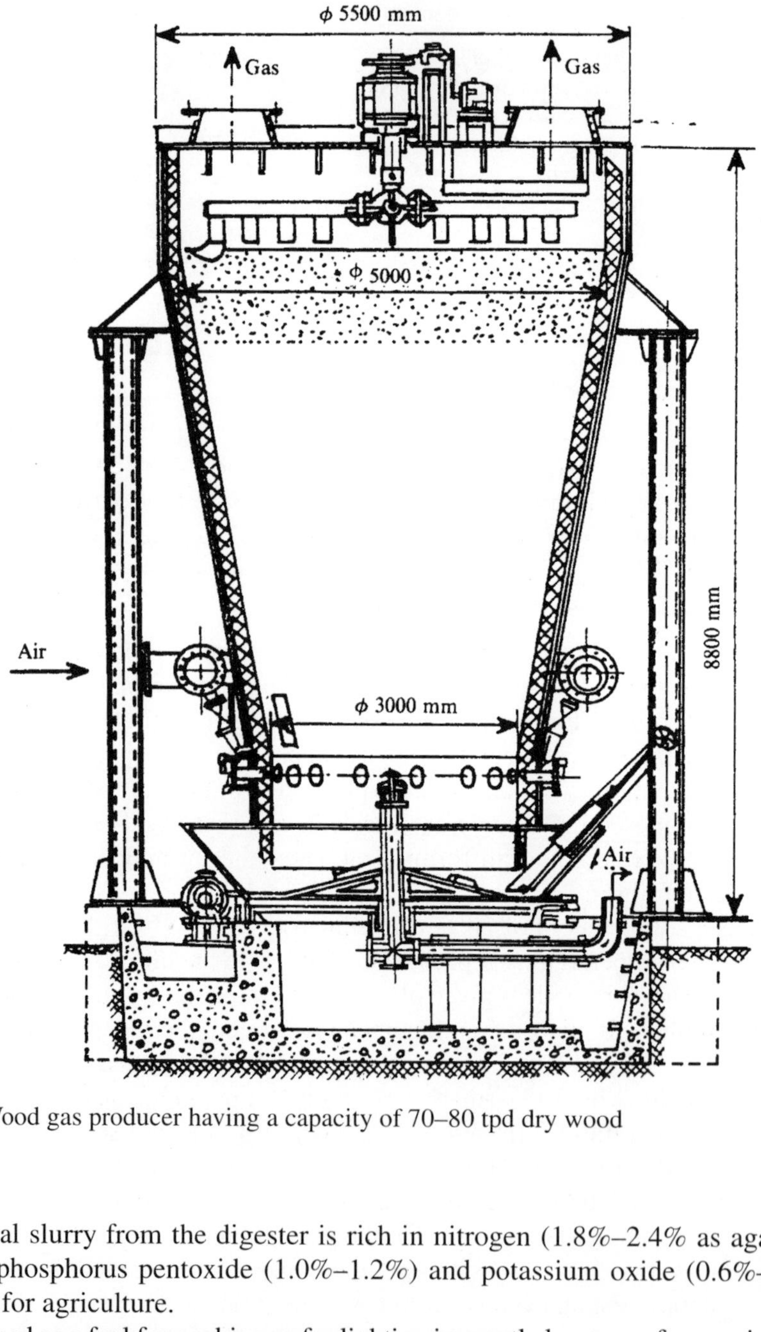

Figure 6.7 Wood gas producer having a capacity of 70–80 tpd dry wood

The residual slurry from the digester is rich in nitrogen (1.8%–2.4% as against 0.75% in fresh dung), phosphorus pentoxide (1.0%–1.2%) and potassium oxide (0.6%–0.8%). It is a good manure for agriculture.

Biogas is used as a fuel for cooking or for lighting in mantle lamps, or for running gasoline and diesel engines. It is also possible to use it as a fuel in industry if available in sufficient quantities.

The future of biogas is very bright in a country like India with a large population of cows, horses and buffaloes.

6.2.10 Refinery Gases

The term *refinery gas* covers a wide range of products and includes gases obtained during distillation, cracking and other processing of petroleum and petroleum fractions. These gases contain paraffins like methane, ethane, propane and butane, olefins like ethylene, propene and butene, hydrogen sulphide and hydrogen. A possible range of composition may be:

Composition, in per cent

H_2S	H_2	CH_4	C_2H_4 + C_2H_6	C_3H_6 + C_3H_8	C_4H_8 + C_4H_{10} (dry)	CV kcal/Nm^3	Specific gravity (air = 1)
6–8	6–8	8–10	8–12	40–50	up to 30	20,500	1.35

The C_4- and C_3-hydrocarbons may be separated by liquefaction and sold as bottled gas. Similarly sulphur may be recovered from the hydrogen sulphide. The hydrogen and lighter hydrocarbons are best utilised in the production of chemicals, fertilisers and plastics. These may be used to enrich town gas also. All these uses and even burnig as a refinery fuel may not completely consume all the refinery gases produced in a big plant. The surplus gas is disposed of by burning it in the open.

6.2.11 Liquefied Petroleum Gases (LPG)

Out of the gasesous hydrocarbons, the C_3- and C_4-compounds can be liquefied at room temperature by the application of moderate pressure and can therefore be conveniently stored and transported as liquids in light pressure vessels. These are known as *liquefied petroleum gas*, or LPG or *bottled gas*. In many countries two grades of this product are marketed: one containing the C_3-hydrocarbons—commercial propane—and the other containing the C_4-hydrocarbons—commercial butane—as the chief components. The latter is safer to handle with storage pressure upto 3 atm and is freely used for domestic purposes. The C_3-hydrocarbons are liquefied and stored at higher pressures upto 10 atm and used in industrial heating. Where only one grade is sold, it is a mixture of the two types containing higher proportion of C_4-hydrocarbons. The lighter grade of LPG contains about 90% propane, below 10% butane–*iso*-butenes and below 2% ethane–ehylene. The heavier grade contains 80% to 90% butane–*iso*-butenes, below 20% propane–propene and below 2% ethane–ethylene.

LPG is prepared from wet natural gas, associated gas and refinery gases. Indian specifications are given in Table 6.11. The combustion characteristics of LPG differ greatly from the usual gaseous fuels. They have high calorific value, high specific gravity, high air requirement and low flame speeds (Table 6.10). Suitable burners are available for using LPG in home and industry. Since LPG is odourless it is necessary to add odorants, such as mercaptans or sulphides, to detect leakage.

Bottled gases are becoming increasingly popular as domestic fuel. The chief point of attraction lies in the fact that a large concentration of thermal energy is contained in a relatively

Table 6.9 Performance of a commercial wood gas producer

Wood type	Properties				Yields (dry basis)			
	Moisture, per cent	Chip size, mm	Blast temp, °C	Blast rate Nm^3/kg	Tar, per cent	Volatile acids, per cent as acetic acid	Liquor, per cent	Gas, Nm^3/kg
Pine	46.8	86	50	1.23	12.3	2.62	18.0	1.94
Birch	44.6	80	51	1.15	10.0	4.02	21.0	1.75

Table 6.10 Properties of liquefied petroleum gases

Type	Gross CV $kcal/Nm^3$ dry	Specific gravity (air = 1)	Theoretical air requirement Nm^3/Nm^3
Pure butane	30,680	2.00	30.94
Pure propane	23,670	1.52	23.80
Coke oven gas	5,020	0.38	4.57

small and portable vessel which provides a gaseous fuel that can be piped or directly sent to a gas burner. This is particularly true in the case of those cities and towns where adequate central gas distribution systems do not exist. When portable fuels are not easily available, LPG is quite popular in industrial heating also. LPG has also found use as a fuel for internal combustion engines of stationary type and also those attached to vehicles, tractors and locomotives.

Propane and butane gases may be used to enrich the low calorific value gases in a town gas supply system upto the limits of permissible specific gravity. Butane may also be diluted with air to get a suitable fuel. A mixture of 20% butane and 80% air has a CV of 6,136 $kcal/Nm^3$ dry and is well outside the limits of inflammability of butane in air. Finally propane and butane may be reformed to get a gaseous fuel or synthesis gas. Propane, butane and their olefinic counterparts are widely used in refinery processes and in the production of petrochemicals.

Oil gasification

Liquid fuels are gasified to produce either gaseous fuels or synthesis gas. Depending upon local conditions, light distillates, middle distillates and heavy oils are all used as raw materials for gasification processes. Some have found commercial application.

The C_3- to C_{15}-hydrocarbons are liquids under ambient conditions, although both lighter and heavier hydrocarbons may remain in solution in them. The transformation of liquid hydrocarbons into the gaseous form involves two basic processes:

1. Reduction in the size of the molecules,
2. Reduction in the C/H ratio.

These can be achieved by cracking, addition of hydrogen and removal of carbon in the form of coke, tar and oxides of carbon.

Table 6.11 Specifications for liquefied petroleum gases (IS: 4576-1999)

Characteristics	Requirements for commercial use		
	Butane	Butane–Propane mixture	Propane
Vapour pressure at 40°C, kPa, gauge, Max	520	1050 (Note 2)	1550
Composition, liquid volume percentage			
(a) C_2-Hydrocarbons	–	Report	Report
(b) C_3-Hydrocarbons	Report	Report	95.0 Min
(c) C_4-Hydrocarbons	Report	Report	4.0 Max
(d) C_5-Hydrocarbons and heavier	2.5 Max	2.5 Max	0.2 Max
(e) Unsaturated hydrocarbons OR	Report	Report	Report
Volatility: Evaporation temperature in °C for 95 per cent by volume at 760 mm Hg pressure, max	2.0	2.0	−38.0
Total volatile sulphur ppm, max	150	150	150
Copper strip corrosion at 38°C for 1 h	Not worse than No.1		
Hydrogen sulphide	Pass	Pass	Pass
Free water content	None	None	None

The C/H ratios by weight of some fuels are: typical coking coal, 15.5:1; heavy fuel oil, 7.8:1; light distillate feedstock, 5.6:1; natural gas and town gas; 3:1. If town gas is manufactured from solid and liquid fuels only by thermal cracking without the addition of hydrogen or oxygen, and if it is assumed that all the hydrogen goes to the formation of the gas, a simple calculation will show that only 24% of the solid fuel, 46% of the heavy oil and as much as 61% of the light oil can be gasified, leaving 76% of the soild fuel, 54% of the heavy oil and only 39% of the light oil as either carbon or carbon-rich tar. The total gasification processes aim at converting the carbon into cabon monoxide and gaseous hydrocarbons by supplying air, oxygen and steam. This clearly bears out the importance of the lower C/H ratio of the oils as a technical point in favour of oil gasification as compared to the gasificaion of solid fuels.

The overall reactions with oxygen and steam in oil gasification leading to hydrogen and carbon monoxide formation are

$$C_xH_y + \frac{x}{2}O_2 \rightleftarrows xCO + \frac{y}{2}H_2 \tag{6.11}$$

$$C_xH_y + xH_2O \rightleftarrows xCO + \left(x + \frac{y}{2}\right) H_2 \tag{6.12}$$

Reaction (6.11) is referred to as partial oxidation and reaction (6.12) as steam reforming. These reactions are the ultimate goals in the production of synthesis gas. However, it is desired in fuel gas manufacture to break down (crack) the oil hydrocarbon molecules to smaller sizes in the gaseous range so that the gaseous product is rich in hydrocarbons and has a high calorific value. The cracking reactions have been discussed in Chapter 5. The resultant gaseous hydrocarbons are predominently paraffinic, chiefly methane, and olefinic in nature while aromatics and naphthenes are also present. The cracking and steam reforming reactions are facilitated by operation under pressure or by the use of catalysts, the temperature remaining an important variable. Cracking and steam reforming reactions (6.13) are highly endothermic while partial oxidation is exothermic.

$$C_nH_{2n+2} + nH_2O \rightleftarrows nCO + (2n+1)\,H_2 \quad \Delta H = +49{,}290 \text{ kcal for } n = 1 \text{ and} \qquad (6.13)$$
$$= +2{,}36{,}000 \text{ kcal for } n = 6$$

The heat requirement of the endothermic reaction is supplied by one of three methods:

1. Cyclic method of dividing the process into blow and run periods as in blue gas generation,
2. Continuous method of burning a part of the feed in the gasifier using oxygen and steam in place of air and steam,
3. External heating by enclosing the gasifier in a furnace.

The oil gasification process may be classified as follows:

1. Non-catalytic cyclic processes, such as Semet–Solvay,
2. Catalytic cyclic processes, such as Segas and Onia–Gegi,
3. Continuous catalytic reforming, such as ICI and Topsoe.
4. Partial combustion processes, such as Koppers–Totzek, Shell and Texaco,
5. Hydrogenation processes.

Since the oil crisis in the 1970s, the oil gasification processes are restricted to the production of syngas for chemicals and fertilisers.

***Continuous steam reforming of naphtha*:** Steam reforming of naphtha has followed the commercial success of steam reforming of natural gas, refinery gases and LPG for the production of fuel gas, ammonia synthesis gas, methanol synthesis gas and hydrogen. An incentive for the development of this process was the availability of cheap surplus naphtha on a very large scale in the world market.

Contrary to hydrocarbon gases, steam reforming of naphtha deposits carbon on the catalysts inspite of the use of excess quantities of steam and operation at comparatively low pressures. However, ICI of Britain have succeeded in developing a suitable nickel catalyst (specially promoted nickel oxide on a ceramic base with hydraulic cement bonding) which allows continuous reforming of desulphurised naphtha at a relativiely high pressure and with a relatively reduced steam consumption.

Since sulphur seriously deactivates the nickel catalyst, naphtha needs desuphurisation before steam reforming. Depending upon the purity of the naphtha, catalytic hydro-desulphurisation

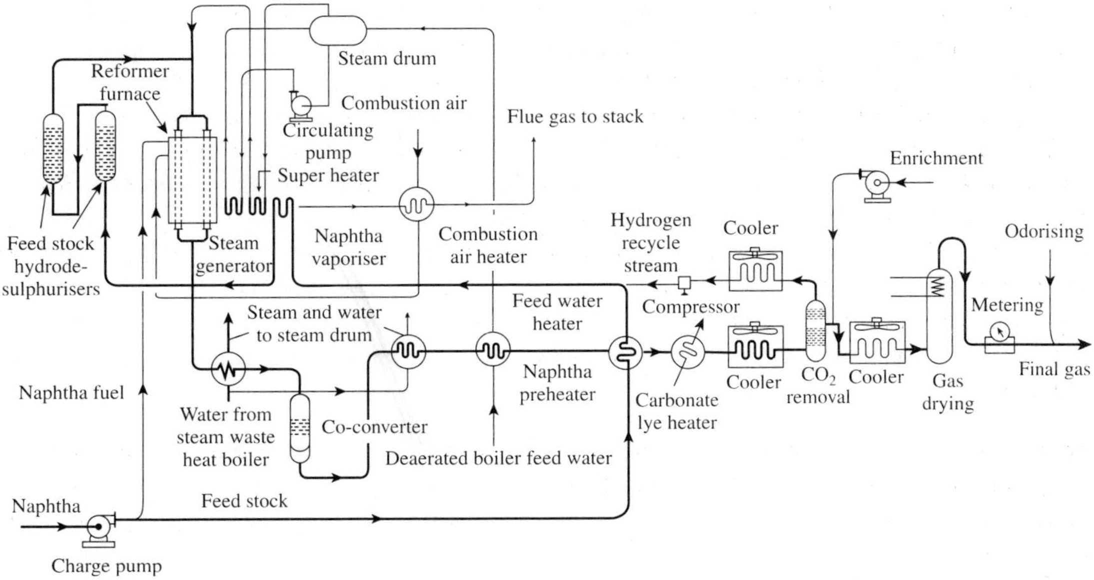

Figure 6.8 Simplified flow-diagram of (ICI's) continuous naphtha reforming plant

is done in a single stage (vapour phase) or in two-stages (liquid phase followed by vapour phase) process.

The main reactor is in the form of three rows of vertical nickel chrome alloy steel tubes of 127 mm internal diameter and 8.4 m length charged with the catalyst. The tubes are heated in a furnace upto 960°C by five rows of burners installed at the furnace top. The flue gases are used to raise waste heat steam at 20 atm, superheat the reaction steam and preheat the combustion air.

A mixture of desulphurised naphtha vapours, surplus hydrogen and superheated steam (3 mol/mol of feedstock carbon) enters the reactor from the top. The reaction is carried out at pressures up to 28 atm and temperatures upto 840°C. The exact temperature and pressure conditions depend upon the ultimate use of the gas. In a plant for a town`s gas supply the gas leaves the reactor at 700°C and 19 atm (Fig. 6.8). It is a medium calorific gas with calorific value of 3,000 kcal/Nm^3 (dry) and specific gravity of 0.46 (air = 1). The sensible heat of the gas is recovered in a waste heat boiler. It is then passed through a shift converter where carbon monoxide is reduced to 3% by catalytic shift conversion reactioin (6.7). The sensible heat of the gas is extracted in the waste heat boiler, boiler water preheater and naphtha with a carbonate lye for carbon dioxide removal. Finally the gas is enriched and dried befoe supplying to consumers. Natural gas, refinery gas, or LPG may be used for enrichment.

When the intention is to make ammonia synthesis gas, the methane content must not exceed 0.2%. On the other hand, it is not possible to lower the methane content below 7% to 8% in the steam reformed gas without using very high temperatures which seriously shorten the reactor tube life. Therefore, a secondary reformer is used to modify the exit gas from the primary reformer. Air is mixed with the primary gas in a quantity which introduces nitrogen in the right proportion to hydrogen in the final synthesis gas. A part of the combustibles burns and autogenously raises the gas temperature to 1,200°C when methane–steam reforming is almost complete over a special nickel catalyst. The gas leaves the secondary reformer at about 900°C. It is then passed through a waste heat recovery system before further treatment for ammonia synthesis.

The chief merits of the ICI steam reforming process are low capital cost, low running cost and well-proved equipment and process design. Its limitation is non-adaptability to middle and heavy oils. Units of capacity exceeding 1,00,000 Nm^3 per hour gas have been constructed. India has also adopted this process for ammonia synthesis for her fertiliser production on quite a large scale for distribution to homes and industries. It is also sutiable for the production of synthesis gas. In addition coke oven gas may also be regarded as the cheapest raw material for ammonia production if the fertilliser plant is located near an integrated iron and steel plant.

***Partial combustion process*:** The Shell gasification process is continuous, autothermic and non-catalytic. It is mainly used in the production of synthesis gas although the product gas can be enriched and supplied as town gas. The feedstock ranges from natural gas to high sulphur heavy oils.

The feedstock, oxygen and steam are separately preheated to about 300°C and then brought together in the reactor at carefully controlled pressures and flow rates. The reactor is essentially a burner where the oxygen and steam are raipidly mixed immediately before partial combustion and a great turbulence is set up in the chamber (Fig. 6.9).

The process operates under pressures of 10–40 atm and at temperatures of 1,000–1,500°C. The reactor, a cylindrical vessel, is internallly lined with layers of highly refractory and insulating materials. The exit gas is a mixture of hydrogen, carbon monoxide and carbon dioxide and contains small quantities of methane and carbon. The carbon produced is less than 0.05% when methane is the feedstock and 2% to 3% by weight when liquid hydrocarbons are gasified. It can be removed by washing with water or scrubbing in light distillates and recovered as a pelleted by-product fuel. The carbon monoxide is shift converted and then the carbon dioxide is removed if the gas is to be used in ammonia synthesis or town gas manufacture. The sensible heat of the gases leaving the reactor at 1,000–1,200°C is extracted in a waste heat boiler.

The typical product gas composition with a naphtha (liquid) feedstock is: CO_2 4.8%, CO 41.8%, H_2 51.7%, CH_4 0.3% and others 1.4%. Its calorific value is 2,870 kcal/Nm^3 dry and specific gravity is 0.52 (air = 1). The material requirements per cubic metre of gas, crude but dry are: naphtha 0.3 kg, steam 1 kg and oxygen 0.25 Nm^3. Cold gas efficiency is 84% of feedstock. The high capital cost of the process makes it suitable only for large capacity installations.

***Hydrogenation processes*:** The process of hydrocracking has been successfully used in the field of oil gasification. Crude oil and heavy fuel oil are hydrocracked to give a gas which can be further steam reformed into synthesis gas.

Preheated streams of the feedstock and a hydrogen-rich gas are fed into a reactor containing a fluidised bed of coke particles. The reactor operates at around 700–900°C at pressures up to 50 atm. The aliphatic contituents are cracked into smaller molecules and the aromatics are dealkylated to give benzene, toluene, naphthalene and polynuclear aromatics. The data on the hydrogenation of a crude oil are: Gas composition in percentage: CO_2 0.2, C_nH_m 0.3, CO 2.9, H_2 36.3, CH_4 45.9, C_2H_6 13.5 and N_2 0.9. Its calorific value is 7,800 kcal/Nm^3 dry and the specific gravity (air = 1) is 0.44. The cold gas efficiency is 73% of the feedstock and benzene, toluene and naphthalene are obtained as by-products.

A characteristic feature of the gas is the high ethane content which makes it more suitable as a town gas than methane alone. Production of ethane makes greater room for hydrogen of the gas which therefore has a higher flame speed.

Light distillates can be subjected to a gas recycle hydrogenation process to get a rich gas similar to that obtained by the hydrogenantion of crude oil and heavy oil.

6.3 CLEANING AND PURIFICATION OF GASEOUS FUELS

The gases need cleaning and purification to a much lesser extent than synthesis gases. Among the fuel gases, greater attention is paid to those distributed by piping to homes and industries than those used in industrial furnaces directly after gas generation. The major problem of purifying gaseous fuels is two-fold: It should not be harmful and obnoxious to the user and it should not create engineering troubles like corrosion and choking. Sometimes the gases are upgraded by removing inerts like carbon dioxide.

The bulk of the steam and tar vapours is removed by condensing in coolers. Further removal of tar mist is achieved in mechanical detarrers, although complete removal requires electrostatic precipitators. The water vapour content of the gases meant for long distancve piping is controlled by cooling down to the desired dew point by the use of refrigerated glycol or calcium chloride solution.

The methods of dust removal may be classified as dry or wet. Dry methods include dust chambers, cyclones, bag filters and electrostatic precipitators; wet methods include packed towers, spray towers and centrifugal washers, such a Thiessen disintegrator. The maximum limits of dust content are: < 0.45 g/m^3 for use in furnaces and <0.01 g/m^3 for use in gas engines. Electrostatic precipitators and centrifugal washers attain the latter specification.

The dry method of hydrogen sulphide removal in iron oxide chambers is widely used in the manufacture of town gas by coal carbonisation. The coke oven gas is usually purified by wet methods which are also employed in the removal of hydrogen sulphide from refinery gases and the gases obtained by various gasificatioin processes using coal and oil.

In the dry method, the gas is passed over moist ferric hydroxide in a suitable form. When bog iron is used hydrogen sulphide is removed by the following reaction:

$$Fe_2O_3 \cdot xH_2O + 3H_2S \rightarrow Fe_2S_3 + (x+3)H_2O$$

The iron compound is regenerated by adding small quantities of air to the gas:

$$2Fe_2S_3 + 3O_2 + 2yH_2O = 2Fe_2O_3 \cdot yH_2O + 6S$$

Elemental sulphur is formed. The regenerated oxide may be repeatedly used till the sulphur content of the spent catalyst reaches 60% to 65%. The sulphur in the spent oxide may be converted to sulphuric acid via sulphur dioxide.

The basic principle of the wet methods consists in scrubbing the gas with a suitable reagent which removes hydrogen sulphide by absorption, with or without a chemical reaction. The spent reagent is regenerated by oxidation (air blowing) or boiling, when sulphur or hydrogen sulphide separates. The reagents of some of the processes are given below:

***Thylox process*:** A solution of sodium or ammonium thioarsenate is used. The reaction during absorption is

$$Na_4As_2S_5O_2 + H_2S \rightleftarrows Na_4As_2S_6O + H_2O$$

The reaction during regeneration (air blowing) is

$$2Na_4As_2S_6O + O_2 \rightarrow 2Na_4As_2S_5O_2 + 2S$$

***Seaboard process*:** A solution of sodium carbonate is used. The reaction during absorption is

$$Na_2CO_3 + H_2S \rightarrow NaHS + NaHCO_3$$

The reaction during desorption (air blowing) is

$$NaHS + NaHCO_3 \rightarrow Na_2CO_3 + H_2S$$

***Girbotal process*:** Aqueous solutions of aliphatic amines, for example, monoethanolamine (MEA) and diethanolamine (DEA) are used. CO_2 is removed along with H_2S. The reaction during absorption is

$$\underset{\text{MEA}}{HO.CH_2.CH_2.NH_2} + H_2S \rightleftarrows \underset{\text{Absorption complex}}{(HO.CH_2CH_2NH_2)H_2S}$$

The reaction during desorption (steam blowing) is

$$(HO.CH_2.CH_2.NH_2)H_2S \rightarrow \underset{\text{MEA}}{HO.CH_2.CH_2.NH_2} + H_2S$$

***Shell phosphate process*:** A solution of tripotassium phosphate is used. The reaction during absorption is

$$K_3PO_4 + H_2S \rightarrow K_2HPO_4 + KHS$$

The reaction during desorption (boiling) is

$$K_2HPO_4 + KHS \rightarrow K_3PO_4 + H_2S$$

The following problems will be better appreciated after the completion of Chapter 7.

Example 1

Calculate the gross calorific value and the Wobbe index of a fuel gas having the following composition, per cent by volume: methane 89.0, ethane 8.0, propane 2.0 and butane 1.0.

Solution

Using the data in Table 7.4 of Chapter 7, the average calorific value (gross) of the fuel is calculated.

$$\begin{aligned}\text{Average CV (gross)} &= 9{,}500 \times 0.89 + 16{,}644 \times 0.08 \\ &\quad + 23{,}688 \times 0.02 + 30{,}714 \times 0.01 \\ &= 10{,}567 \text{ kcal/Nm}^3\end{aligned}$$

$$\begin{aligned}\text{Average molecular weight of the gas} &= 16 \times 0.89 + 30 \times 0.08 \\ &\quad + 44 \times 0.02 + 58 \times 0.01 \\ &= 18.1\end{aligned}$$

$$\begin{aligned}\text{Average molecular weight of air} &= 32 \times 0.21 + 28 \times 0.79 \\ &= 29 \text{ (accepted since } CO_2 \text{ is also present in air)}\end{aligned}$$

$$\text{Therefore, specific gravity of the gas (air = 1)} = \frac{18.1}{29} = 0.62$$

$$\begin{aligned}\text{Therefore, its Wobbe index} &= \frac{\text{CV}}{\sqrt{\text{sp. gr.}}} \\ &= \frac{10{,}571}{\sqrt{0.62}} = 13{,}426 \text{ kcal/Nm}^3\end{aligned}$$

Example 2

In a trial on a producer, the following results were obtained: Gas yield is 3,000 Nm^3/te of coke, CV of coke is 5,800 kcal/kg and CV of gas is 1,600 kcal/Nm^3. Find its cold gas efficiency.

Solution

$$\text{Gas yield} = \frac{3000}{1000} = 3 \text{ Nm}^3\text{kg coke}$$

Cold gas efficiency of the producer

$$= \frac{\text{Gas yield} \times \text{CV of gas}}{\text{CV of coke}} \times 100$$

$$= \frac{3 \times 1160}{5800} \times 100 = 60\%$$

Example 3

From the data given below, calculate the cold gas efficiency of a gas producer using a dry blast.

Coal analysis				Gas analysis, orsat, per cent	
Per cent as charged		Per cent, dmmf			
Moisture	3.0	C	85.2	CO_2	7.0
Ash	14.7	H	5.6	O_2	0.7
Volatiles	35.7	N	2.8	CO	20.3
Fixed carbon	46.6	S	0.4	H_2	12.5
Total	100.0	O	6.0	C_2H_4	0.5
		Total	100.0	Ch_4	3.0
				N_2	56.0
				Total	100.0

Clinker analysis, per cent dry: carbon 15.0 and ash 85.0.
Heat of combustion, $-\Delta H_{c,}$ cal/mol: CO 67,636; H_2 68,317; C_2H_4 337,234 and CH_4 212,798.

Solution

Steps of calculation: (i) Convert the ultimate analysis of the coal into 'as changed'; (ii) Calculate the dry gas yield from the carbon balance. The carbon is distributed over the gas, clinker and tar. Assume 10% loss of carbon with tar. (iii) Calculate the CV of the dry gas from the orsat analysis and the heats of combustion of the components. (iv) Calculate the CV of the coal using Dulong's formula. (v) Compute the cold gas efficiency of producer.

Converting the ultimate analysis on the 'as charged' basis

$$C = 85.2 \times \frac{100 - 3.0 - 1.1 \times 14.7}{100}$$

$$= 85.2 \times 0.808 \qquad = 68.9$$

$$H = 5.6 \times 0.808 \qquad = 4.5$$
$$N = 2.8 \times 0.808 \qquad = 2.3$$
$$S = 0.4 \times 0.808 \qquad = 0.3$$
$$O = 6.0 \times 0.808 \qquad = 4.8$$

Moisture = 3.0
Ash = 14.7
Water of hydration of minerals = 0.1 × 14.7 = 1.5
100.0

For the calculation of gas yield, use 100 kg coal 'as charged' as the basis.

$$\text{By ash balance loss of carbon in the clinker} = \frac{15.0}{85.0} \times 14.7 = 2.6 \text{ kg}$$

$$\text{Loss of carbon in tar} = 68.9 \times 0.1 = 6.89 \text{ kg}$$

Therefore, carbon gasified $= 68.9 - (2.6 + 6.89) = 59.41$ kg
$= 4.95$ kmol

Neglecting the SO_2 reported with CO_2

Carbon in dry gas $= 7.0 + 20.3 + 2 \times 0.5 + 3.0$
$= 31.3$ kmol per 100 kmol dry gas

Therefore, yield of dry gas $= \dfrac{100}{31.3} \times 4.95 = 15.8$ kmol per 100 kg coal
$= 0.158$ kmol/kg coal $= 0.158 \times 22.4$
$= 3.54$ Nm^3/kg

CV of dry gas $= 67{,}636 \times 0.203 + 68{,}317 \times 0.125$
$+ 3{,}37{,}234 \times 0.005 + 2{,}12{,}798 \times 0.03$
$= 30{,}340$ kcal/kmol

CV of the coal $= 80.8 \times 68.9 + 345\left(4.53 - \frac{4.8}{8}\right) + 22.2 \times 0.32$
$= 6{,}930$ kcal/kg

Therefore, cold gas efficiency of the producer

$$= \frac{0.158 \times 30{,}340 \times 100}{6{,}930}$$

$= 69.2\%$

Example 4

Using the data of Example 6.3, calculate the supply of air and steam per kg of coal if BST is 60°C. Given: vapour pressure of water at 60°C = 149.40 mm Hg, and 30% of nitrogen in coal is conveted into ammonia.

Solution

The air supply is calculated from the nitrogen balance. Both air and fuel supply nitrogen, while the gas and ammonia liquor are the recipients of this element. The gas receives 70% of the nitrogen of the coal while the ammonia receives 30%.

N_2 in the gas $= 56$ kmol per 100 kmol dry gas

$$= \frac{56}{100} \times 15.8 \text{ kmol per 100 kg coal}$$

$= 8.85$ kmol per 100 kg coal

$$N_2 \text{ from the coal to the gas} = 2.3 \times 0.7 \times \frac{1}{28}$$
$$= 0.06 \text{ kmol/100 kg coal}$$

$$\text{Therefore, } N_2 \text{ from air} = 8.85 - 0.06$$
$$= 8.79 \text{ kmol per 100 kg coal}$$

$$\text{Therefore, air supplied} = 8.79 \times 22.4 \times \frac{100}{79} \times \frac{1}{100}$$
$$= 2.49 \text{ Nm}^3\text{/kg coal charged}$$

Applying Dalton's law of partial pressures and assuming the blast pressure to be 760 mm,

$$\text{Steam content of the blast} = \frac{149.40}{760 - 149.40}$$
$$= 0.245 \text{ kmol per kmol of dry air}$$

$$\text{Therefore, steam supplied in the blast} = 0.245 \times \frac{2.49}{22.4} \times 18$$
$$= 0.49 \text{ kg/kg coal charged}$$

PROBLEMS

1. Using the gas compositions of Table 6.6 and the calorific values of Table 7.4, calculate the gross calorific value and the Wobbe index of the given gases.

2. When amorphous carbon is used as the fuel in an air-blown producer, the maximum cold gas efficiency is 69.73%. If steam is added in such a way that 0.4 kg steam is decomposed per kilogram of carbon gasified, what will be the maximum cold gas efficiency of the mixed-blast producer? The values of ΔH_c for CO and H_2 are 67,636 and 68,317 kcal/kmol, respectively.

   [Answer: 88.5%]

3. The coal feed to a producer on analysis showed 70% carbon, 3.5% moisture, 7.5% ash and negligible nitrogen and sulphur. The product gas on analysis showed 8.0% CO_2, 16.3% H_2, 20.6% CO, 1.0% CH_4 and 54.1% N_2. Neglecting tar and clinker losses, calculate (a) gas yield (Nm^3/kg), (b) air requirement (Nm^3/kg), (c) steam decomposition (kg/kg), (d) net hydrogen content of coal in per cent and (e) cold gas efficiency of the producer in per cent.

   Note: (i) The steam will include that formed from the oxygen and hydrogen of coal.
   (ii) Calculate the calorific value of coal by using the Dulong formula.

   [Answers: (*a*) 4.41 Nm^3/kg (*b*) 3.02 Nm^3/kg (*c*) 0.253 kg/kg (*d*) 4.0 pecent (*e*) 76.1%]

4. A gas producer gasifies coke containing 10% moisture and 20% ash. If the carbon content of the coke is 95% daf and the gas analysis shows CO_2 5.3%, CO 25%, H_2 12% and CH_4 0.5%, calculate the cold gas efficiency of the producer. (Neglect other elements in the coke in the calculaion and assume N_2 as the remainder in the gas analysis)

   [Answer: 87.6%]

5. Perform the 'simple calculation' refered to in the third paragraph of Oil gasification, section 6.2.11.

6. Using a naphtha feedstock, a steam reformer produces a gas with the following composition, per cent dry: CO_2 4.4, $C_{2.5}H_5$ 4.1, CO 20.0, H_2 49.5, C_2H_6 7.7 and N_2 14.3. If the naphtha is assumed to be C_7H_{16}, calculate the carbon efficienty of the reformer.

   [Answer: 92.8%]

BIBLIOGRAPHY

1. *Gas Engineers Handbook*. New York: Industrial Press, 1966.
2. *Gasmaking*. London: British Petroleum Co, 1965.
3. Gumz W. *Gas Producers and Blast Furnaces*. New York: Wiley, 1950.
4. Lowry HH, (ed.). *Chemistry of Coal Utilization*, Vol 2, 1945;1586:1673; Supplementary Volume, 1963;892:1023; Elliott MA, (ed.). Second Supplementary Volume, 1981;1491: 1599. New York: Willey.
5. Ministry of Petroleum and Natural Gas, Govt. of India, http://petroleum.nic.in/ng.htm.
6. Nonhebel G. *Gas Purification Processes*. London: George Newne, 1964.
7. Pritchard R, Guy JJ and Connor NE. *Industrial Gas Utilization, Engineering Principles and Practice*. Essex, UK: Bowker Publishing Co., 1977.
8. Paul S. LNG option for India. *Ind. Chem. Eng.* Section B, 2004;46:4.

7 Combustion Process
(Stoichiometry and Thermodynamics)

7.1 COMBUSTION STO!CHIOMETRY

The combustible elements in solid and liquid fuels are carbon, hydrogen and sulphur. The combustible components of gaseous fuels are hydrogen, carbon monoxide, methane and unsaturated hydrocarbons. Higher paraffins may be present in some fuels such as LPG, carburetted water gas, fuels obtained from oil gasification and gases from low temperature carbonisation of coal and lignite. The average molecular formula of paraffins present may be assumed as:

- $C_{1.25}H_{4.5}$ for low temperature coal gas,
- $C_{1.1}H_{4.2}$ for carburetted water gas and oil gas.

The composition of the unsaturated hydrocarbons may vary from gas to gas. The following average molecular formulae are representative:

- C_2H_4 for producer gas from coal,
- C_4H_8 for low temperature coal gas and vertical retort coal gas,
- $C_{2.5}H_5$ for coke oven gas, horizontal retort coal gas, carburetted water gas and oil gas.

The combustibles undergo complete combustion by the following well-defined overall chemical reactions.

Solid and liquid fuels

$$C + O_2 \rightarrow CO_2$$
$$2H + ½\, O_2 \rightarrow H_2O$$
$$S + O_2S \rightarrow O_2$$

Gaseous fuels

$$H_2 + ½\, O_2 \rightarrow H_2O$$
$$CO + ½\, O_2 \rightarrow CO_2$$
$$CH_4 + 2O_2 \rightarrow CO_2 + 2H_2O$$
$$C_nH_{2n+2} + \left(\frac{3n+1}{2}\right)O_2 = nCO_2 + (n+1)H_2O$$

Paraffins

$$C_nH_m + \left(\frac{n+m}{4}\right)O_2 = nCO_2 + \frac{m}{2}H_2O$$

Unsaturates

In making use of the above reactions in combustion stoichiometry, it is important to remember the following points:

1. The combustibles in solid and liquid fuels are expressed as the elements in weight per cent. Particular attention should be given to the basis of expression. In Chapter 3 we have seen that the ultimate analysis is expressed on a dmmf basis for the high ash Indian coals. This poses a problem in combustion calculations because the volatile part of mineral matter is included in the combustion products. Under these circumstances it is preferable to express the ultimate analysis in such a way that the following equation holds good:

   coal = carbon + hydrogen + sulphur + nitrogen + oxygen + moisture + ash

   or $100 = C + H_2 + S + N_2 + O_2 + M + A$

   Since the detailed composition of the mineral matter is not generally known, it is assumed for convenience that

   mineral matter – ash = 0.1 ash
   = water of hydration of the minerals

   This assumption is also necessary for the calculation of the net calorific value from the gross. For low-sulphur Indian coals, this assumption is justified; unless carbonate minerals are present in significant quantities.
2. The combustibles in gaseous fuels are given in volume or mole per cent. Usually a dry basis of expression (orsat analysis) is used. If necessary, conversion into a wet basis is done on the assumption that the fuel gas is saturated with water vapour at the given temperature and pressure.
3. The composition of the products of combustion (flue gas) is first calculated as they are produced and then expressed on a dry basis. Further, the sulphur dioxide and carbon dioxide are reported together as carbon dioxide, since both the components are absorbed in the same absorption pipette in the orsat analysis.
4. *Theoretical air* is the amount of air stoichiometrically required for the complete combustion of the combustibles. Similarly, *theoretical products of combustion* refer to the flue gas or exhaust gas obtained by the complete combustion of the fuel using theoretical air.
5. *Excess air* is the amount of air used in excess of the theoretical and is expressed as per cent or fraction of the theoretical air. It is important to remember this definition, particularly when a flue gas contains free oxygen as well as combustibles like carbon monoxide owing to inefficient combustion. In this case excess air will correspond to free oxygen minus oxygen required for the combustion of the residual combustibles.
6. Useful basis of calculation is 100 kg of solid or liquid fuels and 100 kmol of gaseous fuels or flue gas as the case may be.
7. Usually air is assumed dry and CO_2-free in combustion calculations.

8. In dry air, the volume (or mole) ratios of the components are

$$\frac{\text{oxygen}}{\text{nitrogen}} = 0.27 \qquad \frac{\text{air}}{\text{oxygen}} = 4.76$$

$$\frac{\text{nitrogen}}{\text{oxygen}} = 3.76 \qquad \frac{\text{nitrogen}}{\text{air}} = 0.79$$

$$\frac{\text{oxygen}}{\text{air}} = 0.21 \qquad \frac{\text{air}}{\text{nitrogen}} = 1.27$$

9. Average molecular weight of dry, CO_2-free air = 29.0

10. $\text{Excess-air factor} = \dfrac{\text{actual air used}}{\text{theoretical air}}$

At 50% excess air, the value of the factor is 1.5.

7.2 EXAMPLES

The following examples illustrate the principles of combustion stoichiometry:

7.2.1 Flue Gas Analysis from Fuel Analysis and Air Supply

Example 1

Solid fuel: Determine the flue gas analysis, the air–fuel ratio by weight and the volume of combustion products at 250°C when a Wardha valley coal of the following composition burns with 50% excess air.

Proximate analysis, per cent, air-dried				Ultimate analysis , per cent, daf				
moisture	ash	volatile-matter-fixed	carbon	carbon	hydrogen	nitrogen	sulphur	oxygen
8.0	20.0	28.5	43.5	81.0	4.6	1.8	0.6	12.0

If the burning rate of the coal is 3 t/h, what is the capacity of the air blower used? Assume complete combustion.

Solution

Basis of calculation: 100 kg air-dried coal.

The steps are:

1. The ultimate analysis is converted from daf to air-dried basis by multiplying by the

$$\text{factor } \frac{100 - (8 + 20)}{100} = 0.72$$

2. The weights in kilograms are converted into kilomoles by dividing by molecular weight or atomic weight as the case may be.
3. Theoretical (stoichiometric) oxygen requirement is calculated from the combustion equation. The air supply is calculated after deducting the oxygen of the fuel from the total oxygen requirement.
4. The combustion products are also calculated from the chemical equations. Their composition includes excess air. The final analysis of the flue gas is given on dry basis after combining CO_2 and SO_2.
5. The calculated values in each step are given below.

Volume of combustion products (flue gas) at 250°C and 1 atm

$$= \text{kmol} \times 22.4 \times \frac{273 + 250}{273} \text{ m}^3\text{/100 kg air-dried coal}$$

$$= 40.44 \times 22.4 \times \frac{523}{273}$$

$$= 1740 \text{ m}^3 \text{ /100 kg air-dried coal}$$

Flow rate of the flue gas

$$= \frac{1740}{100} \times 3000 \times \frac{273 + 30}{273 + 250} \times \frac{1}{60} = 504 \text{ m}^3\text{/min}$$

Assuming 30°C ambient temperature, capacity of air-blower

$$= 5.44 \times 1.5 \times 4.76 \times 22.4 \times \frac{273 + 30}{273} \times \frac{3000}{100} \times \frac{1}{60}$$

$$\approx 482.8 \text{ m}^3\text{/min}$$

$$\approx 500 \text{ m}^3\text{/min}$$

Flue gas analysis (Problem 1)

Components	**Amount, kmol**	**Volume per cent as produced**	**Volume per cent dry (orsat analysis)**	**Remarks**
CO_2	4.87	12.06	12.7	For converting volume per cent as produced, to volume per cent dry, multiply by $\frac{100}{100-5.18}$ or 1.055
SO_2	0.01	0.03	–	
O_2	2.72	6.73	7.1	
N_2	30.75	76.00	80.2	
H_2O	2.09	5.18	–	
Total	40.44	100.00	100.00	

$$\text{Air–fuel ratio by weight} = \frac{\text{kmol theor.}O_2 \times 4.76 \times \text{excess air factor} \times \text{mol.wt. of air}}{100 \text{ kg air-dried coal}}$$

$$= \frac{5.44 \times 4.76 \times 1.5 \times 29}{100} = 11.27$$

Table for solution of Problem 1

Elements	kg/100 kg daf coal	kg/100 kg air-dried coal	kmol/100 kg air-dried coal	kmol of O_2 theoreticaly required per 100 kg of air-dried coal	kmol /100kg air-dried coal					
					CO_2	H_2O	SO_2	N_2	O_2	Total
C	81.0	58.4	4.87	4.87	4.87	–	–	–	–	4.87
H	4.6	3.3	1.65	0.83	–	1.65	–	–	–	1.65
O	12.0	8.6	0.27	−0.27	–	–	–	–	–	–
N	1.8	1.3	0.05	–	–	–	–	0.05	–	0.05
S	0.6	0.4	0.01	0.01	–	–	0.01	–	–	0.01
Moisture	–	8.0	0.44	–	–	0.44	–	–	–	0.44
Ash	–	20.0	–	–	–	–	–	–	–	–
Total	100.0	100.0	–	5.44	4.87	2.09	0.01	0.05	0.00	7.02
Excess O_2 at 50% of the theoretical = 5.44 × 0.5 = 2.72					–	–	–	–	2.72	2.72
N_2 from total air supplied = Theoretical O_2 × 3.76 × excess air factor = 5.44 × 3.76 × 1.3 = 30.7					–	–	–	30.70	–	30.70
					4.87	2.09	0.01	30.75	2.72	40.44

Example 2

Liquid fuel: Determine the flue gas analysis and air–fuel ratio by weight when a medium viscosity fuel oil with 84.9% carbon, 11.4% hydrogen, 3.2% sulphur, 0.4% oxygen and 0.1% ash is burned with 20% excess air. Assume complete combustion.

Solution

Basis of calculation: 100 kg fuel oil. No change in the basis of expression is necessary. The data of the ultimate analysis are converted into kilomoles (see table on page 256) and the procedure given in the earlier problem is followed. The results are tabulated below.

Flue gas analysis (Problem 2)

Components	kmol	Volume per cent as produced	Volume per cent dry (orsat analysis)	Remarks
CO_2	7.08	11.78	13.0	Convert vol. per cent as produced to vol. per cent dry by multiplying by $\frac{100}{100-9.47}$ or 1.105
SO_2	0.10	0.17	–	
O_2	2.00	3.33	3.7	
N_2	45.25	75.25	83.3	
H_2O	5.70	9.47	–	
Total	60.13	100.00	100.00	

$$\text{Air fuel ratio by weight} = \frac{10.02 \times 4.76 \times 1.2 \times 29}{100} = 16.0(\,$$

Example 3

Gaseous fuel: A furnace burns producer gas with 10% excess air at the rate of 7200 km³/h and discharges flue gases at 400°C and 760 mm Hg. Calculate the flue gas analysis, air requirement and volume of flue gases per hour. The gas is supplied from a gas holder and its orsat analysis is: CO_2 4.0, C_nH_m 0.4, CO 29.0, H_2 12.0, CH_4 2.6 and N_2 52.0. Ambient temperature is 30°C and pressure is 760 mm Hg. Assume complete combustion.

Solution

Basis of calculation: 100 kmol of producer gas as supplied. The steps are:

1. Assume the gas is saturated with water vapour at 30°C and 760 mm Hg. Vapour pressure of water (tension of aqueous vapour or aqueous tension) at 30°C = 31.89 mm Hg. Content of water vapour in the saturated gas

$$= \frac{100 \times 31.89}{760} = 4.2 \text{ kmol}$$

2. Convert the orsat analysis data into volume per cent wet (saturated) by multiplying by

$$\frac{100 - 4.2}{100} \text{ or } 0.958.$$

3. Use the combustion equations and tabulate the results as shown on page 257.

Table for solution of Problem 2

Elements	kg/100 kg fuel oil	kmol/100 kg fuel oil	kmol of O_2 theoretically required per 100 kg of fuel oil	Products, kmol/100 kmol fuel oil					
				CO_2	H_2O	SO_2	N_2	O_2	Total
C	84.9	7.08	7.08	7.08	–	–	–	–	7.08
H	11.4	5.70	2.85	–	5.70	–	–	–	5.70
O	0.4	0.01	−0.01	–	–	–	–	–	–
S	3.2	0.10	0.10	–	–	0.10	–	–	0.10
Ash	0.1	–	–	–	–	–	–	–	–
Total	100.0		10.02	7.08	5.70	0.10	–	–	12.88
Excess O_2 at 20% of the theoretical $= 10.02 \times 0.2 = 2.00$				–	–	–	–	2.00	2.00
N_2 from total air supplied $= 10.02 \times 3.76 \times 1.2 = 45.25$				–	–	–	45.25	–	45.25
				7.08	5.70	0.10	45.25	2.00	60.13

Combustion calculations (Problem 3)

Components	kmol/ 100 kmol dry gas	kmol/ 100 kmol wet gas	kmol of O_2 theoretically required	kmol of combustion products				
				CO_2	O_2	N_2	H_2O	Total
CO_2	4.0	3.8	–	3.8	–	–	–	3.8
C_nH_m (assume C_2H_4)	0.4	0.4	1.20	0.8	–	–	0.8	1.6
O_2		–	–	–	–	–	–	–
CO	29.0	27.8	13.90	27.8	–	–	–	27.8
H_2	12.0	11.5	5.75	–	–	–	11.5	11.5
CH_4	2.6	2.5	5.00	2.5	–	–	5.0	7.5
N_2	52.0	49.8	–	–	–	49.8	–	49.8
H_20	–	4.2	–	–	–	–	4.2	4.2
Total	100.0	100.0	25.85	34.9	–	49.8	21.5	106.2
Excess O_2 at 10% of the theoretical = 25.85 × 0.1 = 2.6				–	2.6	–	–	2.6
N_2 from total air supplied = 25.85 × 3.76 × 1.1=106.9				–	–	106.9	–	106.9
				34.9	2.6	156.7	21.5	215.7

Flue gas analysis (Problem 3)

Components	kmol	Vol. per cent as produced	Vol. per cent dry (orsal analysis)	Remarks
CO_2	34.9	16.2	18.0	Convert vol. per cent as produced to vol. per cent dry by multiplying by $\frac{100}{100-9.9}$ or 1.110
O_2	2.6	1.2	1.3	
N_2	156.7	72.7	80.7	
H_2O	21.5	9.9	–	
Total	215.7	100.0	100.0	

Air requirement = 25.85 × 4.76 = 1.1 kmol/100 kmol gas or $Nm^3/100\ Nm^3$ ga

$$= 25.85 \times 4.76 \times 1.1 \times \frac{7200}{100}\ Nm^3/h$$

$$= 9{,}750\ Nm^3/h$$

Flue gas gases produced = 215.7 kmol/100 kmol gas or $Nm^3/100\ Nm^3$ gas

$$= 215.7 \times \frac{7200}{100}\ Nm^3/h$$

$$= 15{,}530\ Nm^3/h$$

$$= 15{,}530 \times \frac{673}{273}\ m^3/h \text{ at } 400°C \text{ and } 760\,mm$$

$$\approx 38{,}290\ m^3/h \text{ at } 400°C \text{ and } 760\ mm$$

7.2.2 Excess Air Calculation from Flue Gas Analysis

Example 4

Using the calculated orsat analysis of the flue gas in Problem 1, determine the per cent excess air used in the combustion. No other data are available.

Solution

The orsat analysis of the flue gas is CO_2 12.7%, O_2 7.1% and N_2 80.2%. There is no combustible in the gas. Hence, the oxygen entirely corresponds to the excess air. We know

$$\text{Excess air, per cent} = \frac{\text{excess air}}{\text{theoretical air}} \times 100$$

Since the orsat analysis is on a dry basis, the content of water vapour in the flue gas is not available. Furthermore, the oxygen available from the fuel is also not known. Therefore, the total air cannot be calculated from the oxygen balance of the flue gas analysis. The only other course left is to use the nitrogen balance. In the absence of fuel analysis, one does not know how much nitrogen is contributed by the fuel itself. From all these considerations it may be seen that the problem defies any straightforward solution. However, the contribution of solid and liquid fuels and some gaseous fuels to the nitrogen content of the respective flue gases is negligible. The present problem is solved below with the assumption that the nitrogen of the flue gas is entirely contributed by air and that the fuel does not contain any oxygen.

Basis of calculation: 100 kmol dry flue gas

Components	kmol	kmol C accounted for	kmol O_2 accounted for
CO_2	12.7	12.7	12.7
O_2	7.1	–	7.1
N_2	80.2	–	–
Total	100.0	12.7	19.8

$$\text{Excess } O_2 = 7.1\,\text{kmol}$$

$$\text{Total } O_2 = \frac{\text{Total } N_2}{3.76} = \frac{80.2}{3.76}$$

$$= 21.33\,\text{kmol}$$

$$\text{Hence, excess air, per cent} = \text{excess } O_2\text{, per cent}$$

$$= \frac{7.1}{21.33 - 7.1} \times 100$$

$$= 49.9$$

This result is close to the given condition (50% excess air) of Problem1.

7.2.3 Dew Point of Flue Gases

Flue gases flow through ducts and various other parts of the furnace system. It is detrimental to have water deposited on them. A knowledge of dew point of flue gases is therefore desirable.

Example 5

Determine the dew point of the gases obtained in Problems 1, 2 and 3.

Solution

By definition, the dew point of a gaseous mixture is the temperature at which its water vapour starts condensing. It is therefore the temperature at which the saturation pressure for water equals the partial pressure of water vapour in the gaseous mixture.

From the wet flue gas analysis of the three problems, the content of water vapour is: 5.18%, 9.47% and 9.9%, respectively.

The partial pressure of water vapour in the three gaseous mixtures (assume gas pressure = 760 mm Hg):

$$\text{For Problem 1: } \frac{5.18}{100} \times 760 = 39.4 \text{ mm Hg}$$

$$\text{For Problem 2: } \frac{9.47}{100} \times 760 = 72.0 \text{ mm Hg}$$

$$\text{For Problem 3: } \frac{9.9}{100} \times 760 = 75.2 \text{ mm Hg}$$

Thermodynamic tables give the following temperatures at which water has the above saturation pressures (aqueous tension):

For Problem 1: 33.7°C
For Problem 2: 45.0°C
For Problem 3: 45.9°C

These are the required dew points.

7.2.4 Combustion Calculations Involving Loss of Combustibles with Flue Gases and Ash

Example 6

The flue gas from a coal-fired furnace gives the following orsat analysis: CO_2 11.5%, CO 1.5%, and O_2 6.5%. A coal from the Dishergarh seam of the Raniganj field containing carbon 68.0%, nitrogen 1.9%, sulphur 0.3% and ash 14.7% is used. The dry refuse removed from the furnace has a combustible content of 40% as carbon. Calculate (a) the volume of air per kilogram of coal, (b) kilomoles of dry flue gases per 100 kg coal, (c) the content of surplus hydrogen and water in coal, (d) the volume of flue gases per kilogram coal and (e) per cent of excess air. Assume dry air, ambient temperature 30°C, pressure 760 mm and flue gas exit temperature 300°C.

Solution

The steps are

1. Use 100 kg coal as basis of calculation.
2. Calculate the quantity of dry refuse from ash balance between coal and refuse. Hence, determine the carbon lost in the refuse and the carbon burnt to CO_2 and CO.
3. Calculate the kilomoles of dry flue gas from carbon balance. Take due care for the sulphur in coal which is burnt to SO_2 and reported as CO_2 in the flue gas.
4. Use nitrogen balance for calculating the quantity of total oxygen supplied. Calculate the oxygen (sum of free and combined) accounted for in the flue gas. The deficit in oxygen is due to the combustion of hydrogen of coal into water.
5. The quantity of hydrogen calculated from the oxygen deficit is the surplus hydrogen in coal. This hydrogen is in excess of that required to combine with the oxygen of coal to form water (vide Dulong's formula for calculation of CV).
6. The water content of coal is calculated by the method of difference, that is, 100 per cent of (C + S + N_2 + ash + surplus H_2). This water is the total of actual and *potential water content* of coal.The latter term means the sum of oxygen in coal and the hydrogen required to combine with it to form water.
7. Volume of the flue gas is calculated from the dry flue gas, the water content and the surplus hydrogen content of coal.
8. The quantity of excess air is finally calculated by the method used in 7.2.2.

Basis of calculation: 100 kg coal as fired

Components	kg	kmol
C	68.0	5.67
N	1.9	0.07
S	0.3	0.01
Ash	14.7	–
	84.9	
Surplus hydrogen	?	
Water	?	
	100.00	

$$\text{From ash balance, C in refuse} = 14.7 \times \frac{40}{60} = 9.8 \text{ kg}$$

$$= 0.82 \text{ kmol}$$

C burnt = 5.67 − 0.82 = 4.85 kmol

Total C + S burnt = 4.85 + 0.01 = 4.86 kmol

Dry flue gas analysis

Components	kmol/ 100 kmol dry flue gas	kmol (C + S)/ 100 kmol dry gas	kmol /37.5 kmol dry flue gas	Total O_2 accounted for 37.5 kmol dry flue gas
CO_2	11.5	11.5	37.5 × 0.115 = 4.32	4.32
CO	1.5	1.5	37.5 × 0.015 = 0.56	0.28
O_2	6.5	–	37.5 × 0.065 = 2.44	2.44
N_2 (by difference)	80.5	–	37.5 × 0.805 = 30.82	
Total	100.0	13.0	37.50	7.04

1. Basis: 100 kg coal as fired

Total N_2 = 80.5 kmol per 100 kmol dry flue gas × $\frac{37.5}{100}$ kmol dry flue gas per 100 kg coal fired

= 30.18 kmol per 100 kg coal as fired

From (C + S) balance

$$\text{Quantity of dry flue gas} = \frac{100}{13.0} \times 4.86$$

$$= 37.5 \text{ kmol/100 kg coal}$$

2. Air contains 79% N_2

$$\text{Hence } \frac{\text{air}}{\text{nitrogen}} = \frac{100}{79} \approx 1.27$$

The number of kilomoles of components in 37.5 kmol dry flue gas and the total O_2 accounted for in it are given in the table above.

N_2 from air = total N_2 – N_2 from coal

= 30.18 − 0.07 = 30.11 kmol

Total air supplied = 30.11 × 1.27 kmol/100 kg coal

$$= 30.11 \times 1.27 \times 22.4 \times \frac{303}{273} \times \frac{1}{100}$$

= 9.5 m^3/kg coal at 30°C and 760 mm Hg

$$\text{Total } O_2 \text{ from the air} = \frac{30.11}{3.76} = 8.02 \text{ kmol}$$

O_2 deficit = 8.02 − 7.04 = 0.98 kmol

Hence, surplus H in coal = 0.98 × 2 = 1.96 kmol/100 kg coal

= 3.9 kg/100 kg coal

Water content of coal = 100 − (84.9 + 3.9) = 11.2 kg

This includes the oxygen of coal and the hydrogen required to combine with it.

Now O_2 required for burning CO in 37.5 kmol dry flue gas

$$= 0.56 \times \frac{1}{2} = 0.28 \text{ kmol}$$

$$\text{Excess } O_2 \text{ in the flue gas} = 2.44 - 0.28 - 0.82 = 1.34 \text{ kmol}$$

$$\text{Per cent excess air} = \frac{1.34}{8.02 - 1.34} \times 100 = 20.1$$

Flue gases produced = dry flue gases + water from coal
+ water from surplus hydrogen of coal.

$$= 3.75 + \frac{11.2}{18} + 1.96 \text{ kmol/100 kg coal}$$

$$= 37.5 + 0.62 + 1.96 \text{ kmol/100 kg coal}$$

$$= \frac{37.5 + 0.62 + 1.96}{100} \times 22.4 \times \frac{573}{273}$$

$$= 18.8 \text{ m}^3\text{/kg coal at 300°C and 760 mm}$$

7.3 RAPID METHODS OF COMBUSTION STOICHIOMETRY

Detailed methods of combustion stoichiometry should be followed only for an acquaintance with the subject. Once the principles of combustion stoichiometry are known, industrial problems are worked out more quickly by other means. A few of them are presented below.

7.3.1 Use of Fundamental Formulae

The formulae are easily derived from combustion equations. They are useful in a comparatively quick solution of combustion problems.

Solid and liquid fuels

H_2O, C, H_2, O_2, N_2 and S represent the weight per cent of moisture, carbon, hydrogen, oxygen, nitrogen and sulphur in the fuel, respectively.

Total O_2 required for complete combustion of the fuel

$$= \left[\frac{C}{12} + \frac{H_2}{4} + \frac{O_2}{32} + \frac{S}{32}\right] \text{kmol/100 kg fuel} \tag{7.1}$$

$$= 0.224\left[\frac{C}{12} + \frac{H_2}{4} - \frac{O_2}{32} + \frac{S}{32}\right] \text{Nm}^3\text{/kg fuel} \tag{7.2}$$

Volume of theoretical air required

$$A_o = 4.76\left[\frac{C}{12} + \frac{H_2}{4} + \frac{O_2}{32} + \frac{S}{32}\right] \text{kmol/100 kg fuel}$$

$$= \frac{4.76 \times 22.4}{100}\left[\frac{C}{12} + \frac{H_2}{4} - \frac{O_2}{32} + \frac{S}{32}\right] \text{Nm}^3\text{/kg fuel} \quad (7.3)$$

$$= (0.089C + 0.267H_2 - 0.039O_2 + 0.033S)\ \text{Nm}^3\text{/kg fuel}$$

Volume of theoretical flue gas wet, that is as produced

$$V_o = (\alpha + \beta + \gamma)\ \text{Nm}^3\text{/kg fuel} \quad (7.4)$$

where,

$$\alpha = \text{volume of } CO_2 \text{ produced, Nm}^3\text{/kg fuel}$$

$$= \frac{22.4}{100}\left[\frac{C}{12} + \frac{S}{32}\right] \quad (7.5)$$

$$= (0.01866C + 0.007S)\ \text{Nm}^3\text{/kg fuel}$$

$$\beta = \text{volume of } H_2O\text{, Nm}^3\text{/kg fuel}$$

$$= \frac{22.4}{100}\left[\frac{H}{2} + \frac{H_2O}{18}\right] \quad (7.6)$$

$$= (0.112H + 0.012H_2O)\ \text{Nm}^3\text{/kg fuel}$$

$$\gamma = \text{volume of } N_2\text{, Nm}^3\text{/kg fuel}$$

$$= 0.79A_o + \frac{N_2}{28 \times 100} \times 22.4 \quad (7.7)$$

$$= (0.79A_o + 0.008N_2)\ \text{Nm}^3\text{/kg fuel}$$

Volume of theoretical dry flue gas

$$V_o = (\alpha + y)\ \text{Nm}^3\text{/kg fuel} \quad (7.8)$$

Analysis of theoretical flue gas wet, per cent:

$$CO_2 = \frac{100\alpha}{\alpha + \beta + \gamma} \quad (7.9)$$

$$H_2O = \frac{100\beta}{\alpha + \beta + \gamma} \quad (7.10)$$

$$N_2 = \frac{100\gamma}{\alpha + \beta + \gamma} \quad (7.11)$$

Orsat analysis of theoretical flue gas dry, per cent:

$$CO_2 = \frac{100\alpha}{\alpha + \gamma} \tag{7.12}$$

$$N_2 = \frac{100\gamma}{\alpha + \gamma} \tag{7.13}$$

Excess air (EA) is calculated from the flue gas analysis by using one of the following formulae:

1. Case of complete combustion

$$\text{per cent EA} = \frac{100O_2}{0.27N_2 - O_2} \tag{7.14}$$

2. Case of incomplete combustion

$$\text{per cent EA} = \frac{100\,(O_2 - 0.5CO - 0.5H_2 - 2CH_4)}{0.27N_2 - (O_2 - 0.5CO - 0.5H_2 - 2CH_4)} \tag{7.15}$$

where, O_2, CO, H_2, CH_4 and N_2 represent the per cent of the respective components in the flue gas. These equations neglect nitrogen in the fuel and are valid for solid and liquid fuels for all practical purposes.

Excess air factor is defined as

$$n = \frac{A - A_o}{A_o} = \text{per cent EA}/100 \tag{7.16}$$

Air supplied

$$A = (1 + n)A_o \ Nm^3/kg \text{ fuel} \tag{7.17}$$

Flue gas produced

$$V = (V_o + nA_o) \ Nm^3/kg \text{ fuel} \tag{7.18}$$

Dry flue gas

$$V = (V_o^1 + nA_o) \ Nm^3/kg \text{ fuel} \tag{7.19}$$

Flue gas analysis, per cent wet:

$$CO_2 = \frac{100\alpha}{V} \tag{7.20}$$

$$H_2O = \frac{100\beta}{V} \tag{7.21}$$

$$N_2 = \frac{100\,[\gamma + 0.79\,(A - A_o)]}{V} \tag{7.22}$$

$$O_2 = \frac{100\ nA_o}{V} \times 0.21 = \frac{21nA_o}{V} \tag{7.23}$$

Orsat analysis of flue gas, that is per cent dry, is obtained by multiplying wet per cent of CO_2, O_2 and N_2 by 100/(100 − H_2O per cent wet) to get per cent dry of the respective components.

Gaseous fuels

Let CO_2, CH_4, C_nH_m, H_2, CO, O_2 and N_2 represent the volume or mole per cent of the respective components in the fuel gas as fired, that is, wet or saturated with water vapour at the ambient temperature and pressure.

Volume of theoretical air required

$$A_o = 0.0476[(n+\frac{m}{4})C_nH_m + 0.5CO + 2CH_4 + 0.5H_2 - O_2]\ Nm^3/Nm^3 \text{ fuel gas} \tag{7.24}$$

Volume of theoretical wet flue gas produced

$$V_o = (\alpha + \beta + \gamma)\ Nm^3/Nm^3 \text{ fuel gas} \tag{7.25}$$

where

α = volume of CO_2 produced, Nm^3/Nm^3 wet fuel gas

$$= 0.01\ (CO_2 + nC_nH_m + CO + CH_4) \tag{7.26}$$

β = volume of H_2O produced, Nm^3/Nm^3 wet fuel gas

$$= 0.01\left(\frac{m}{2}C_nH_m + 2CH_4 + H_2 + H_2O\right) \tag{7.27}$$

γ = volume of N_2 from air and fuel, Nm^3/Nm^3 wet fuel gas

$$= 0.01N_2 + 0.79A_o \tag{7.28}$$

Volume of theoretical dry flue gas

$$V'_o = \alpha + \gamma\ Nm^3/Nm^3 \text{ wet fuel gas} \tag{7.29}$$

Excess air ratio, total air supplied and flue gas produced (wet or dry) may be calculated by the formulae (7.14) to (7.23). If the gaseous fuel contains significant quantities of nitrogen, for example, producer gas and blast furnace gas, the formulae for excess air calculation are modified as below

1. Case of complete combustion

$$\text{per cent EA} = \frac{100 \times O_2'}{0.27\left(N_2' - \frac{N_2}{V}\right) - O_2'} \tag{7.30}$$

2. Case of incomplete combustion

$$\text{Per cent EA} = \frac{100\,(O_2' - 0.5CO' - 0.5H_2' - 2CH_4')}{0.27\,(N_2' - N_2/V') - (O_2' - 0.5CO' - 0.5H_2' - 2CH_4')} \quad (7.30A)$$

In the earlier formulae the symbols with prime denote the content (per cent) in dry flue gas. The term N_2/V' in the denominator gives the contribution of the gaseous fuel to the nitrogen in the flue gas.

Example 7

Solve Problem 1 using the formulae given above.

Solution

Analysis of the coal as fired, per cent: C 58.4, H 3.3, O 8.6, N 1.3, S 0.4, moisture 8.0 and ash 20.0.

Theoretical air

$$\begin{aligned} A_o &= 0.089C + 0.267H_2 + 0.033S - 0.033O \\ &= 0.089 \times 58.4 + 0.267 \times 3.3 + 0.033 \times 0.4 - 0.033 \times 8.6 \\ &= 5.80\ \text{Nm}^3/\text{kg coal} \end{aligned}$$

Carbon dioxide

$$\begin{aligned} \alpha &= 0.01866C + 0.007S = 0.01866 \times 58.4 + 0.007 \times 0.4 \\ &= 1.09\ \text{Nm}^3/\text{kg coal} \end{aligned}$$

Water

$$\begin{aligned} \beta &= 0.112H + 0.0125H_2O = 0.112 \times 3.3 + 0.0124 \times 8.0 \\ &= 0.47\ \text{Nm}^3/\text{kg coal} \end{aligned}$$

Nitrogen

$$\begin{aligned} \gamma &= 0.79A_o + 0.008N = 0.79 \times 5.80 + 0.008 \times 1.3 \\ &= 4.60\ \text{Nm}^3/\text{kg coal} \end{aligned}$$

Theoretical flue gas (wet)

$$\begin{aligned} V_o &= \alpha + \beta + \gamma = 1.09 + 0.47 + 4.60 \\ &= 6.16\ \text{Nm}^3/\text{kg coal} \end{aligned}$$

At n = 0.5,

Air supplied

$$\begin{aligned} A &= (1 + n)\,A_o = (1 + 0.5)\,5.8 = 8.7\ \text{Nm}^3/\text{kg coal} \\ &= \frac{8.7 \times 29}{22.4} = 11.27\ \text{kg/kg coal} \end{aligned}$$

Flue gas produced (wet)

$$V = V_o + nA_o = 6.16 + 0.5 \times 5.8 = 9.06 \text{ Nm}^3\text{/kg coal}$$

$$= 9.06 \times \frac{523}{273} = 17.4 \text{ m}^3\text{/kg coal at 250°C and 760 mm}$$

Flue gas analysis, per cent wet

$$CO_2 = \frac{1.09}{9.06} \times 100 = 12.0$$

$$H_2O = \frac{0.47 \times 100}{9.06} = 5.2$$

$$N_2 = \frac{[4.60 + 0.79\,(8.7 - 5.8)] \times 100}{9.06} = 76.1$$

$$O_2 = \frac{21 \times n \times A_o}{V} = \frac{21 \times 0.5 \times 5.8}{9.06} = 6.7$$

$$\text{Capacity of air blower} = 8.7 \times 3000 \times \frac{1}{60} \times \frac{273 + 30}{273} = 483 \text{ m}^3\text{/min}$$

$$\approx 500 \text{ m}^3\text{/min}$$

Example 8

Calculate excess air from the flue gas analysis of Problem 6 by using the applicable formula.

Solution

Neglecting carbon loss with clinker

$$\text{per cent EA} = \frac{100(O_2 - 0.5CO)}{0.27N_2 - (O_2 - 0.5CO)}$$

$$= \frac{100(6.5 - 0.5 \times 1.5)}{0.27 \times 80.5 - (6.5 - 0.5 \times 1.5)}$$

$$= \frac{100 \times 5.75}{21.40 - 5.75} = \frac{575}{15.99}$$

$$= 36.0$$

7.3.2 Use of Empirical Correlations

Statistical relationships exist between the net calorific value of all industrial fuels, their theoretical air requirement and the volume of theoretical flue gases they produce. The net calorific values are themselves related to the gross calorific values of the respective fuels. Therefore a knowledge of gross calorific value can be used to calculate A_o and V_o. The empirical correlations are given in Tables 7.1 and 7.2.

Table 7.1 Conversion of gross CV(C_G) into net CV (C_N)

Solid fuels, kcal/kg	
Coal: $C_N = 0.954C_G + 110$	(7.31)
Coke: $C_N = C_G - 40$	(7.32)
Liquid fuels, kcal/kg	
$C_N = 0.75C_G + 2000$	(7.33)
Gaseous fuels, kcal/Nm³	
Blast furnace gas: $C_N = 0.99C_G$	(7.34)
Coke oven gas: $C_N = 0.895C_G$	(7.35)
Water gas : $C_N = 0.915C_G$	(7.36)
Producer gas: $C_N = 0.95C_G$	(7.37)

Table 7.2 Calculation of volumes of theoretical air (A_o) and theoretical flue gases (V_o)

Type of fuel	A_o Nm³/kg for solid and liquid fuels; Nm³/Nm³ for gaseous fuels	V_o Nm³/kg for solid and liquid fuels; Nm³/Nm³ for gaseous fuels
Solid fuels		
C_N<5,500 kcal/kg	$1.01 \times \frac{C_N}{1000} + 0.5$	$0.89 \times \frac{C_N}{1000} + 1.65$
C_N>5,500 kcal/kg	$1.01 \times \frac{C_N}{1000} + 0.55$	$1 \times \frac{C_N}{1000} + 0.9$
Liquid fuels	$0.85 \times \frac{C_N}{1000} + 2.0$	$1.11 \times \frac{C_N}{1000}$
Gaseous fuels		
Blast furnace gas	$0.795 \times \frac{C_N}{1000}$	$0.634 \times \frac{C_N}{1000} + 1$
Producer gas	$0.875 \times \frac{C_N}{1000}$	$0.640 \times \frac{C_N}{1000} + 1.11$
Mixtures of blast furnace gas and coke oven gas	$1.07 \times \frac{C_N}{1000} - 0.16$	$1.04 \times \frac{C_N}{1000} + 0.7$
Water gas, carbureted water gas, coke oven gas	$1.09 \times \frac{C_N}{1000} - 0.25$	$1.35 \times \frac{C_N}{1000} - 0.67$
Hydrogen, carbon monoxide	2.38	2.88

Simple empirical relationships are also available for calculating excess air from a knowledge of the type of fuel gas. From CO_2-balance between dry theroritical and dry actual flue gases, it can be seen that excess air ratio.

$$n = \frac{V_o'}{A_o}\left\{\frac{(CO_2)_{theor}}{(CO_2)_{actual}} - 1\right\} \tag{7.38}$$

For solid and liquid fuels

$$\frac{V_o'}{A_o} = 1 \tag{7.39}$$

Therefore,

$$n = \frac{(CO_2)_{theor}}{(CO_2)_{actual}} - 1 \tag{7.40}$$

For gaseous fuels, the ratio V_o'/A_o varies between 0.9 and 1.95 in regularity with the variation in net calorific value

C_N, kcal/Nm^3	900	1800	2700	3600	4500
$\frac{V_o'}{A_o}$	1.95	1.25	1.0	0.9	0.9

Again the content of CO_2 in theoretical dry flue gas is approximately constant for a given class of fuels (Table 7.3). This is the maximum CO_2-content of dry flue gas obtained from that class of fuel. The value of $(CO_2)_{max}$ that is $(CO_2)_{theor}$ may also be calculated from the flue gas orsat analysis.

1. Case of complete combustion

$$(CO_2)_{max} = \frac{100CO_2}{100-(O_2+3.76O_2)} = \frac{100CO_2}{100-4.76O_2} \tag{7.41}$$

2. Case of incomplete combustion

$$(CO_2)_{max} = \frac{100(CO_2+CO+CH_4)}{100-4.76(O_2-0.5\,CO-0.5\,H_2-2CH_4)} \tag{7.42}$$

These relationships form the basis of Ostwald charts. The CO_2 and O_2 contents of the flue gas from the complete combustion of a particular class of fuel should vary with the amount of excess air in a regular manner. This regularity is shown in a very simple and clear manner in Ostwald charts. Each class of fuel has its own Ostwald chart. The two limiting points in the chart are the $(CO_2)_{max}$ and the $(O_2)_{max}$ which is 21% (oxygen content of air). The composition of flue gas containing excess air is shown by a point in the line joining the two limiting points (Fig. 7.1), perfect combustion line.

The Ostwald chart also contains lines of incomplete combustion, represented by contents of H_2 and CO in the flue gas (several experiments have indicated that when CO is present, it is

Table 7.3 CO_2 content of theoretical dry flue gas

Fuel	$(CO_2)_{max}$, that is, $(CO_2)_{theor}$, per cent
Bituminous coal	18.7
Lignite, semi-anthracite, anthracite, semi-coke	19.5
Wood, charcoal	20.5
Fuel oil	15.8
Town gas	
C_N = 4,900 kcal/Nm^3	11.0
4,450 kcal/Nm^3	12.0
4,000 kcal/Nm^3	13.0
Coke oven gas	11.0
Producer gas from coal	19.0
Producer gas from coke	20.5
Blast furnace gas	24.5

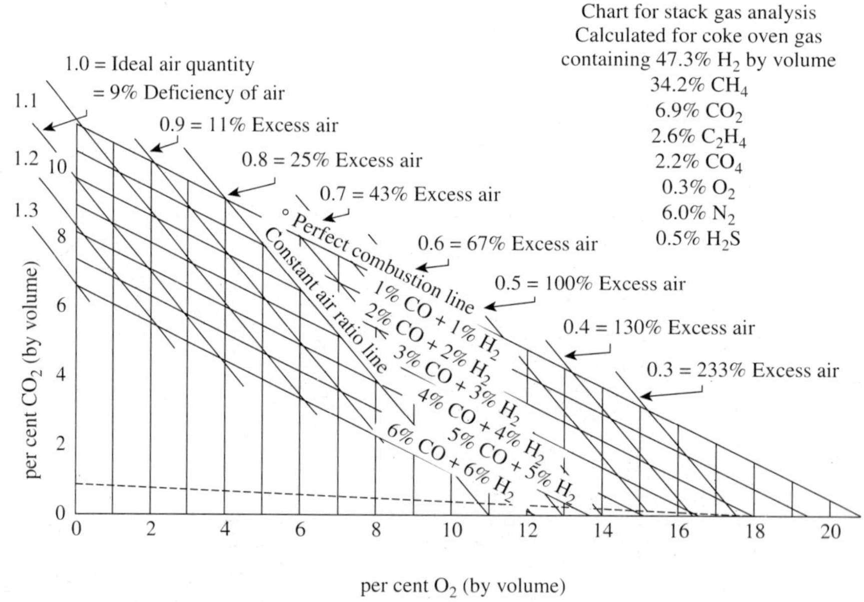

Figure 7.1 Ostwald chart for coke-oven gas

accompanied by an approximately equal volume of H_2). The chart also includes constant air-ratio lines. Air-ratio, R, is defined as

$$R = \frac{\text{theoretical air}}{\text{actual air}} = \frac{1}{1+n} \tag{7.43}$$

where n = excess air factor

Ostwald charts are useful in checking flue gas in combustion calculations. Each flue gas is represented by a single point in the chart meant for the particular fuel. The Ostwald charts for various industrial fuels are available in Reference 5.

Once the excess air factor, n, is known, the volume of actual air supply, A, and the actual flue gas produced, V, can be calculated.

Example 9

The CO_2 content of the products of complete combustion of a coal of 7,200 kcal/kg CV is 12.0% by orsat analysis. Calculate (a) the per cent excess air, (b) the volume of air supply at 30°C and 760 mm and (c) the volume of flue gas leaving the furnace at 400°C and 760 mm per kilogram of coal fired.

Solution

From Table 7.1

$$\begin{aligned} C_N &= 0.954C_G + 110 \\ &= 0.954 \times 7200 + 110 \\ &= 6980 \text{ kcal/kg} \end{aligned}$$

From Table 7.3

$$(CO_2)_{max} = 18.7\%$$

$$\text{Excess air factor} = \frac{(CO_2)_{max} - 1}{(CO_2)} = \frac{18.7 - 1}{12.0} = 0.56$$

Therefore,

Excess air = 56%

From Table 7.2

$$A_o = 1.01 \times \frac{C_N}{1000} + 0.55 = 1.01 \times 7 + 0.55$$

$$= 7.62 \text{ Nm}^3/\text{kg coal}$$

$$V_o = \frac{C_N}{1000} + 0.9 = 7.9 \text{ Nm}^3/\text{kg coal}$$

Air supply

$$A = (1 + n)\, A_o = 1.56 \times 7.62 \text{ Nm}^3/\text{kg coal}$$

$$= 1.56 \times 7.62 \times \frac{303}{273} = 13.2 \text{ m}^3/\text{kg coal at 30°C and 760 mm}$$

Volume of flue gas,

$$V = V_o + nA_o = 7.9 + 0.56 \times 7.62$$
$$= 12.17 \text{ Nm}^3\text{/kg coal}$$
$$= 12.17 \times \frac{673}{273} = 30 \text{ m}^3\text{/kg coal at 400°C and 760 mm}$$

7.3.3 Use of Humid Air in place of Dry Air for Combustion Calculation

If necessary, humidity of atmospheric air may be taken into account in combustion calculation.

Let the humidity of air = H g/kg of dry air

Volumetric content of water vapour in air

$$v_w = \frac{H/1000 \times 22.4/18}{22.4/29}$$
$$= 0.0016H \text{ Nm}^3\text{/Nm}^3 \text{ dry air} \tag{7.44}$$

Theoretical humid air consumption for combustion

$$A'_o = A_o (1 + v_w) \text{ Nm}^3 \text{/kg or Nm}^3 \text{ fuel} \tag{7.45}$$

Actual humid air supply

$$A' = (1 + n) A_o (1 + v_w) \text{ Nm}^3 \text{/kg or Nm}^3 \text{ fuel} \tag{7.46}$$

Product of combustion with humid air

$$\text{Volume of } CO_2 = \alpha' = \alpha \text{ Nm}^3\text{/kg or Nm}^3 \text{ fuel} \tag{7.47}$$
$$\text{Volume of } H_2O = \beta' = \beta + v_W (1 + n) A_o \text{ Nm}^3\text{/kg or Nm}^3 \text{ fuel} \tag{7.48}$$
$$\text{Volume of } N_2 = \gamma' = \gamma, \text{ Nm}^3\text{/kg or Nm}^3 \text{ fuel} \tag{7.49}$$

Volume of combustion products (wet)

1. With theoretical humid air

$$(V_o)_{ha} = \alpha + \beta' + \gamma = (V_o + v_w A_o) \text{ Nm}^3\text{/kg or Nm}^3 \text{ fuel} \tag{7.50}$$

2. With excess air (humid)

$$V_{ha} = (V_o)_{ha} + v_w nA_o + nA_o$$
$$= V_o + nA_o + v_w (1 + n) A_o$$
$$= V + v_w (1 + n) A_o \text{ Nm}^3\text{/kg or Nm}^3 \text{ fuel} \tag{7.51}$$

Volume of dry combustion products

1. $(V'_o)_{ha} = \alpha + \gamma = V'_o$ Nm^3/kg or Nm^3 fuel (7.52)

2. $(V'_o)_{ha} = V'_o + nA_o = V'$ Nm^3/kg or Nm^3 fuel (7.53)

Therefore, volume of dry combustion products and their orsat analysis remain unchanged by the use of humid air in place of dry air.
If the humidity is expressed as per cent relative humidity, RH, volume of water vapour

$$v_w = \frac{\frac{RH}{100} \times p_s}{P - \frac{RH}{100} \times p_s} \ Nm^3/orNm^3 \text{ dry air} \quad (7{:}54)$$

where, P = atmospheric pressure, mm Hg

p_s = vapour pressure of water at the ambient temperature, mm Hg.

Alternatively, humidity, H, is read from psychometric charts at the given RH and temperature, and then v_w is calculated from H.

Example 10

Solve Problem 7 using air of 60% relative humidity at 30°C and 760 mm.

Solution

Water vapour pressure at 30°C = 31.886 mm

$$v_w = \frac{0.63 \times 1.886}{760 - 0.6 \times 31.886} = 0.0258 \ Nm^3/Nm^3 \text{ dry air}$$

$$\text{Air supplied} = A(1 + v_w) = 8.7(1 + 0.0258)$$
$$= 8.92 \ Nm^3/kg \text{ coal}$$
$$= \frac{8.92 \times 29}{22.4} = 11.55 \text{ kg/kg coal}$$

Flue gas produced

$$V_{ha} = V + v_w(1 + n)A_o = 9.06 + 0.0258 \times 1.5 \times 5.8$$
$$= 9.29 \ Nm^3/kg \text{ coal} = 9.29 \times \frac{523}{273} = 17.8 \ m^3/kg \text{ at } 250°C$$

Water vapour from air

$$V_{ha} - V = 9.29 - 9.06$$
$$= 0.23 \ Nm^3/kg \text{ coal}$$

Flue gas analysis, per cent, wet

$$CO_2 = 12.0 \times \frac{9.06}{9.29} = 11.7$$

$$N_2 = 76.1 \times \frac{9.06}{9.29} = 74.2$$

$$O_2 = 6.7 \times \frac{9.06}{9.29} = 6.6$$

$$H_2O = \frac{0.47 + 0.23}{9.29} = 7.5$$

Total $= \overline{100}$

Orsat analysis will remain unchanged by the use of humid air in place of dry air.

7.4 COMBUSTION THERMODYNAMICS

Heat of combustion, equilibrium constant of reactions occurring during combustion, enthalpy of combustion systems and flame temperature are the thermodynamic functions of the combustion processes which fundamentally influence the utilisation of fuels in diverse appliances. A brief account of each of these functions is given below.

7.4.1 Heat of Combustion

By applying Hess's law the heat of combustion of any substance can be calculated from the heats of formation of reactants and products. Table 7.4 presents the heats of combustion of fuel gas components and carbon at 25°C and 1 atm pressure. The heat of combustion of carbon depends upon its allotropic form. In physics and thermochemistry, β-graphite is used as a basis for heat of formation ($\Delta H_f = 0$) since it is a well-defined body material. In technical processes, however, it is preferable to use amorphous carbon, say, coke carbon as the basis. The data of Table 7.4 relate to solid amorphous carbon. In relation to β-graphite, the other allotropic forms of carbon have the following heats of formation: β-graphite = 0, α-graphite = 270 kcal/kmol, diamond = −220 kcal/kmol, amorphous carbon = + 2948.2 kcal/kmol (assumed to get a round figure of 97,000 for $-\Delta H_c$).

CV can be easily derived from the heat of combustion.

$$\text{For carbon CV} = \frac{97000}{12} = 8083 \text{ kcal/kg}$$

$$\text{For methane} = \frac{212 \times 798}{22.4} = 9500 \text{ kcal/Nm}^3$$

Table 7.4 Heat of combustion of carbon and some gases at 298°K and 1 atm pressure

Substance	Heat of combustion (ΔH_c) kcal/kmol	Calorific value, kcal/Nm^3 dry
Carbon (solid amorphous)	97,000	8,083 (kcal/kg)
Hydrogen	68,317.4	3,050 (33,887.6 kcal/kg)
Carbon monoxide	67,636.1	3,020
Methane	212,798	9,500
Ethane	372,820	16,644
Ethylene	337,234	15,055
Acetylene	310,615	13,867
Propane	530,605	23,688 (11,947 kcal/kg liquid)
n-Butane	687,982	30,714 (11,749 kcal/kg liquid)
Propane	491,987	21,964 (11,692 kcal/kg liquid)
1-Butene	649,757	20,000 (11,581 kcal/kg liquid)
Benzene	789,080	35,227 (9,999 kcal/kg liquid)
Toluene	943,580	42,124 (10,143 kcal/kg liquid)
Hydrogen sulphide	134,744	6,015

The calorific values of other gases are also given in Table 7.4. When water is a product of combustion, it is considered in the liquid state for arriving at the heat of combustion of Table 7.4. Therefore, the figures in the last column of this table are the gross calorific values of the substances concerned. The calorific values of commercial fuels are discussed in Chapters 3, 5 and 6.

The heat of combustion of a fuel is also known as its potential heat. If in a combustion system there is undeveloped heat, that is loss of combustible with the flue gas and or solid refuse, the process is obviously inefficient. The efficiency of combustion is calculated in the following way.

$$\eta\text{, per cent} = 1 = \frac{(\text{potential heat in flue gas} + \text{potential heat in refuse})}{\text{potential heat in fuel}} \times 100$$

The combustion efficiency is different from the furnace efficiency. The latter is much more involved. Any meaningful discussion of this term is beyond the scope of this book.

7.4.2 Enthalpy of Combustion System

In most practical cases of combustion, heat transfer from the combustion gases takes place at a fairly constant pressure, not far from the atmospheric. The heat absorbed by the surroundings from a combustion system at constant pressure is equal to the decrease in the enthalpy of the system.

$$Q = -\Delta H \tag{7.55}$$

where Q = heat absorbed by surroundings

$-\Delta H$ = enthalpy decrease of the system in heat units

In other words, enthalpy change is the change of heat content of the system at constant pressure. It is not necessary to know the absolute enthalpy of a system. A knowledge of enthalpy in relation to a reference state, say, 0°C and 760 mm is sufficient. The enthalpy of a gas at a temperature t°C is thus

$$\Delta H_t = \bar{C}_{p_{0-t}} \times t_1 \tag{7.56}$$

where $\bar{C}_{p_{0-t}}$ is the mean specific heat between the reference temperature, 0°C, and the given temperature, t°C, expressed as kcal/Nm^3 °C (volume basis) or kcal/kg°C (weight basis). Like C_p, the units of ΔH, may be kcal/Nm^3 or kcal/kg. If molal heat capacity, MC_p, is used, ΔH_t will be in kcal/mol.

The values of mean specific heat and enthalpy of gases at different temperatures with reference to 0°C are given in the Appendices 1–4. The enthalpy change between two temperatures, $t_1$and t_2 can be determined from these tables.

$$\Delta H_{t_1} - \Delta H_{t_2} = \left(\bar{C}_{p_{0-t_1}} \times t_1\right) - \left(\bar{C}_{p_{0-t_2}} \times t_2\right) \tag{7.57}$$

Depending upon the particular problem, either the mean specific heat values, or the enthalpy values are directly used.

The specific heat of a gas is independent of pressure at low pressures and the average specific heat of a gas mixture is

$$(\bar{C}_{p_{0-t}})\text{ mixture} = x_a (\bar{C}_{p_{0-t}})_A + x_b (\bar{C}_{p_{0-t}})_B = \text{mean specific heat of components, A, B} \ldots$$

where $(\bar{C}_{p_{0-t}})_A, (\bar{C}_{p_{0-t}})_B, \ldots$ = mean specific heat of components, A, B ... etc. $x_a, x_b \ldots$ = volume fraction or weight fraction of components as the case may be.

The enthalpy of the mixture.

$$(\Delta H_{0\text{-}t})_{mix} = (\bar{C}_{p_{0-t}})_{mix} \times t \tag{7.59}$$

Alternatively the enthalpy of the mixture is calculated from the enthalpy of its components.

$$(\Delta H_{0\text{-}t})_{mix} = x_a (\Delta H_{0\text{-}t})_A + x_b (\Delta H_{0\text{-}t})_B + \ldots \tag{7.60}$$

Although normally enthalpy is the product of specific heat and temperature, it is necessary to take into account some other factors at very high temperatures above 1500°C for arriving at correct enthalpy values.

The most significant of these factors is the storage of energy by dissociation reactions — Appendix 3 and Appendix 4 neglect this effect. Dissociation reactions of combustion systems are discussed below. Appropriate corrections are to be applied in using the data of Appendix 3 and Appendix 4 above 1,500°C.

The specific heat and enthalpy of combustion gases are used in calculating flame temperature, heat loss with flue gases and furnace efficiency. Since the enthalpy data do not include latent heat of condensation of water vapour, net calorific value of fuel is to be used for calculating combustion efficiency and furnace efficiency.

Example 11

From the data of Problem 6, calculate the total heat loss with the flue gas and the potential heat loss with the solid refuse, both expressed as a per cent of the potential heat (calorific value) in coal. Hence, determine the combustion efficiency.

Solution

Analysis of coal, per cent as fired: C 68.0, surplus H_2 3.9, S 0.3, water (including water from the oxygen of coal) 11.2, ash 14.7.

By Dulong's formula

$$\begin{aligned}\text{CV (gross)} &= 80.8\text{C} + 345\text{H}_2 + 22.2\text{S}\\ &= 80.8 \times 68.0 + 345 \times 3.9 + 22.2 \times 0.3\\ &= 6842 \text{ kcal/kg coal fired}\end{aligned}$$

$$\begin{aligned}\text{CV (net)} &= 6842 - 53\left[\text{net H}_2 + \frac{\text{water}}{9}\right]\\ &= 6842 - 53\left[3.9 + \frac{11.2}{9}\right]\\ &= 6569 \text{ kcal/kg}\end{aligned}$$

$$\text{Carbon in refuse} = 0.098 \text{ kg/kg coal}$$

Potential heat loss with the refuse

$$= 0.098 \times 8080 = 792 \text{ kcal/kg coal fired}$$

$$= \frac{792}{6569} \times 100 = 12.1\% \text{ of the potential heat in the coal}$$

$$\begin{aligned}\text{Flue gas produced} &= 37.50 + 0.62 + 1.96\\ &= 40.08 \text{ kmol/100 kg coal}\\ &= \frac{40.08 \times 22.4}{100} \text{ Nm}^3\text{/kg coal}\\ &= 8.98 \text{ Nm}^3\text{/kg coal}\end{aligned}$$

Flue gas analysis, per cent

$$\left.\begin{aligned}\text{H}_2\text{O} &= \frac{0.62 + 1.96}{40.08} \times 100 = 6.4\\ \text{CO}_2 &= 11.5 \times \frac{100 - 6.4}{100} = 10.8\end{aligned}\right\} \text{calculated from orsat analysis}$$

$$\left.\begin{aligned} CO &= 1.5 \times \frac{100 - 6.4}{100} = 1.4 \\ N_2 &= 80.5 \times \frac{100 - 6.4}{100} = 75.3 \\ O_2 &= 6.5 \times \frac{100 - 6.4}{100.0} = 6.1 \end{aligned}\right\} \text{calculated from orsat analysis}$$

Enthalpy of flue gas at 300°C

$$= 0.064 \times 110.0 + 0.108 \times 134.7 + 0.014 \times 94.4 + 0.753 \times 94.2 + 0.061 \times 97.2$$
$$= 99.9 \text{ kcal/Nm}^3$$

Let mean specific heat of flue gas at 30°C (datum temperature)

$$= 0.064 \times 0.357 + 0.108 \times 0.400 + 0.014 \times 0.311 + 0.753 \times 0.311 + 0.061 \times 0.314$$
$$= 0.324 \text{ kcal/ Nm}^{3}{}^{\circ}\text{C}$$

Therefore, enthalpy at 30°C $= 0.324 \times 30$

$$= 9.72 \text{ kcal/Nm}^3$$

Therefore, sensible heat loss with the flue gas

$$= 99.9 - 9.72$$
$$= 90.18 \text{ kcal/Nm}^3 \text{ gas}$$

Again, potential heat loss with the flue gas

$$= (CV)_{co} \times \frac{1.4}{100}$$
$$= 3020 \times 0.014$$
$$= 42.28 \text{ kcal/Nm}^3\text{gas}$$

Therefore, combustion efficiency

$$= \left[1 - \frac{792 + 42.28 \times 8.98}{6569}\right] \times 100$$
$$= 82.2\%$$

But, total heat loss with the flue gas

$$= 90.18 + 42.28$$
$$= 132.46 \text{ kcal/Nm}^3 \text{ gas}$$
$$= 132.46 \times 8.98 = 1190 \text{ kcal/kg coal fired}$$
$$= \frac{1190}{6569} \times 100 = 18.1\% \text{ of the potential heat in coal}$$

7.4.3 Equilibrium Constants of Combustion Reactions

There are a number of reversible reactions which determine the fate of the combustion process. A proper evaluation of the latter needs consideration of the equilibrium constants of these reactions. Appendix 5 gives the equilibrium constants of some of the reactions at the wide temperature range of 300−4000 K. Out of these, the following reactions are of particular significance (ΔH values are at 25°C):

Dissociation of water vapour

$$H_2O \rightleftarrows H_2 + ½O_2 \qquad \Delta H = +\ 57{,}798\ \text{kcal}$$

Dissociation of water vapour into hydroxyl group

$$H_2O \rightleftarrows OH + ½H_2 \qquad \Delta H = +\ 67{,}858\ \text{kcal}$$

Dissociation of carbon dioxide

$$CO_2 \rightleftarrows CO + ½O_2 \qquad \Delta H = +\ 67{,}636\ \text{kcal}$$

Dissociation of hydrogen

$$H_2 \rightleftarrows 2H \qquad \Delta H = +\ 104{,}178\ \text{kcal}$$

Dissociation of oxygen

$$O_2 \rightleftarrows 2O \qquad \Delta H = +\ 118{,}318\ \text{kcal}$$

Formation of nitric oxide

$$½O_2 + ½N_2 \rightleftarrows NO \qquad \Delta H = +\ 21{,}600\ \text{kcal}$$

These reactions are endothermic and convert a part of the sensible heat of the combustion gases into the potential heat in the form of heat of combustion of the combustible components formed and the heat of formation of atomic hydrogen and atomic oxygen. The combustion of fuels is therefore rendered incomplete at high temperatures. This has two effects: first, the combustion efficiency is lowered and second, the temperature of the system falls. The second point is discussed under flame temperature. Another effect of dissociation is an increase in the volume and number of moles of the gases.

When the equilibrium constants at different temperatures are known, the degrees of dissociation at those temperatures can be calculated.

The temperature dependence of the equilibrium constants for CO_2 dissociation is given in the following formula, applicable to temperatures above 1,600K.

$$4.571 \log K_{p_{CO_2}} = \frac{69894.4}{T} + 41470 \log T + 0.378 \times 10^{-3} T - 0.0972 \times 10^{-3} T^2 - 36.04838 \qquad (7.61)$$

A similar formula for H_2O dissociation, applicable for all temperatures is given below

$$4.571 \log K_{p_{H_2O}} = \frac{57111.1}{T} - 2.6135 \log T - 0.84834 \times 10^{-3} T + 0.19602 \times 10^{-6} T^2 - 2.96716 \quad (7.62)$$

where T is in K.

If x is the fraction of CO_2 dissociated at any temperature, the equilibrium constant at that temperature:

$$K_{p_{CO_2}} = \frac{1-x}{x} \times \frac{2+x}{x} \times \frac{1}{\sqrt{P}} \quad (7.63)$$

Similarly,

$$K_{p_{H_2O}} = \frac{1-y}{y} \times \frac{2+y}{y} \times \frac{1}{\sqrt{P}} \quad (7.64)$$

where, y is the fraction of H_2O dissociated and P is the total pressure in atmosphere. In the usual cases of combustion, P = 1. Equations 7.63 and 7.64 are also valid for mixtures containing carbon dioxide and water vapour. In such cases, partial pressures of these components are to be used for P.

The fraction dissociated is called degree of dissociation, α. Table 7.5 gives the degree of dissociation of carbon dioxide and water vapour in the temperature range, 1,400–2,900°C. Out of the six reactions written above, these two reactions are the most significant in usual combustion with air.

Example 12

Combustion products (composition of undissociated gas): 11% CO_2, 9.5% H_2O, 3.3% O_2 and 75.3% N_2, leave a combustor at 1,600°C. Calculate the heat loss due to dissociation of CO_2 and H_2O per Nm^3 of the gas.

Solution

Initial partial pressures

$$P_{CO_2} = 0.119 \text{ atm}$$

$$P_{H_2O} = 0.095 \text{ atm}$$

Degrees of dissociation at 1,600°C (Table 7.5)

$$\alpha_{CO_2} = 1.54\% \text{ at } 0.119 \text{ atm}$$

$$\alpha_{H_2O} = 0.6\% \text{ at } 0.095 \text{ atm}$$

Table 7.5 Dissociation of carbon dioxide and water vapour

Temp°C	$K_{P_{CO_2}}$	Carbon dioxide pressure, atm				$K_{P_{H_2O}}$	Water vapour pessure, atm			
		1	0.5	0.2	0.1		1	0.5	0.2	0.1
		per cent, dissociated					per cent, dissociated			
1400	18060	0.2	0.25	0.37	0.4	61030	0.08	0.10	0.14	0.17
1500	5820	0.4	0.5	0.73	0.85	22260	0.15	0.20	0.3	0.35
1600	2120	0.8	1.0	1.3	1.6	9007	0.3	0.4	0.5	0.6
1700	858.4	1.5	1.8	2.4	3.0	3995	0.5	0.6	0.8	1.0
1800	379.2	2.5	3.0	4.0	5.0	1916	0.8	1.0	1.4	1.8
1900	180.8	3.9	4.9	6.5	8.1	984.0	1.2	1.6	2.2	2.7
2000	92.08	6.0	7.5	10.0	12.4	536.1	1.9	2.4	3.3	4.0
2100	49.67	8.9	11.1	14.7	18.1	307.8	2.8	3.4	4.7	5.8
2200	28.16	12.7	15.7	20.6	25.2	185.1	3.8	4.8	6.5	8.0
2300	16.69	17.5	21.4	27.6	33.4	116.0	5.1	6.4	8.7	10.8
2400	10.29	23.2	28.1	35.8	42.4	75.32	6.8	8.5	11.4	14.1
2500	6.565	29.8	35.7	44.4	51.5	50.58	8.8	10.9	14.6	17.9
2600	4.321	37.0	43.7	53.1	60.3	35.00	11.1	13.8	18.1	22.2
2700	2.923	44.5	51.7	61.2	68.2	24.87	13.7	16.9	22.2	27.0
2800	2.028	52.0	59.3	68.6	74.9	18.11	16.6	20.4	26.5	32.0
2900	1.438	59.3	66.3	74.9	80.5	13.50	19.8	24.2	31.1	37.2

Note: *The equilibrium constants hold good for mixtures as well as pure gases.

Heat of dissociation

$$\Delta H_{CO_2} = +\ 67{,}636 \text{ kcal/kmol} = +\ 3{,}020 \text{ kcal/Nm}^3$$
$$\Delta H_{H_2O} = +\ 57{,}798 \text{ kcal/kmol} = +\ 2{,}580 \text{ kcal/Nm}^3$$

Heat absorbed due to dissociation

$$= 3020 \times 0.0154 \times 0.119 + 2580 \times 0.006 \times 0.095$$
$$= 7.02 \text{ kcal/Nm}^3 \text{ gas (neglecting the small rise in the volume of the gas by dissociation)}$$

This is the required heat loss and represents the potential heat of CO and H_2 formed by the dissociation of CO_2 and H_2O.

7.4.4 Enthalpy–Temperature Diagrams

When combustion is complete, the enthalpy of a flue gas is actually the summation of the enthalpy of theoretical flue gas and that of excess air. For a given fuel, the composition of its theoretical flue gas is fixed. Therefore, the enthalpy of its flue gas is a function of two variables only, namely, temperature and excess air. At temperatures above 1,600°C, the endothermic effect of dissociation reactions must be taken into account. These considerations form the

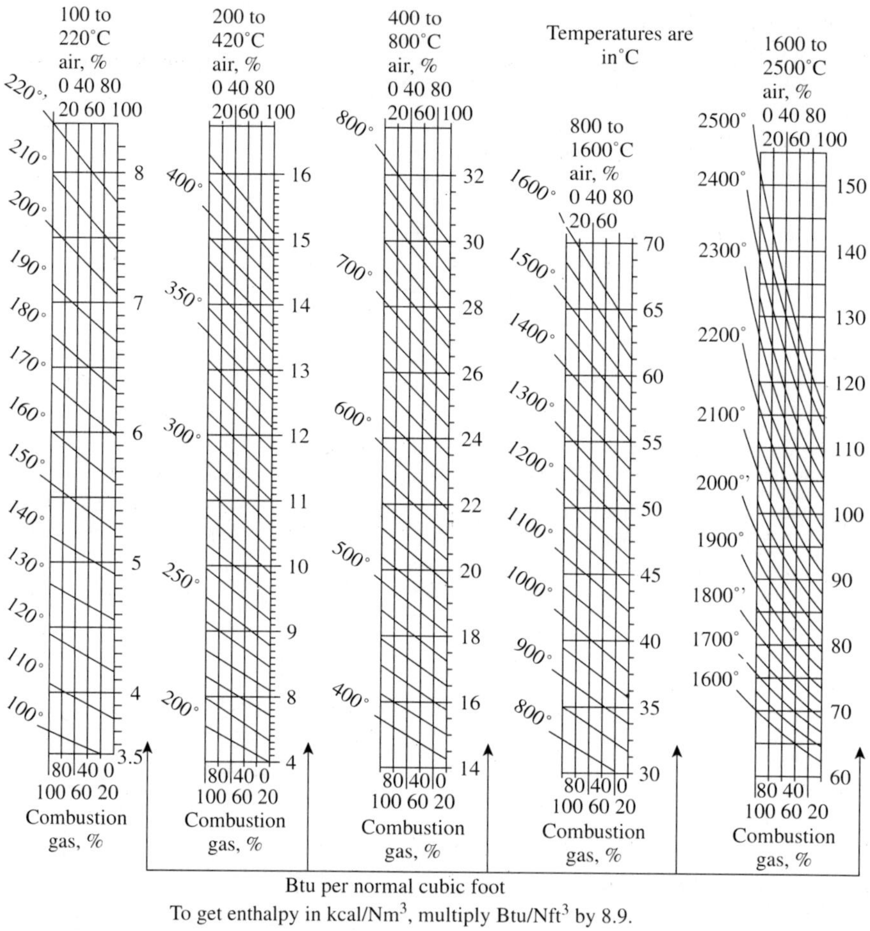

Figure 7.2 Ht-diagram for combustion gases

basis of the construction of enthalpy–temperature diagrams. Each class of fuel has its own enthalpy–temperature diagrams. However it has been found that the theoretical wet flue gases obtained from all industrial fuels have approximately the same enthalpy (kcal/Nm^3), at the same temperature.

The equilibrium constants hold good for mixtures as well as for the pure gases. This situation lays the foundation for the well-known Ht- or It-diagram of Rosin and Fehling (Fig. 7.2). The diagram covers the range from 100 to 2,500°C and shows the relationship between enthalpy and temperature for different air content of the flue gas. The air content of flue gas,

$$v_A = \frac{nA_n}{V} \times 100\% \tag{7.65}$$

The diagram takes into account various significant dissociation reactions occurring above 1,600°C.

Enthalpy–temperature diagrams in general and Ht-diagrams in particular are very useful in rapidly working out the problems concerning heat loss with the flue gas and flame temperature.

The full benefits of the Ht-diagram are derived by using a few auxiliary tables given in Reference 4 of the Bibliography.

Example 13

Using Ht-diagram calculate the enthalpy of the flue gas of Problem 12.

Solution

Air content of flue gas

$$v_A = 3.3 \times 4.76 = 15.7\%$$

Therefore, from the Ht-diagram, enthalpy at 1,600°C

$$= 70 \times 8.9 = 623 \text{ kcal/Nm}^3$$

7.4.5 Flame Temperature

The average temperature attained by the combustion products of a mixture of fuel and oxidant is known as the flame temperature of the inflammable mixture. Four different types of flame temperature are recognised: theoretical flame temperature, adiabatic flame temperature, actual flame temperature and maximum adiabatic flame temperature, or simply maximum flame temperature. It will be evident that the theoretical flame temperature is not a tangible concept, while the others are.

When the combustion of the fuel goes to completion and the entire heat of combustion goes to heat the products of combustion, the resultant temperature is the theoretical flame temperature. In reality, combustion is never complete at high temperatures owing to dissociation reactions. When the endothermic effect of dissociation reactions is taken into account, the adiabatic flame temperature is obtained. Obviously, this is lower than the theoretical flame temperature.

In a practical combustion process, adiabatic conditions do not exist. There is always some heat loss to the surroundings of the system. The resultant average temperature of the combustion products is termed the actual flame temperature.

All the three flame temperatures defined above depend upon the composition of the fuel–oxidant mixture. If the quantity of oxidant is to small, it will lead to incomplete combustion; if the quantity is too large, it will dilute the products and take away heat. Both the factors lower the flame temperature. Theoretical flame temperature has its maximum value at the stoichiometric composition of fuel and oxidant. Since the degree of dissociation reactions markedly increases with temperature, the endothermic effect also rises with it. Heat loss due to this factor increases with the rise in temperature at an accelerated rate. Therefore, the maximum adiabatic flame temperature is realised when the fuel is slightly in excess of the stoichiometric composition (Fig 7.3). Maximum adiabatic flame temperatures of some fuel gases in air and of a few in oxygen are given in Table 7.6. It may be seen that with the exception of acelyene which is somewhat above the range and with a slight exception for hydrogen and carbon monoxide, all

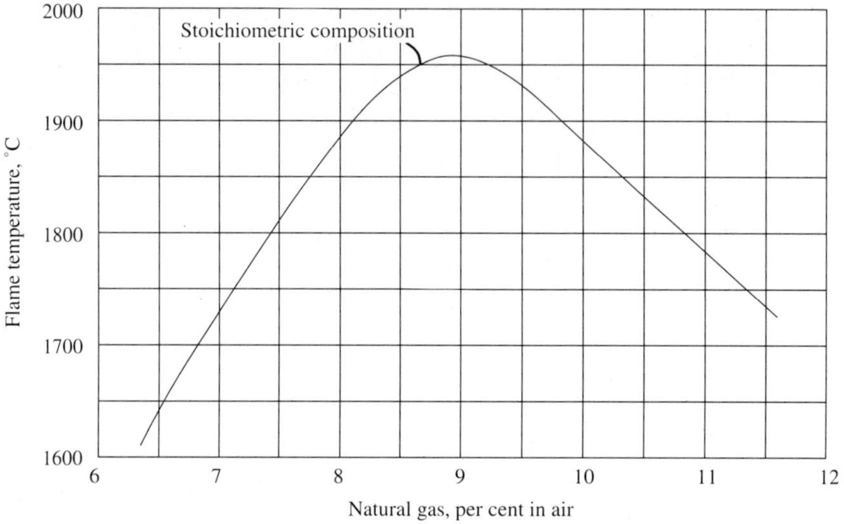

Figure 7.3 Variation of adiabatic flame temperature with composition of fuel–air mixture

Table 7.6 Maximum adiabatic flame temperature (MAFT)

Sr. No.	Gas	In air per cent combustibles in gas–air mixture giving MAFT	MAFT °C	In oxygen per cent combustibles in gas–O_2 mixture giving MAFT	MAFT °C
1.	Hydrogen	31.6	2,045	78.0	2,660
2.	Carbon monoxide	32.0	2,100	70.0	2,925
3.	Methane	10.0	1,875	–	–
4.	Ethane	5.8	1,895	–	–
5.	Propane	4.15	1,925	–	–
6.	Butane	3.2	1,895	–	–
7.	Isobutane	3.2	1,900	–	–
8.	Ethylene	7.0	1,975	–	–
9.	Propane	4.5	1,935	–	–
10.	Butane	3.4	1,930	–	–
11.	Acetylene	9.0	2,325	44.0	3,137
12.	Natural gas	9.0	1,930	–	–
13.	Coal gas	17.6	1,918	–	–
14.	Carturetted water gas	18.8	1,930	–	–
15.	Producer gas	46.7	1,663	–	–
16.	Blast furnace gas	–	1,460	–	–

gases listed from 1 to 14 in Table 7.6 have their maximum flame temperatures around 1,900–1,950°C. This is because an increase in the calorific value of a fuel is counter-balanced by the additional air requirement and hence by a greater quantity of products of combustion. The last two gases, that is, producer gas and blast furnace gas have quite a low flame temperature owing to very high content of inerts.

From the few data given for combustion in oxygen, it will be seen that the flame temperature of a fuel is much higher in oxygen than in air. This is because of the high nitrogen content of air, which takes away a significant quantity of heat. It is noteworthy that the depression effect of dissociation on flame temperature is many times more in oxygen combustion than in air combustion. For example, methane–oxygen flame, with a stoichiometric ratio of 1:2 has an adiabatic flame temperature of 2,700°C as against its theoretical flame temperature of 5,050°C. The drop in flame temperature due to dissociation is thus as high as 2,310°C. The corresponding values for stoichiometric combustion in air are: theoretical flame temperature, 2,060°C, adiabatic flame temperature, 1,870°C and the drop due to dissociation, 190°C, that is, 1/12 of the drop in oxygen. If the products of combustion of a system are difficult to dissociate at high temperatures the resultant adiabatic flame temperature should be very high. A cyanogens–oxygen flame has such products, CO and N_2, and gives a temperature of 4,580°C. Similarly, a hydrogen–fluorine flame has a maximum adiabatic temperature of 4,300°C

In the above discussion and for getting the values of Table 7.6, fuel–oxidant mixture is taken at room temperature and combustion is assumed to occur at atmospheric pressure. Flame temperatures can be appreciably increased if additional heat is supplied to the system in the form of, say, preheated air and if possible, preheated fuel.

Flame temperatures are calculated by striking a heat balance between the fuel and air on the one hand and the products of combustion on the other. Let us take the case of actual flame temperature in air. Heat balance is given by

$$C_N + \Delta H_{fuel} + A\Delta H_{air} = V\Delta H_{wg} + q_{diss} + q_{loss}$$

where,

C_N = net CV of fuel, kcal/Nm^3 fuel

ΔH_{fuel} = enthalpy of the fuel above the reference temperature, kcal/Nm^3

ΔH_{air} = enthalpy of air above reference temperature, kcal/Nm^3 air.

ΔH_{wg} = enthalpy of combustion products above reference temperature, kcal/Nm^3 combustion gases.

A = air supplied, Nm^3/Nm^3 fuel

V = combustion gases producted, Nm^3/Nm^3 fuel

q_{diss} = heat loss by dissociation reactions, kcal/Nm^3 fuel

q_{loss} = heat loss to surroundings, kcal/Nm^3 fuel

But,

$$\Delta H_{wg} = t_f \bar{C}_{P_{wg,0-t_f}} - t\bar{C}_{P_{wg,0-t}}$$

t_f = flame temperature, °C

t = reference temperature, say 25°C

$\bar{C}_{P_{wg,0-t_f}}$ = mean specific heat of combustion products between t_f and 0°C

$\bar{C}_{P_{wg,0-t}}$ = that between t and 0°C

Therefore,

$$t_f = \frac{C_N + \Delta H_{fuel} + A\,\Delta H_{air} + q_{diss} - q_{loss} + Vt\,\bar{C}_{P_{wg,\,0-t}}}{V\,\bar{C}_{P_{wg,\,0-t_f}}} \quad (7.66)$$

If the fuel and air are supplied at the reference temperature, $\Delta H_{fuel} = 0$ and $\Delta H_{air} = 0$. In the case of theoretical flame temperature, $q_{diss} = 0$ and $q_{loss} = 0$. In the case of adiabatic temperature, $q_{loss} = 0$.

Actual flame temperature may be 250–400°C, lower than the theoretical temperature.

Flame temperature has a special significance as it governs the thermal efficiency of the transference of heat from the flames to the heating surface. From the second law of thermodynamics, this efficiency is given by

$$\eta = \frac{T_f - T_s}{T_f} = 1 - \frac{T_s}{T_f}$$

Where T_f is the flame temperature and T_s the surface temperature in absolute scale. The higher the flame temperature, the greater the efficiency. Hence the special significance of preheating air/ fuel before combustion.

In the calculation of t_f, the difficulty lies in its appearance on both sides of the formulae. The specific heat of waste gases depends upon the unknown t_f. A trial and error method is therefore used. Two approaches are possible: one through specific heat and another through enthalpy determinations, as illustrated in the following problems.

Adiabatic flame temperature may be calculated rapidly though less accurately by using a Ht-diagram.

Example 14

Calculate (a) the theoretical and (b) the adiabatic flame temperatures in the combustion of methane in stoichiometric air.

Solution

Basis of calculation: 1 Nm^3 methane

Stoichiometric combustion

$$CH_4 + 2\,(O_2 + 3.76N_2) \rightarrow CO_2 + 2H_2O + 7.52N_2$$

1 Nm^3 2 × 4.76 Nm^3 1 Nm^3 2 Nm^3 7.52 Nm^3

Volume of combustion products,

$$V = 1 + 2 + 7.52 = 10.52\ Nm^3$$

Theoretical flame temperature

$$t_f = \frac{C_N + V_{25}\,\bar{C}_{P_{wg,\,0-25}}}{V\bar{C}_{P_{wg,\,0-t_f'}}}$$

$$\text{Net CV, } C_N = \text{Gross CV} - 4.7\,V_{H_2}$$
$$= 9500 - 4.7 \times 200$$
$$= 8{,}560 \text{ kcal/Nm}^3$$

$$\bar{C}_{P_{wg,\,0-25}} = (0.390 + 2 \times 0.356 + 7.52 \times 0.311) \times \frac{1}{10.52}$$
$$= \frac{3.442}{10.52} \text{ kcal/Nm}^3\,°\text{C} = 0.3272 \text{ kcal/Nm}^3\,°\text{C}$$

Assume $t_f = 2100°C$

Neglecting the change in the gas composition by dissociation reactions, we get

$$\bar{C}_{P_{wg,\,0-2100}} = (0.583 + 2 \times 0.469 + 7.52 \times 0.357) \times \frac{1}{10.52}$$
$$= \frac{4.206}{10.52} \text{ kcal/Nm}^3\,°\text{C}$$

Therefore,

$$t'_f = \frac{8560 + 10.52 \times 25 \times \frac{4.206}{10.52}}{10.52 \times \frac{4.206}{10.52}}$$
$$= \frac{8665.15}{4.206} = 2060°\text{C}$$

Second trial

Assume $t'_f = 2{,}070°C$. Neglecting the change in the gas composition by dissociation reactions, we get

$$\bar{C}_{P_{wg,\,0-2070}} = \frac{1}{10.52}(0.582 + 2 \times 0.468 + 7.52 \times 0.357)$$
$$= \frac{4.196}{10.52} \text{ kcal/Nm}^3\,°\text{C} = 0.398$$

$$t'_f = \frac{8646}{4.198} = 2060\,°\text{C}$$

This temperature is the same as on the previous trial and may be taken as the theoretical temperature.

Adiabatic flame temperature

$$t_f = \frac{C_N + V25\,\bar{C}_{P_{wg,\,0-25}} - q_{diss}}{\bar{C}_{P_{wg,\,0-25}}}$$

Partial pressures of CO_2 and H_2O in the undissociated combustion products.

$$P_{CO_2} = \frac{1}{10.52} = 0.095 \approx 1 \text{ atm}$$

$$P_{H_2O} = \frac{2}{10.52} = 0.19 \approx 0.2 \text{ atm}$$

Assume $t'_f = 1900°C$. Neglecting the change in the gas composition by dissociation reactions,

$$\bar{C}_{P_{wg,\,0-1900}} = \frac{1}{10.52}(0.576 + 2 \times 0.460 + 7.5 \times 0.354$$

$$= \frac{4.151}{10.52} \text{ kcal/Nm}^3\ {}^{o}\text{C}$$

At 1,900°C, degrees of dissociation

$$\alpha_{CO_2} = 8.1\% \text{ at } 0.1 \text{ atm}$$

$$\alpha_{H_2O} = 2.2\% \text{ at } 0.2 \text{ atm}$$

Neglecting the volume rise by dissociation

$$q_{diss} = \frac{1}{10.52}(3020 \times 0.081 \times 1 + 258 \times 0.22 \times 2) \text{ kcal/Nm}^3 \text{ products}$$

$$= \frac{358}{10.52} \text{ kcal/Nm}^3 \text{ products}$$

$$= 34 \text{ kcal/Nm}^3 \text{ fuel}$$

$$\text{Therefore, } t'_f = \frac{8646 - 34}{4.156} = 2{,}072°C$$

Second trial

Assume $t'_f = 1{,}980°C$

$$\alpha_{CO_2} = 11.5\%$$

$$\alpha_{H_2O} = 3.1\%$$

$$\bar{C}_{P_{wg,\,0-1980}} = \frac{1}{10.52}(0.579 \times 2 \times 0.463 + 7.52 \times 0.356$$

$$= \frac{3.213}{10.52} \text{ kcal/Nm}^3\text{°C} = 0.305$$

$$q_{diss} = 3020 \times 0.115 + 2580 \times 0.031 \times 2$$

$$= 508 \text{ kcal/Nm}^3 \text{ fuel}$$

$$\text{Therefore, } t'_f = \frac{8646 - 507}{4.185} = 1945°C$$

Third trial

Assume $t'_f = 1960°C$

$$\alpha_{CO_2} = 10.6\%$$

$$\alpha_{H_2O} = 2.9\%$$

$$\bar{C}_{P_{wg,\,0-1960}} = \frac{4.185}{10.52} \text{ kcal/Nm}^3\,°C$$

$$q_{diss} = 3020 \times 0.116 + 2580 \times 0.029 \times 2 = 350.32 + 149.64$$

$$= 500 \text{ kcal/Nm}^3 \text{ fuel}$$

$$\text{Therefore, } t'_f = \frac{8146 - 500}{4.185} = 1{,}827°C$$

This checks well against the value. However, this temperature is much higher than the experimental value of the maximum adiabatic flame temperature 1875°C for methane in air (Table 7.6). The chief reason is the simplification of the method of calculation.

Example 15

What will be the theoretical flame temperature in the above case if the air is preheated to 700°C?

Solution

Datum temperature = 25°C

Enthalpy of air at 700°C above 0°C

$$\Delta H_{air} = \bar{C}_{P_{0-700}} \times 700$$

$$= 0.32 \times 700$$

$$= 229.6 \text{ kcal/Nm}^3 \text{ air}$$

$$\Delta H_{air} \text{ at } 25°C = 0.310 \times 25$$

$$= 7.8 \text{ kcal/Nm}^3$$

Therefore,

$$t_f = \frac{C_N + A\Delta H_{air} + V \times 25 \times \bar{C}_{P_{wg,\,0-25}}}{V \times \bar{C}_{P_{wg,\,0-tf}}}$$

$$= \frac{8560 + (2 \times 4.76)\,(229.6 - 7.8) + 10.52 \times 25 \times 3.442/10.52}{10.52 \times \bar{C}_{P_{wg,\,0-tf}}}$$

$$= \frac{10.758}{10.52 \times \bar{C}_{P_{wg,\,0-tf}}}$$

Therefore, $\Delta H_{wg,\,0-t_f}$ = 10,826 kcal/Nm³ of fuel or 10.52 Nm³ of waste gas

Assume t_f = 2500°C

$$\Delta H_{wg,\,0-2500} = \frac{1}{10.52}(1483.6 + 6 + 2 \times 1212.5 + 7.52 \times 908.0)\ 10.52$$
$$= 10{,}742.8 \text{ kcal/Nm}^3 \text{ of fuel}$$

It is slightly less than the required value of enthalpy (10,825 kcal/Nm³ of fuel) and may be used for further calculation.

Second trial

Assume t_f = 2,550°C

$\Delta H_{wg,0-2550}$ = 1515.6 + 2 × 1242.0 + 7.52 × 927.9

= 10,977.6 kcal/Nm³ of fuel

By interpolation

$$t_f = 2500 + \frac{2550 - 2500}{10977.6 - 10736.6} \times (10826 - 10736.6)$$
$$= 2500 + 18.6$$
$$= 2519°\text{C}$$

Example 16

Using a Ht-diagram, calculate the adiabatic flame temperature obtained by burning coke oven gas (gross CV = 4,800 kcal/Nm³) with 20% excess air preheated to 600°C. Use appropriate tables in Reference 4.

Solution

Net CV,

$$C_N = 4800 \times 0.895 = 4300 \text{ kcal/Nm}^3$$

Theoretical air,

$$A_o = 1.09 \times \frac{4300}{1000} - 0.25$$
$$= 4.44 \text{ Nm}^3/\text{Nm}^3 \text{ fuel}$$

Theoretical waste gas

$$V_o = 1.35 \times \frac{4300}{1000} - 0.67 = 5.14 \text{ Nm}^3/\text{Nm}^3 \text{ fuel}$$

Therefore, actual wg $= V_o + nA_o = 5.14 + 0.2 \times 4.44$
$= 6.03$ Nm³/Nm³ fuel

Actual air $= (1 + n) A_o = 1.2 \times 4.44 = 5.33$ Nm³/Nm³ fuel

Enthalpy of inlet air above 25°C

$$\Delta H_{air,\,600-25} = (194.6 - 7.8) \times 5.33 \text{ kcal/ Nm}^3 \text{ fuel}$$

$$= \frac{186.8 \times 5.33}{6.03} = 165 \text{ kcal/Nm}^3 \text{ waste gas}$$

Heat input of fuel above 25°C = heat of combustion + sensible heat (nil)

$$= \frac{4300}{6.03} = 713 \text{ kcal/ Nm}^3 \text{ waste gas}$$

Therefore, enthalpy of waste gas above 25°C = 165 + 713
= 878 kcal/ Nm³ waste gas
= 98.7 Btu/Nft³

Air content of waste gas

$$v_A = \frac{nA_o \times 100}{V} = \frac{0.2 \times 4.44 \times 100}{6.03} = 14.7\%$$

From the Ht-diagram, enthalpy of waste gas at 25°C above 0°C

= 0.9 Btu/Nft³ waste gas

Therefore, total enthalpy of waste gas at t_f °C above 0°C

= 98.7 + 0.9 = 99.6 Btu/Nft³

From the Ht-diagram, t_f = 2060°C

Above 1,600°C, the Ht-diagram takes into account the effect of heats of dissociation reactions. Therefore, this is the required adiabatic flame temperature.

PROBLEMS

1. A liquid hydrocarbon fuel having a H–C ratio of 0.174 by weight is burnt with 20% excess air. Calculate (a) the volume of air at 30°C and 740 mm, (b) the volume of flue gas at 300°C and 740 mm and (c) the weight of water formed per kilogram of fuel.
[Answer: (a) 15.75m³/kg, (b) 31.57 m³/K, (c) 1,332 kg/kg]

2. A natural gas (CO_2 0.5%, N_2 3.0%, CH_4 92.5%, C_2H_6 4.0%) is burnt with 15% excess air. Determine (a) the volume of air, (b) the volume of flue gas per cubic metre of the

gas and also (c) the dew point, (d) composition and (e) average molecular weight of the combustion products.

[Answer: (a) 12.09 Nm^3/Nm^3, (b) 11.91 Nm^3/Nm^3, (c) 563°C, (d) CO_2 10.2%, O_2 3.0%, N_2 86.8%, (e) 27.8)]

3. A Raniganj coal (carbon 88%, hydrogen 4.8%, nitrogen 2.0%, sulphur 0.6% and oxygen 4.6% daf; moisture 3% and ash 14.5% as received) is burnt at the rate of 300 kg/h. Air supply at 30°C and 740 mm is in 25% excess of theoretical. The flue gas leaves at 300°C and 740 mm. Compute (a) the capacity of air blower, (b) the volume of exit gas per minute, (c) the orsat analysis of flue gas and (d) the dew point of flue gas.

[Answer: (a) 55 m^3/min, (b) 313 m^3/min, (c) CO_2 14.8%, O_2 4.3%, N_2 80.9% (d) 32.7ºC]

4. A gas has the following composition: H_2 31%, CO 24% and $N_2$45%. If the orsat analysis of the combustion products in a furnace is CO_2 12.1%, O_2 2.8% and CO 0.0%, calculate the amount of excess air used in the combustion.

[Answer: 20.2%]

5. A propane–butane mixture (1:4) is burnt in a Burnsen burner. If the fuel–air mixture in the mixing tube contains 5.0% fuel, calculate the primary air in the mixture as a percentage of the theoretical air.

[Answer: 64.4%]

6. A furnace is fired with a Singareni coal at the rate of 200 kg/h. The coal analyses as: carbon 60.5%, hydrogen 3.2%, nitrogen 1.4%, sulphur 0.5%, oxygen 9.4%, moisture 5.0% and ash 20.0%, air dried. The dry refuse contains 20% combustibles. The flue gas has the following orsat analysis: CO_2 13.6%, O_2 5.5% and CO 1.2%. Calculate (a) the hourly heat loss due to unburnt combustibles, (b) the capacity of the air blower and (c) volume flow-rate (m^3/h) of the flue gas at 250°C and 740 mm.

[Answer: (a) 132,264 kcal/h (b) 30 m^3/mm (c) 2,923 m^3/h]

7. If octane is burnt with 90% theoretical air, calculate the CO content of the exhaust gas. (Assume the reactivity to oxygen in the order of $H_2 > CO$).

[Answer: 5.0% dry]

8. A heater burnts fuel oil of specific gravity 0.90 and composition: carbon 85.5%, hydrogen 12.0% and sulphur 2.5%, at the rate of 170 l/h . The combustion is supplemented by a gaseous fuel of orsat analysis: H_2 13.5%, N_2 62.5%, CH_4 2.0%, CO 14.0% and CO_2 8.0%. Oil to gas ratio is 2:1 by weight. If 20% excess air is used, calculate the capacity of the fan delivering the air. Assume air at 30°C and 760 mm, and with humidity of 18.8 g moisture per kilogram dry air.

[Answer: 40 m^3/min]

9. A flue gas has the following orsat analysis: CO_2 8.2%, O_2 9.9%, CO 0.1%, H_2 0.4% and N_2 81.4%. Calculate (a) the CO_2 content of the dry theoretical flue gas of the given fuel and (b) the per cent excess air used in the combustion.

[Answer: (a) 15.5% (b) 80.4%]

10. The products of combustion of a fuel gas contain CO_2 14.2%, H_2O 5.8%, O_2 4.0% and the rest N_2. If it is at a temperature of 1,000°C and mixed with atmospheric air at 30°C in the volume ratio of 1: 2, calculate the temperature of the mixture.

[Answer: 398°C]

11. Calculate the heat loss with the flue gas in Problems 1 and 3. Express the result in kcal/kg fuel.

[Answer: 1329 kcal/kg, 882 kcal/kg]

12. The ultimate analysis of a coal tar fuel is C 90.0%, H_2 6.0%, S 0.3%, N_2 1.2% and O_2 2.5%. It is burned with 20% excess air and the flue gas leaves the combustion chamber at 300°C. Calculate the heat loss with the flue gases per kg of dry fuel. (Neglect the ash of coal tar fuel.)

[Answer: 1.041 kcal/kg]

13. Calculate (a) the theoretical and (b) the adiabatic flame temperatures of pure benzene burnt in stoichiometric air.

[Answer: (a) 2,813°C (b) 2,556°C]

14. Determine the amount of heat lost by the dissociation of CO_2 and H_2O in the preceding problem.

[Answer: 1.99 kcal/kg benzene]

15. Compute the enthalpy change of the theoretical products of combustion of methane in cooling from 1,500°C to 300°C. Express the result in (a) kcal/mol of methane and (b) kcal/Nm^3 of the combustion products

[Answer: (a) 1.138×10^5 kcal/mol (b) 482.9 kcal/Nm^3]

16. Calculate (a) the theoretical and (b) the adiabatic flame temperatures in the combustion of natural gas with 20% excess air. The orsat analysis of the gas is: CH_4 94.0%, C_2H_6 1.2%, C_2H_2 0.7%, C_4H_{10} 0.4%, C_5H_{12} 0.2%, CO_2 0.2% and N_3 3.3%. The air is preheated to 800°C. It has a humidity of 18.8 g moisture per kilogram of dry air.

[Answer: (a) 2,384°C (b) 2,148°C]

17. A coal-fired furnace emits flue gas at the rate of 52,200 m^3/h at 250°C. If the coal analyses as carbon 58.4%, hydrogen 3.3%, nitrogen 1.3%, sulphur 0.4%, oxygen 8.6%, moisture 8.0% and ash 20.0%, and its burning rate is 3 t/h, calculate the orsat analysis of the flue gas.

[Answer:$CO_2$12.7%, O_2 7.1%, N_2 80.2%]

18. A gaseous fuel contains CO_2 3.4%, C_2H_4 3.7%, $C_6H_6$1.5%, O_2 0.3%, CO 17.4%, H_2 36.8%, CH_4 24.9% and N_2 12.0%. It is burnt with air in a furnace. The CO_2 analyser shows 10.0% in the flue gases. Calculate (a) the per cent excess air used and (b) the complete orsat analysis of the flue gases.

[Answer: (a) 37.1% (b) CO_2 10.0%, O_2 5.9%, N_2 84.1%]

19. Burning of carbon monoxide in air at atmospheric pressure gives a gas in which the molal ratio of CO to CO_2 is unity. The process takes place in an insulated chamber, and the

products of combustion exit at a temperature of 2,430°C. Carbon monoxide enters the furnace at 30°C while the air is preheated to the required temperature. Calculate (a) the composition of the exit gas, (b) the per cent theoretical air used and (c) the temperature of the preheated air.

[Answer: (a) CO_2 25.8%, CO 25.8%, N_2 48.4%, (b) 50% (c) 2,282°C]

20. A coal analyses carbon 80% and hydrogen 5%. It is burnt with air to yield flue gases containing CO_2 12.5% and O_2 3.5% (orsat analysis). Some carbon monoxide is suspected in the flue gases. Estimate (a) the complete orsat analysis of the flue gases and (b) the air/coal ratio by weight.]

[Answer: (a) CO_2 12.5%, CO 4.2%, O_2 3.5%, N_2 79.8%, (b) 11.7]

21. The flue gases from a furnace burning a hydrocarbon fuel enter the stack at normal pressure and 200°C and show the following orsat analysis: CO_2 12.0%, CO 1.4%, O_2 3.4% and N_2 83.2%. Estimate (a) the per cent of excess air, (b) the volume of stack gases, expressed as cubic metres per kilogram of carbon burnt, (c) the dew point of the stack gases, (d) the atomic ratio of hydrogen to carbon in fuel and (e) the mole of stack gases per mole of carbon burnt.

[Answer: (a) 13.9% (b) 27.1m^3/kg (c) 47.5°C (d) 1.80 (e) 8.36 mol/mol]

Bibliography

1. Hougen OA and Watson KM. Material and Energy Balance, *Chemical Process Principles,* Vol. 1, New York: Wiley,1959.
2. Hougen OA, Watson KM and Ragatz RA. Thermodynamics, *Chemical Process Principles,* Vol. 2, New York: Wiley,1959.
3. Lewis B and Von Elbe G. *Combustion, Flames and Explosions of Gases*, Third edition, Orlando: Academic Press, 1987.
4. Lewis WK, Radasch AH and Lewis HC. *Industrial Stoichiometry*, Second edition. New York: McGraw-Hill, 1954.
5. Rose JW and Copper JR. *Technical Data on Fuel*, Seventh edition, World Energy Conference, London, 1976.
6. Trinks W and Mawhinney MH. *Industrial Furnaces,* Fifth edition, New York: Wiley, 1961.

8 Combustion Process
(Kinetics)

8.1 NATURE OF COMBUSTION PROCESS

The potential energy stored in fuels is released by combustion and made available in the form of heat and power. The process is essentially chemical in nature and involves the reaction of fuel components with oxygen. Heat of reaction is evolved and the temperature of the products is raised. The latter are in the gaseous and vaporous form and tend to occupy a large volume at the raised temperature attained, or generate high pressure if the volume is not allowed to expand. The heat can be directly used for domestic and industrial purposes, or it can raise steam to

1. Generate electricity (steam turbines),
2. Heat,
3. Supply mechanical power (steam engines).

The energy of the hot combustion products can also be directly used in gas turbines to do mechanical work or generate electricity.

There are two methods by which the combustion of fuels can generate electricity without using an alternator: by using fuel cells or a magneto-hydrodynamic (MHD) generator. In a fuel cell, which is essentially an electrochemical cell, the free energy of the combustion reaction is directly converted into electrical energy at a theoretical efficiency of 100%. In a MHD power generator, an exceedingly high temperature of combustion products is realised between 2,000°C and 3,000°C, at which temperature the latter are ionised and the charge of the ions is used to produce electrical power. The heat remaining in the combustion products is utilised in further generating electricity through the steam cycle and in preheating the air for combustion. The total efficiency of such a plant is expected to be 55% as against only 30% of a thermal power station. Vigorous work is going on in developing the fields of fuel cells and MHD generators to make them commercially feasible.

All applied combustion processes are characterised by the formation of a *flame*. The term 'flame' refers to the heat- and light-emitting gaseous zone which separates the combustion products from the fuel that is undergoing combustion. A flame appears just after the fuel is ignited and a sufficiently high temperature is created which tends to persist slightly beyond the burning zone till a drop in temperature renders the gases completely non-luminous. A flame thus covers luminous burning and after-burning zones and is confined between the ignition zone and the completely non-luminous combustion gases. A non-luminous Bunsen flame is actually a flame of low luminosity. The *flame front* is the surface between the luminous region and the unburnt fuel. A well-defined flame front is obtainable with gaseous fuels. Flame thickness or flame length depends upon many factors, namely, types of fuel, presence of diluent (nitrogen in air, or recirculated flue gases), intensity of mixing of fuel and oxidant. A fast-burning combustion and luminescence occur almost simultaneously. Since a flame separates the fuel from the combustion products, it may be significantly defined as the zone of very steep concentration and temperature gradients. It is this characteristic which distinguishes flames from chemical reactions in general.

Because of the luminosity of a flame, some fuels like kerosene are used as illuminants. Certain devices such as gas mantle are used to augment the flame luminosity.

8.2 TYPES OF COMBUSTION PROCESSES

There are five general types of combustion processes depending upon the means of contacting fuel and oxidant and the conditions of the flame. These are briefly described below.

8.2.1 Combustion with Stationary Flames

This is the usual combustion practiced in ovens and furnaces. The resultant flame front is more or less stationary in space. A stationary flame may be further sub-divided as *premixed* and *diffusion*. In the former case fuel and oxidant are premixed before they enter the burning zone. When they are separately delivered to the burning zone the flame is called a diffusion flame. In actual practice a part of the total air may be premixed with the fuel and the remaining may be directly supplied to the combustion area. The premixed air is known as *primary air* and the rest is called *secondary air*. While solid, liquid and gaseous fuels can all give a premixed flame, a true diffusion flame is obtained only with gaseous fuels.

A stationary flame may be laminar or turbulent. The velocity of a premixed system is responsible for its flame being laminar or turbulent. In the case of a diffusion flame, both the direction of flow and the velocity of fuel and oxidant determine the nature of the flame. A laminar diffusion flame is favoured by low and equal velocity and parallel flow of gas and air. When they flow at an angle to one another with higher and unequal velocity, a strongly turbulent diffusion flame results.

Generally speaking, diffusion flames are considerably larger than premixed flames, the other conditions of burning remaining similar. Again laminar flames are longer than turbulent flames. A highly turbulent flame is unsteady and irregular in shape. When the air hole of a Bunsen burner is completely closed, a laminar diffusion flame is obtained. It is long, steady, quiet, noiseless, well-shaped and luminous. If the big air-hole is fully opened, a turbulent premixed flame is realised. It is short, unsteady, flickering, noisy, irregular-shaped and of low luminosity.

The main advantage of the diffusion flame over the premixed one lies in the fact that one or both of the fuel and air can be preheated before combustion. On the other hand, when the aim in usual thermal appliances is an intense combustion in a comparatively short space, a truly laminar, diffusion flame is rarely used. However, long flames are desired in a coke oven and open hearth furnace and these are two typical examples where it is desired to have a long flame, so that an extended zone of uniform temperature is obtained.

8.2.2 Surface Combustion or Flameless Combustion

At high temperatures all refractory solid surfaces have the property of accelerating the rate of combustion of fuel gases and air. Some solids, for example, platinum, can accelerate the process even at low temperatures. Combustion with a stationary flame is limited by, among other factors, a range of velocity and concentration conditions of the gas and air. Stable combustion is possible even outside this range if the reaction is allowed to proceed in contact with solid surfaces. This phenomenon is known as *surface combustion*. Its industrial application is to achieve rapid combustion of a large quantity of fuel in a comparatively small space with the production of high temperature and enhanced heat transfer rate. A tunnel burner is a type of gas burner operating on the surface combustion principle (see Chapter 9). Well-defined normal flames are absent in surface combustion.

Submerged combustion

This is a special case of the surface combustion process. The burner is partly or fully submerged in a liquid and the hot combustion gases are allowed to bubble through it in an agitated condition. A high rate of heat transfer results from the direct contact of hot gases with the liquid and leads to a high evaporation rate. Up to 95% of the potential heat of the fuel may be the useful heat for the process. Submerged combustion finds commercial application particularly in the evaporation of severely scaling liquids and highly corrosive solutions.

8.2.3 Combustion with Explosion Flame

This occurs in a homogeneous mixture of fuel and air and is characterised by the flame front progressing rapidly through the mixture. The process may be either under constant pressure, for example, mine explosion, or constant volume, for example, combustion in a gasoline engine. Detonation is a special type of explosion where the extremely high reaction rate generates ultra-high velocity pressure waves of 1–4 km/s and an abnormal rate of pressure rise.

Pulsating combustion

When one end of a long tube is open and the other closed by a non-return valve, and the air and fuel are introduced at the closed end, pulsating combustion takes place. On ignition, the pressure in the system tends to rise sharply at near-constant volume; the rise in pressure prevents the flow of the air and fuel momentarily. When the exhaust gases leave the tube through the open end, a fresh supply of air and fuel arrives at the hot zone and combustion is repeated in the form of pulsations. The frequency of these pulsations corresponds to the resonant frequency of the combustion unit. There is no stationary flame in the system. Pulsating combustion is essentially one specific type of explosion flame. During the operation of a pulsating combustion unit, a standing pressure wave is produced in the pipe with a pressure node (velocity antinode) at the exit end, and a pressure antinode (velocity node) at, or near, the combustion zone. The main interest in pulsating combustion systems is in obtaining higher rates of combustion and heat transfer than in a conventional burner. The noise levels of the pulsating combustors can be lowered by adopting suitable silencers.

8.2.4 Slow Combustion

This is the general term given to the combustion processes taking place at sub-flame temperatures below 400°C, at slow and conveniently determinable rates. Slow combustion of higher hydrocarbons has been useful in determining chain-reaction rates. In a premixed system of fuel vapour and air, slow combustion proceeds simultaneously at a number of points in the whole system. No reaction zone or flame-front is visible. This process is called *homogeneous combustion*. It is often characterised by the appearance of *cool flames* in succession. These flames emit small quantities of heat and pale bluish light usually seen only in the dark. Several cool flames—up to four—may follow one another. Cool flames are a feature of gas phase oxidation of organic substances below 400°C. Slow combustion has no direct application in industry. It is indirectly useful in studying the combustion mechanism

8.2.5 Combustion of Solid Fuels on Grate

The process of burning solid fuels in suspension, for example, pulverized fuel firing or cyclone firing, can more or less fit into one or more of the combustion processes described above. However, the combustion of lumps on grate is in a class by itself. An industrial grate essentially consists of assemblies of firebars mounted parallel to one another. *Primary air* is passed through the gap between the neighbouring bars. A glowing fuel bed with an ash layer underneath is maintained on the grate and fresh charge of fuel is fed into it from time to time. The rate of feeding corresponds to the combustion rate. The insulating ash layer serves to protect the firebars from intense heat and also helps in the distribution of air.

The fresh charge on heating undergoes drying and decomposition. The combustible volatiles are burnt in the *secondary combustion* space above the fuel bed. An additional supply of air—*secondary air*—is made for this purpose. This combustion more or less corresponds to the stationary flame process. The residual solid fuel undergoes true primary combustion on grate. It is a surface process involving heterogeneous reactions between carbon and oxygen. The liberated heat converts the fuel into a glowing mass which radiates heat to the heating surface and also to the fresh charge of fuel. Carbon monoxide is a product of combustion on grate and takes part in the secondary combustion.

The above method of burning solid fuels is known as *overfeed firing*. In *underfeed firing*, fresh fuel is fed to the glowing bed from below. In *crossfeed firing*, the fuel is fed transversely across the front of the grate.

In down-jet combustion, the grate is dispensed with. A jet of air is impinged on the surface of the bed of fuel. The sides and bottom of the bed are enclosed. The combustion products leave the bed through the upper surface. Fresh fuel is fed from the top while the refuse is periodically removed from the bottom. The principle of down-jet combustion has not met with much success in practice. Difficulties in ash removal appear to be the main hurdle.

8.3 MECHANISM OF COMBUSTION REACTIONS

The most studied combustion reactions are those involving homogeneous combustion of gases and vapours. Here again the reaction of the hydrogen–oxygen system has received by far the greatest attention of researchers in the field. The other systems include carbon monoxide–oxygen, methane–oxygen and higher hydrocarbons–oxygen. Reaction mechanisms in high temperature combustion processes differ from homogeneous combustion in many respects.

Low boiling liquid fuels are burnt in the vapour state. High boiling liquids undergo cracking on heating and their combustion is essentially a combination of two processes: combustion of cracked vapour and combustion of solid carbonaceous material.

Solid fuels also involve two stages; first, the combustion of inflammable gases and vapours emitted by its heating and oxidation, and secondly, the combustion of residual coke or char. The solid carbonaceous material from liquid fuels and the coke and char from solid fuels are mainly composed of carbon. The mechanism of the carbon–oxygen reaction which has also been studied in detail is therefore significant in understanding the combustion of solid and liquid fuels.

The more clearly defined combustion reactions are discussed below with an initial reference to the characteristics of chain reactions.

8.3.1 Chain Reactions

The most remarkable feature of rapid combustion and explosion is that, under certain conditions, the reaction of an almost iso-thermal system can switch over from one of a slow and finite rate to an infinitely fast and uncontrollable violent reaction. For usual chemical reactions under isothermal conditions, the rate depends on the concentration of the reactants. Since the latter fall with time, the rate also should fall with the progress of the reaction under isothermal conditions. Herein lies the basic distinction of combustion reactions.

Combustion reactions of gases and vapours are all chain reactions. Any chain reaction of finite rate has three consecutive phases: the *initiation* phase, the *propagation* phase and the *termination* phase. *Chain carriers* are formed in the first phase by thermal or other means. These may be atoms such as H or O, free radicals such as OH, CHO or CH, or even molecules such as formaldehyde. The chain carriers react with the molecules of the reactants and produce new chain carriers together with the molecules of products. The newly formed chain carriers further participate in reactions and thereby promote the propagation phase. The process is terminated by chain-breaking reactions in which the chain carriers are eliminated from the reaction zone by chemical reactions or adsorption on another substance, for example, on a container surface. When the intermediate reactions in the propagation phase produce several new chain carriers in one step, each of the new carriers can initiate a new series of reactions and thus the process is branched out in several directions. This is known as a *chain-branching* reaction. If the carriers are not terminated as fast as they are produced, the result is the progress of the overall process at a violent rate till all the reactants are exhausted. Rapid combustion and explosion are brought about by branched-chain reactions.

$$\text{Rate} = \frac{F(c)}{f_s + f_g + A(1-\alpha)} \tag{8.1}$$

where,

$F(c)$ = a function depending on the rate of initial production of chain carriers

f_s and f_g = factors depending on the rates of destruction of the carriers at the surface and in the gas, respectively

α = branching factor, that is the number of chain carriers produced from one initial carrier

A = a constant or a function of concentration

The condition for overall branching is

$\alpha > 1$, that is $1 - \alpha$ = a negative quantity.

If $A(\alpha - 1)$ equals $(f_s + f_g)$, the reaction rate becomes infinite and brings about an explosion.

The terms f_s and f_g are pressure dependent in the opposite way: while f_s increases by a decrease in pressure, the term $(f_s + f_g)$ will pass through a maximum. Under certain conditions of diminution of either f_s (pressure rise) or f_g (pressure fall), the term $f_s + f_g$ may become lower than $A(I - \alpha)$. This suggests that the branched-chain reactions will be terminated and explosion made impossible at both lower and higher pressures than a certain limiting value. An explosion may again occur at comparatively high pressures, particularly at increased temperatures by the formation of new chain carriers, by other reactions and also by thermal effect.

Besides the branched-chain mechanism of explosion, thermal mechanism is also recognised in the study of explosion reactions. Heat liberated by slow combustion may raise the temperatures of the system, whereupon the reaction rate is accelerated. If the heat generated exceeds the heat lost to the surroundings, the reaction rate may steadily increase to a final explosive velocity. Theoretically, a branched-chain explosion can take place isothermally. A clear-cut distinction of the two mechanisms is not practically possible.

Combustion is often characterised by an *induction period*, or *delay period*, or *ignition lag*, or *time lag*, during which the concentration of chain carriers builds up. The reaction rate increases exponentially during this period and ends up in either a steady reaction or an explosion, depending upon the prevailing conditions.

8.3.2 Thermal Mechanism

Non-isothermal systems can be accelerated to an explosive violence by a thermal mechanism under favourable conditions. The heat liberated by exothermic reactions is partly lost to the surroundings and the rest raises the temperature of the system. At the higher temperature, the rate is accelerated if the opposite effect of the decreased concentration of the reactants is of lesser significance. This is possible when the reaction is highly exothermic and the heat loss to the surroundings is much less than the heat generated. Under adiabatic, or almost adiabatic conditions, the system may attain a certain temperature (ignition temperature) when the rate of reaction approaches infinity and the system explodes. The intital period during which the reaction rate slowly rises with time, and after which the rate is suddenly accelerated, is the induction period of the system. A simplified mathematical treatment of the thermal mechanism is given later in the chapter in Section 8.4.

The thermal mechanism does not exclude the chain mechanism. An increase in temperature also increases the number of chain carriers in the system. Thus a thermal mechanism can be complemented by a chain mechanism. It is also possible that a system is initially of a chain type and with an increase in temperature gets converted into the thermal type. It is futile to attempt to explain combustion and explosion by purely thermal or purely chain mechanisms.

8.3.3 Hydrogen–Oxygen Reaction

Active chain carriers may be formed by the following reactions which may result from intermolecular collision, thermal dissociation or excitation by electric spark.

$$H_2 \rightarrow 2H$$
$$O_2 \rightarrow 2O$$
$$H_2O \rightarrow OH + H$$

These also may be diffused into the unburned gaseous mixture from an adjacent flame.

It is generally agreed that the low-pressure explosion in the H_2–O_2 system is due to the following chain-propagating reactions

$$OH + H_2 \rightarrow H_2O + H \tag{8.2}$$
$$H + O_2 \rightarrow OH + O \tag{8.3}$$
$$O + H_2 \rightarrow OH + H \tag{8.4}$$

The last two reactions are of the chain-branching type where each carrier generates two new carriers. Reaction (8.2) develops one carrier for each molecule of H_2O produced. The new carriers formed can start their own new chains. If they are not removed as fast as they are produced the system ends up with an explosion. Chemical calculations show that as little as 1% branching in the H_2–O_2 reaction would bring about an explosion.

At low pressures, chain carriers are removed by adsorption on the container surface. There is a critical pressure below which chain-termination is predominant and no matter what is its temperature the system does not explode. This is known as the *first explosion limit* of the H_2–O_2 system. The nature of the surface influences the adsorption process, hence the first explosion limit depends upon the material of the vessel. Salts such as iodides of cesium and potassium are powerful chain-breakers and a vessel coated with one of these salts can raise the first limit by a factor of 10 or even 100. On the other hand addition of inert gases, or increase in the vessel diameter, hinders the adsorption on the surface and thus gives lower values of the first pressure limit.

At pressures above the first limit the chains may be terminated by a termolecular reaction in the gas phase

$$H + O_2 + M \rightarrow HO_2 + M \qquad (8.5)$$

where, M represents a molecule present in the system. This molecule takes away the excess energy produced in the formation of the unstable HO_2 whose heat of formation exceeds its heat of decomposition. There is a critical pressure called *second explosion limit* above which the reaction (8.5) is quite predominant; the chain carriers are fast removed from the system and explosion gives way to slow combustion. HO_2 is destroyed on the surface by the reactions.

$$2HO_2 \xrightarrow{\text{surface}} H_2O_2 + O_2 \qquad (8.6)$$

$$H_2O_2 \xrightarrow{\text{surface}} H_2O + \tfrac{1}{2} O_2 \qquad (8.7)$$

The nature of the surface has markedly less effect on the second limit than on the first limit. The second limit is almost independent of the vessel diameter. These indicate the validity of the chain-breaking reaction (8.5) taking place in the gas phase.

At still higher pressures, and particularly with increased temperatures, another set of chain-branching reactions occur. The process is accelerated until a *third explosion limit* is reached.

$$HO_2 + H_2 \rightarrow H_2O_2 + H \qquad (8.8)$$

$$HO_2 + H_2O_2 \rightarrow H_2O + O_2 + OH \qquad (8.9)$$

$$H + O_2 + H_2O_2 \rightarrow H_2O + O_2 + OH \qquad (8.10)$$

Explosion in this region of pressure is due to a combustion of chain and thermal mechanisms. The heat given out by the chain reactions cannot be dissipated sufficiently fast, so that self-heating occurs, followed by an explosion.

The three explosion limits, P_1, P_2 and P_3 are schematically shown in Fig. 8.1 on the plots of pressure versus reaction rate and temperature versus pressure.

Slow combustion takes place below P_1 and between P_2 and P_3, while the system explodes between P_1 and P_2 and above P_3. P_1 falls slightly with temperature while P_2 rises with it. The effect of temperature on P_3 is remarkable, the limiting pressure falling sharply with a temperature rise. All these observations are well explained by the mechanism described above.

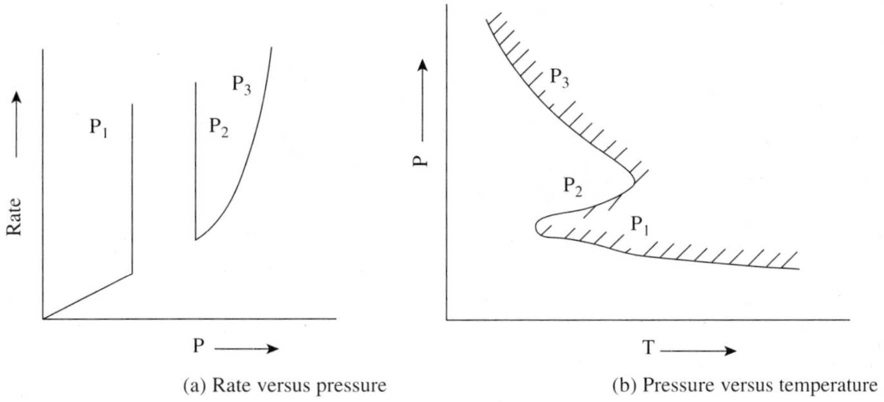

Figure 8.1 Explosion limits for H_2–O_2 reaction, P_1, P_2 and P_3

The chain mechanism is predominant in the process below the second limit. A combination of chain and thermal mechanism is operative above it.

Depending upon various conditions of the system the values of the three explosion limits are: P_1 0.5 to 9.0 mm, P_2 18 to 250 mm and P_3 240 to 699 mm Hg. Above atmospheric pressure the system is always explosive. Figure 8.1b shows the so-called *explosion peninsula*, the peninsular shape of the pressure temperature diagram. The system is explosive in the region of the right-hand side of the curve.

8.3.4 Carbon Monoxide–Oxygen Reaction

The low-pressure ignition of carbon monoxide and oxygen is characterised by an explosion peninsula (Fig. 8.2). There are two explosion limits. No third limit has been reported. A pale blue glow is often observed outside the explosion peninsula. There are two limits of glow, marking what may be called a glow peninsula.

The mechanism of CO–O_2 reaction has been the subject of great controversy. According to Lewis and von Elbe, the chain propagating reactions are

$$O_2 + O + M \rightarrow O_3 + M \tag{8.11}$$

$$CO + O_3 \rightarrow CO_2 + 2O \tag{8.12}$$

The oxygen atom and ozone molelcule serve as chain carriers. Reaction (8.12) is chain-branching.

According to Semenov, the chain-branching is due to

$$CO + T_o \rightarrow T^*_{CO_2}$$

$$T^*_{CO_2} + O_2 \rightarrow CO_2 + 2T_o$$

(* = excited state; T_o = triplet state)

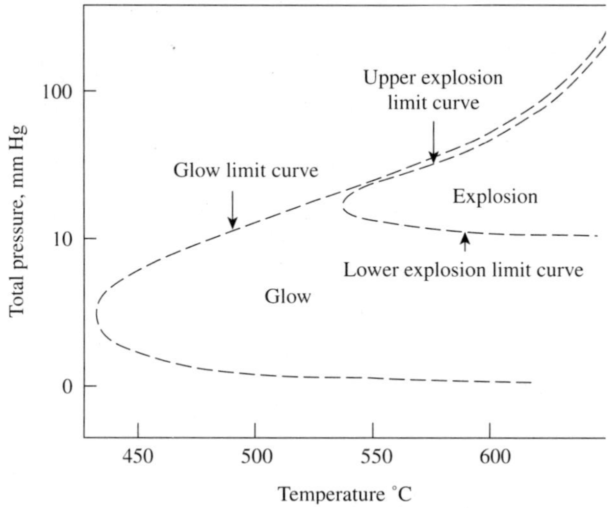

Figure 8.2 Glow and explosion limits for $CO–O_2$ system

At very low pressures, chain carriers are adsorbed by the container surface. When carrier concentration is built up with rising pressure, a glow is observed. This is the result of rapid oxidation and formation of CO_2 molecules in the gas phase. Explosion is inhibited by products. With further rise in pressure the heat generated exceeds the heat lost to the wall and explosion takes place. The lower glow and explosion take place. The lower glow and explosion limits do not change much with temperature. This may be due to the small activation energies of both branching and termination reactions.

The upper glow and explosion limits virtually coincide. The most likely chain-breaking reaction is

$$CO + O + M \rightarrow CO_2 + M \tag{8.13}$$

The other reactions are:

$$T^*_{CO_2} + CO \rightarrow CO_2 + T^*_{CO} \tag{8.14}$$

$$T^*_{CO} + O_2 \rightarrow T^*_{CO_3} + h\gamma \tag{8.15}$$

$$T^*_{CO_3} \rightarrow CO_2 + T_O + h\gamma \tag{8.16}$$

$$T^*_{CO_3} \rightarrow CO + T_{O_2} + h\gamma \tag{8.17}$$

($h\gamma$ = radiation emitted)

and also
$$CO + O_3 + M \rightarrow CO_2 + O_2 + M \tag{8.18}$$

Minute traces of water have a powerful catalytic effect on the combustion of carbon monoxide*. Several times as much electrical energy is required to ignite a fully dry $CO–O_2$ mixture as is necessary when traces of water vapour are present. Water lowers the ignition temperature and

*In fact $CO–O_2$ mixture without any trace of hydrogeneous impurity may be virtually non-explosive.

raises the upper explosion limit. The effect is remarkable with only traces of water vapour. Further addition of water has little effect and sometimes even leads to a negative effect.

In fact $CO–O_2$ mixture without any trace of hydrogenous impurity may be virtually non-explosive. The mechanism of a wet $CO–CO_2$ reaction may involve the following additional chain-branching reactions.

$$H + O_2 \rightarrow OH + O \tag{8.19}$$

$$O + H_2 \rightarrow OH + H \tag{8.20}$$

$$CO + OH \rightarrow CO_2 + H \tag{8.21}$$

$$CO + O + H_2O \rightarrow CO_2 + H + OH \tag{8.22}$$

H may be eliminated by

$$H \xrightarrow{\text{wall}} \text{inert} \tag{8.23}$$

$$H + O_2 + M \longrightarrow HO_2 + M \tag{8.24}$$

$$H + CO + M \longrightarrow HCO + M \tag{8.25}$$

$$HCO + O_2 \longrightarrow CO + HO_2 \text{ (very fast)} \tag{8.26}$$

HO_2 may react as

$$H_2 + HO_2 \longrightarrow H_2O_2 + H \tag{8.27}$$

$$CO + HO_2 \longrightarrow CO_2 + OH \text{ (very fast)} \tag{8.28}$$

$$HO_2 \xrightarrow{\text{wall}} \text{inert} \tag{8.29}$$

The upper limit markedly rises with increasing temperature. This is partly explained by the reaction energies.

Some values of the two limits are: P_1 2.5 to 12.5 mm and P_2 40 to 500 mm Hg.

8.3.5 Hydrocarbons–Oxygen Reaction

The pressure–temperature diagram (Fig. 8.3) for a 1:1 propane–oxygen mixture shows some of the characteristics of the reaction of hydrocarbons with oxygen in the gas phase. The system ignites in the right-hand zone of the thick line, while the left-hand zone is generally for slow oxidation. However, in a certain boundary region the system shows faint luminosity, cool flame, increased luminosity, bright blue flame and finally the characteristic yellow flame of true ignition—all in succession on either increasing pressure at a constant temperature or increasing temperature at a constant pressure.

The oxidation of hydrocarbons involves branched-chain reactions. Long induction periods (about a minute), slow acceleration and cool flames indicate that the branching reaction is infrequent and markedly differs from those in the $H_2–O_2$ and $CO–O_2$ systems. The situation is known as degenerate or delayed branching. In the $H_2–O_2$ and $CO–O_2$ systems the branching is due to reactions of atoms and radicals with fuel or oxygen molecules. In the hydrocarbon–O_2 systems

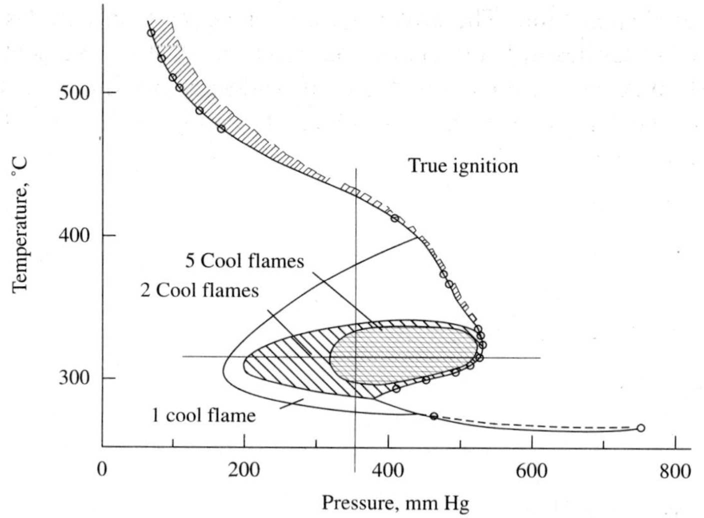

Figure 8.3 Pressure–temperature diagram of propane–oxygen (1:1) mixture

a relatively stable intermediate product, M, is first formed by a non-branching process. This undergoes further reactions to give either inert products or new radicals to initiate the primary chain and produce more carriers.

If RH denotes hydrocarbons, the following give the general mechanism of the oxidation process.

Formation of R radicals

$$RH + O_2 \rightarrow R^{\bullet} + HO_2 \quad (8.30)$$

$$^{\bullet}OH + RH \rightarrow R^{\bullet} + H_2O \quad (8.31)$$

$$HO_2^{\bullet} + RH \rightarrow R^{\bullet} + H_2O_2 \quad (8.32)$$

Formation of peroxide radical RO_2

$$R^{\bullet} + O_2 \rightarrow RO_2^{\bullet} \quad (8.33)$$

Reactions of RO_2 radicals

$$RO_2^{\bullet} \rightarrow R'CHO + {}^{\bullet}OH \quad (8.34)$$

$$RO_2^{\bullet} \rightarrow R''CHO + CH_3O^{\bullet} \quad (8.35)$$

aldehyde (methoxyl radical)

($R' = R$ minus one CH_2; $R'' = R$ minus two CH_2)

$$\begin{matrix} R_1 \diagdown \\ \quad CO_2 \\ R_2 \diagup \end{matrix} \rightarrow R_3\text{–}CO\text{–}R_4 + R_2O^{\bullet} \quad (8.36)$$

ketone

$(R_1 = R_3 + R_4)$

$$R'CHO \rightarrow \underset{\text{carbonyl radical}}{R'CO^{\bullet}} + H_2O \tag{8.37}$$

$$RCO^{\bullet} + O_2 \rightarrow \underset{\text{peracid radical}}{R'CO_3} \tag{8.38}$$

$$R'CO_3^{\bullet} + R'CHO \rightarrow \underset{\text{peracid}}{R'CO_3H} + R'CO^{\bullet} \tag{8.39}$$

$$RO_2 \xrightarrow{\text{isomerisation}} \underset{\text{peroxide}}{{}^{\bullet}R_5OOH} \tag{8.40}$$

$(R_5 = R \text{ minus } H)$

$$^{\bullet}R_5OOH \rightarrow \underset{\text{olefin}}{R_5''CH = CH_2} + HO_2^{\bullet} \tag{8.41}$$

$$^{\bullet}R_5OOH + CH_3 \rightarrow R_5O_2^{\bullet} + CH_4 \tag{8.42}$$

$$\underset{\text{peroxide}}{^{\bullet}R_5OOH} \rightarrow \underset{\text{alkoxy radical}}{R_5O^{\bullet}} + {}^{\bullet}OH \tag{8.43}$$

$$R_5O^{\bullet} \rightarrow \underset{\text{alkyl radical}}{R_5^{\bullet}} + \underset{\text{formaldehyde}}{CH_2O} \tag{8.44}$$

$$R_5O^{\bullet} \rightarrow R_5''^{\bullet} + \underset{\text{acetaldehyde}}{CH_3CHO} \tag{8.45}$$

Various intermediate products and radicals are formed from the initial peroxide radical.
Formation of CO and CO_2

$$R'CHO \xrightarrow{-H} R'CO^{\bullet} \begin{cases} \rightarrow R' + CO \\ \xrightarrow{O_2} CO_2 + \text{other products} \end{cases} \tag{8.46}$$

Termination reaction

$$HO_2 \xrightarrow{\text{wall}} \text{inerts} \tag{8.47}$$

$$RO_2 \xrightarrow[\text{phase}]{\text{gas}} \text{inerts} \tag{8.48}$$

In the case of oxidation of methane, the simplest scheme satisfying the experimental facts is

$$CH_4 + O_2 \rightarrow CH_3^{\bullet} + HO_2^{\bullet} \tag{8.49}$$

$$CH_3^{\bullet} + O_2 \rightarrow \underset{\text{formal-dehyde}}{CH_2O} + CH^{\bullet} \tag{8.50}$$

$$HO_2^{\bullet} + CH_4 \rightarrow \underset{\text{hydrogen peroxide}}{H_2O_2} + CH_3^{\bullet} \tag{8.51}$$

$$^{\bullet}OH + CH_4 \rightarrow CH_3^{\bullet} + H_2O \tag{8.52}$$

$$CH_2O + O_2 \rightarrow CHO^{\bullet} + HO_2^{\bullet} \tag{8.53}$$

$$HO_2^{\bullet} + CH_2O \rightarrow CHO^{\bullet} + H_2O_2 \tag{8.54}$$

$$^{\bullet}OH + CH_2O \rightarrow CHO^{\bullet} + H_2O \tag{8.55}$$

$$CHO^{\bullet} + O_2 \rightarrow CO + HO_2^{\bullet} \tag{8.56}$$

$$HO_2^{\bullet} \xrightarrow{\text{wall}} \text{inert} \tag{8.57}$$

$$H_2O_2 \rightarrow H_2O + O_2 \tag{8.58}$$

Formaldehyde is mainly responsible for the degenerate branching at least in the low temperature, 400 to 450°C oxidation.

In the degenerate branching the number of chains is so low that the reaction rate of a truly isothermal system cannot be very high. Thermal factors are involved in the production of cool flames and ignition. When the heat generated far exceeds the heat lost to the wall, the conditions favour cool flame and ignition. This explains why cool flames always originate near the centre of the vessel and spread outwards.

Ozone, peroxides, nitrogen dioxide, organic nitrates and nickel carbonyl vapour are promoters of ignition of hydrocarbons. Halogen compounds, such as carbon tetrachloride ($CC1_4$) and bromochloromethane ($BrClCH_2$) are the retarders.

The phenomenon of 'knock' in internal combustion engines is connected with the chain mechanism of hydrocarbon oxidation. Many substances enhance the knocking tendency of fuels, for example, ozone, organic nitrates and nitrites, alkyl peroxides and aldehydes, particularly formaldehyde. Their pro-knock effect is due to their property to initiate chain reactions in the process. The anti-knock compounds on the other hand, for example, tetraethyl lead and iron carbonyl are capable of terminating the chains. It is well known that the knock-resistance property of fuels directly varies with the relative rates of their slow combustion.

A knowledge of the reaction mechanism and the effect of various factors including the container wall permits the prediction of conditions for an explosion of a reactive mixture of gases in a glass vessel. A different situation is encountered when the same gas mixture approaches a flame front. The mixture is first warmed by conduction and convection from the hot flame gases. It also mixes with new molecules and radicals which diffuse from the flame towards the cold gas. Thus some reactions will be initiated in the pre-flame region and further accelerated with rise in temperature and formation of chain-branching radicals. The bulk of the overall reaction will take place at the high temperatures in the flame in a very short time. The gaseous products will finally move away from the flame and the excess radicals will recombine comparatively slowly with the fall in temperature. The entire process is complicated by the simultaneous operation of reaction kinetics, transport phenomena and laws of conservation.

Pyrolysis is an important pre-flame process in the combustion of hydrocarbons. Except in pyrolysis of several aliphatic compounds at low temperatures, free radicals play a part in all other pyrolysis processes. The pyrolyses of methane and acetylene are particularly important, for methane is a major fuel component while acetylene is formed in high yields by the decomposition of other hydrocarbons above 1,000°C. In fact acetylene is the essential precursor to the formation of carbon particles in flames.

Flames are characterised by the presence of ions and such radicals as CH, C_2 and C_n. None of these is found in slow combustion experiments.

8.3.6 Combustion of Elementary Carbon

The primary reaction of solid carbon with oxygen is a heterogeneous reaction in which a complex of indefinite composition is formed at the surface of carbon by the process of chemisorption. The surface complex then breaks up into CO and CO_2. In the presence of free oxygen, CO then undergoes homogeneous oxidation into CO_2.

$$xC + yO_2 \rightarrow \underset{\text{(Surface complex)}}{C_xO_{2y}} \rightarrow mCO + nCO_2 \tag{8.59}$$

$$CO + \tfrac{1}{2}O_2 \rightarrow CO_2 \tag{8.60}$$

The CO_2 formed can react with hot carbon to produce CO by the well-known heterogeneous reaction called the *Boudouard reaction*

$$CO_2 + C \rightarrow 2CO \tag{8.61}$$

Depending upon the concentration of oxygen in the system, the product will be rich in CO_2 or CO. It has also been shown that certain chlorinated compounds, for example, $POCl_3$ and CCl_4, which are well-known chain breakers, inhibit the oxidation of CO while water vapour catalyses it. Therefore, an addition of these chlorinated compounds results in the product gas being rich in CO. An addition of excess of water vapour can resume the oxidation of CO and lead to an increase in CO_2 concentration.

The reaction mechanism essentially remains the same whether the carbon is in the form of a bed of lumps or a suspension of powder in air. Since, however, heterogeneous reactions are involved, diffusion plays a significant role in the overall kinetics of the process.

8.4 SPONTANEOUS IGNITION TEMPERATURE (SIT)

The minimum temperature to which a combustible system is to be raised for bringing about rapid combustion or explosion is known as its spontaneous ignition temperature, or simply ignition temperature. It depends upon the inherent property of the system and also upon extraneous factors. By keeping the latter constant, the ignition temperature is obtained as a characteristic property of the combustible. An expression for SIT of a gas or vapour is derived below on the basis of the thermal mechanism of combustion.

The rate of a chemical reaction in gaseous phase is:

$$K = \frac{dn}{dt} = f(p) \times e^{-E/RT} \tag{8.62}$$

where,

n = mole number of product per unit volume
t = reaction time
f(p) = a function of pressure
E = energy of activation
R = gas constant
T = temperature of the system in absolute scale.

The rate of heat liberation by a chemical reaction is

$$q_I = QKV = QVf(p)\, e^{-E/RT} \tag{8.63}$$

where,

Q = heat liberated per mole of product

V = volume of reaction vessel

The rate of heat loss to the vessel wall is

$$q_{II} = B\,(T - T_0) \tag{8.64}$$

where,

B = a constant = heat transfer coefficient × area
T_0 = temperature of vessel wall, K.

Equations [8.63] and [8.64] show that the rate of heat liberation increases exponentially with temperature while the rate of heat loss does so in a linear relationship. Figure 8.4 shows the variation of q_I and q_{II} with temperature. While there is one curve for q_I for the given combustible, there is a set of several parallel lines for q_{II} depending on the vessel wall temperatures T_0, T_0', T_0'' and so on. The lines at T_0' and T_0'' cut the curve at two points. The lower point 'a' indicates a stable equilibrium. It is easily seen that any shifting of the system to either side of the point, will bring it back to this point. The upper point 'b' represents an unstable equilibrium practically impossible to achieve. The line at T_0 touches the curve tangentially at c where the system has unilateral stability. If the system is shifted slightly to a lower temperature, it comes back to c. On the other hand, any shifting towards a higher temperature will move the system further away from it. The point 'c' corresponds to the ignition temperature, T_i.

The two conditions for c are

1. Thermal equilibrium, $q_I = q_{II}$
2. q_{II} is tangential to q_I

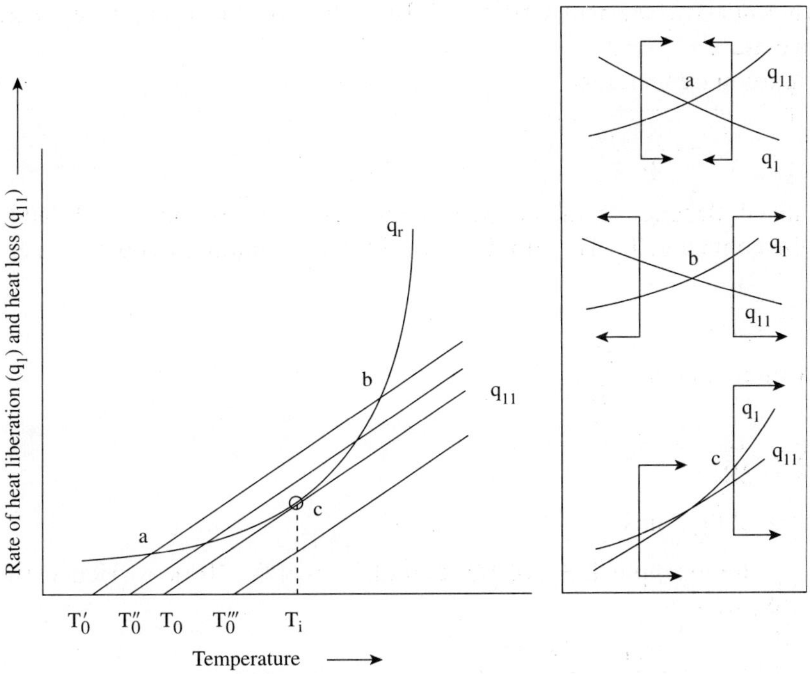

Figure 8.4 Thermal mechanism of ignition

that is,

$$\frac{d_{q_1}}{dT} = \frac{d_{q_{11}}}{dT}$$

Therefore,

$$QVf(p)e^{-E/RT} = B\,(T - T_0) \tag{8.65}$$

and also

$$\frac{Vf(p)e^{-E/RT}}{RT^2} \times E = B \tag{8.66}$$

From these two equations (8.65) and (8.66):

$$\frac{E}{RT^2}(T - T_0) = 1$$

$$\text{or } T^2 - \frac{ET}{R} + \frac{ET_0}{R} = 0$$

Therefore,

$$T = \frac{E}{2R}\left[1 \pm \sqrt{1 - \frac{4RT_0}{E}}\right]$$

This gives two values of the temperature, T. By definition the ignition temperature is the minimum temperature.

Therefore, ignition temperature

$$T_i = \frac{E}{2R}\left[1 - \sqrt{1 - \frac{4RT_0}{E}}\right] \tag{8.67}$$

It is thus seen that T_i depends on the inherent properties of the combustible, E, and also on the conditions of the container, T_0. The limiting values of the container temperature are

$$0 \le T_0 \le \frac{E}{4R} \tag{8.68}$$

If T_0 is maximum, that is, $\frac{E}{4R}$,

$$(T_i)\max = \frac{E}{2R}$$

that is, $(T_i)\max = 2\,(T_0)\max$ (8.69)

Usually, $E \gg T_0$, (for example, E + 50,000 cal/mol, $T_0 = 500 - 1000$ K) Hence, by expanding the series in (8.68) we get

$$T_i = \frac{E}{2R}\left[1 - \left(1 - \frac{2RT_0}{E} - \frac{2R^2T_0^2}{E^2} + \ldots\right)\right]$$

$$= T_0 + \frac{RT_0^2}{E}$$

Therefore, $T_i - T_0 = \frac{RT_0^2}{E}$ (8.70)

Substituting E = 50,000 cal/mol and T_0 = 1,000°K,

$$\frac{RT_0^2}{E} = 40$$

that is $T_i - T_0 = 40$°C.

This shows that for all practical purposes, T_0 may be taken as the ignition temperature without great error. This analysis justifies the usual methods of determination of ignition temperature for all practical purposes.

Taking $T = T_i$, and $T_i - T_0 = \frac{RT_0^2}{E}$ in equation (8.65)

$$QVf(p)e^{-E/RT_0}\left(1 + \frac{RT_0}{E}\right) = B\frac{RT_0^2}{E}$$

Since, $RT_0 \ll 1$, we get the following simplified relation

$$QVf(p)e^{-E/RT_0} = \frac{BRT_0^2}{E}$$

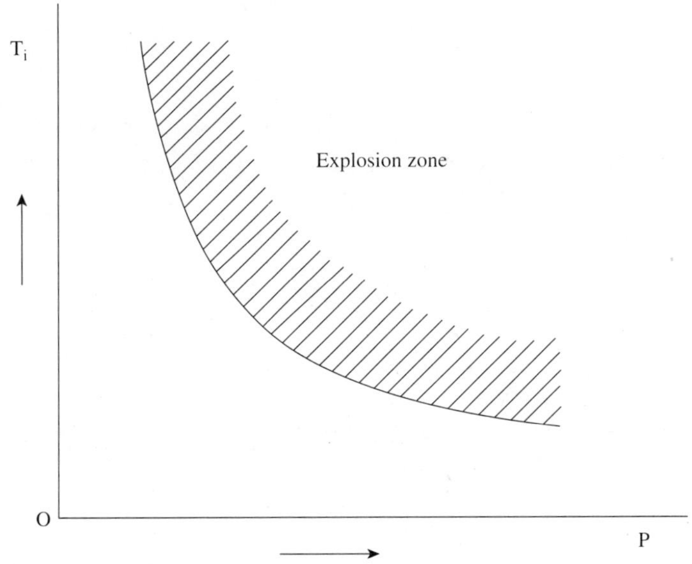

Figure 8.5 Ignition temperature versus pressure

Taking logarithm

$$\ln \frac{f(p)}{T_0^2} = \frac{E}{RT_0} - \ln \frac{QVE}{BR} \tag{8.71}$$

For a bimolecular reaction, $F(p) \propto p^2$. Writing critical pressure p_{cr} for p, we finally get

$$\ln \frac{p_{cr}}{T_0^2} = \frac{E}{2R_0T_0} + \text{constant} \tag{8.72}$$

Equation (8.72) is known as Semenov's equation, giving the relation between pressure and ignition temperature for successful thermal explosion of combustible mixtures (Fig. 8.5). Putting $T_i = T_0$ in equations (8.71) and (8.72), it can be shown that at constant pressure.

$$\ln Q = \text{constant} + \frac{E}{RT_i} \tag{8.73}$$

Q is the heat liberated by combustion. If the fuel–oxidant ratio in the system is varied between 0 and 1, Q will vary with the maximum near the stoichiometric composition. Hence, equation (8.73) will lead to Fig. 8.6 which indicates the existence of two limiting compositions of fuel–oxidant mixture for the system to ignite and explode. These are known as limits of inflammability (see Section 8.6).

Equation (8.70) gives an expression for the induction period. The time required by the system to rise by a temperature difference of RT_0^2/E gives the induction period.

It is now clear that the ignition temperature of a fuel is not a characteristic constant but depends upon circumstances. Some of the influencing factors are the fuel–oxidant ratio, nature

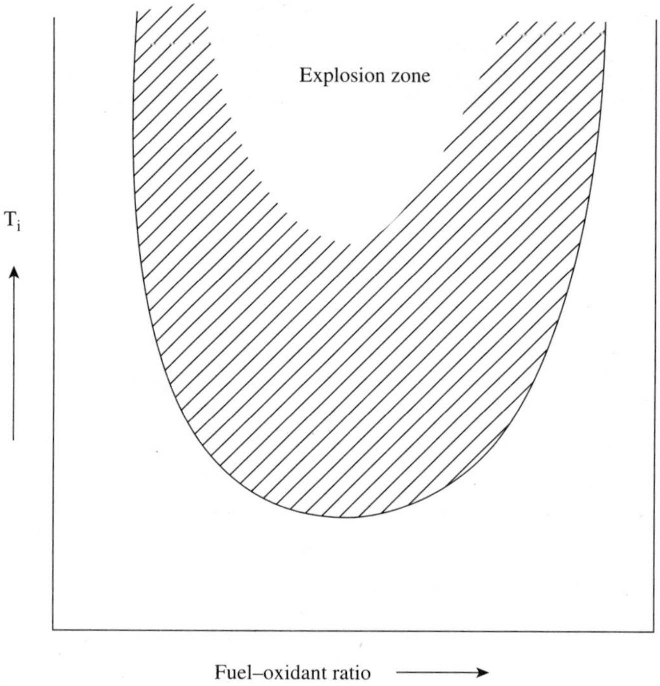

Figure 8.6 Ignition temperature versus mixture composition

of oxidant, induction period, vessel wall and presence of promoters or inhibitors. Obviously the method of determination is of great importance in obtaining the ignition temperature of substances.

Three principal methods have been used for the determination of SIT of gases and vapours:

1. Admission of an explosive mixture to an evacuated, heated vessel;
2. Heating of the fuel and oxidant separately and then mixing them in a heated vessel;
3. Adiabatic compression of the premixed system until ignition occurs.

The SITs of some gases and vapours are given in Table 8.1. These relate mainly to the first method.

The ignition temperature of liquid fuels is determined by dropping oil drops into a heated crucible and observing the minimum temperature at which a glow occurs followed by explosion or flame. A similar method is used in the case of coal also. About 5 to 6 g of coal powder is thrown into the bottom of the vessel heated at a constant rate. The ignition temperature of coke is determined by a slow method in which a sample of coke is heated at a constant rate. The temperature at which the rise of coke mass temperature is suddenly accelerated is recorded as its ignition temperature. Table 8.1 includes the data on solid and liquid fuels.

Table 8.1 Spontaneous ignition temperature in air and oxygen

Substance	In air, °C	In oxygen, °C	Substance	In air, °C	In oxygen, °C
Acetone	561	485	Hexane	248	–
Acetylene	305	296	Hexane, iso	–	284
Ammonia	651	–	Hydrocyanic acid	538	–
Amyl benzene	–	255	Hydrogen	572	560
Amylene	273	–	Hydrogen sulphide	292	220
Anthracene	472	–	Isodiodecane	500	–
Benzene	580	566	Isoprene	440	–
1,3-Butadiene	418	335	Kerosene	254	–
Butane	408	283	Methane	632	556
Bytylene	443	–	Methyl alcohol	470	461
Carbon disulphide	120	107	2-Methyl butane		
Carbon monoxide	609	588	(isopentane)	420	294
Cetane	235	–	Methyl chloride	632	–
Creosote	356	–	a-Methyl naphthalene	566	–
Cresole,o	599	–	2-Methyl propane	462	319
Cresol, m	626	–	Naphtha	277	–
Cyanogen	850	–	Napthalene	587	560
Cyclohexane	–	296	Nonane	285	–
Cyclopropane	498	454	Octane	218	208
Cymene	466		Pentane	290	258
Decalin	262	–	Petroleum ether	329	–
Decane,n	250	–	Phenol	715	500
Dimethylether	350	252	Propane	493	468
Dodecane	534	–	Propylene	458	–
Ethane	472	–	Styrene	490	450
Ether, diethyl	193	182	Tolune	552	516
Ethyl alcohol	392	–	Turpentine	252	–
Ethyl benzene	477	468	Xylene, o	496	–
2-ethyl butane	–	273	Wood	300	–
Ethylene	490	485	Wood charcoal	200–350	–
Gasoline(regular)	280	–	Peat	220–250	–
Gasoline	299	–	Lignite	250–350	–
73 octane					
Gasoline (92 octane)	390	–	Bituminous coal	380–420	270–290
Gasoline (100 octane)	429	–	Anthracite	500	340
Heptane	230	214	Soft coke	425–500	–
			Hard coke	500–650	–

8.5 VELOCITY OF FLAME PROPAGATION

A simple demonstration of flame propagation is furnished by an explosion in a glass tube. When the combustible mixture is ignited at one end by an external source of heat, say, a spark, the flame front can be seen to travel to the other end until the whole mixture is burnt off. Flame propagation is in fact a characteristic feature of all burning processes. A flame becomes

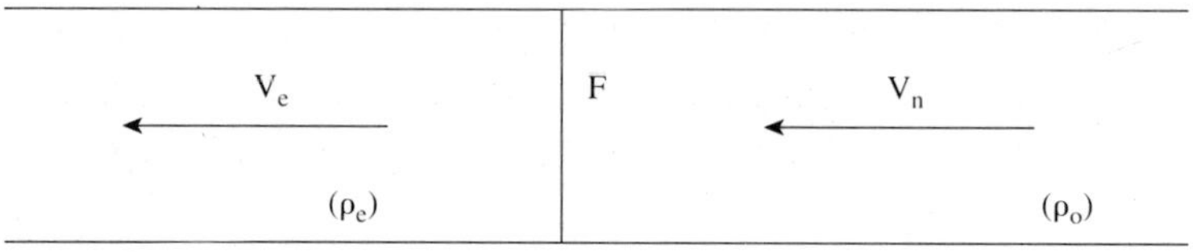

Figure 8.7 Stationary flame in a tube
ρ_e = density of initial mixture
ρ_n = density of exhaust gases
F = flame front
V_n = normal velocity of flame propagation
V_e = velocity of exhaust gases

stationary under favourable circumstances. Let us consider a system with the combustion surface (flame front) perpendicular to the axis of a cylindrical tube (Fig. 8.7). The combustible mixture is flowing from the right to the left. The flame will therefore tend to move from the left to the right towards the oncoming fresh gas. If the gas velocity is equal to the flame velocity, the combustion surface will remain stationary. The direction of the flame velocity is normal to the surface F. It is called the normal velocity of flame propagation, v_n, to differentiate it from the much faster propagation of flame in detonation (several km/s). The term burning velocity is often used instead of velocity of flame propagation. In short, it is the velocity, relative to the unburnt gas, with which a suitably defined flame front travels along the normal to its surface.

Since the mass flow-rate per unit cross-sectional area of the tube is constant under a steady state, we have

$$v_n \rho_o = v_e \rho_e$$

where,

v_n = initial gas velocity
v_e = velocity of exhaust gas
ρ_o = density of unburnt gas
ρ_e = density of exhaust gas

Therefore

$$v_e = v_n \frac{\rho_o}{\rho_e} \tag{8.74}$$

As $\rho_o >> \rho_e$, therefore $v_e >> v_n$

It will thus be seen that there is a sharp gradient of gas velocity across the flame front which is also characterised by the sharp gradients of gas density, temperature and composition.

In an explosive flame system the flame front moves with a velocity which is the vector sum of the gas velocity and the normal velocity of flame propagation. If the mass flow of gases could be prevented in the container then the actual rate of advance of the flame front into the unburned gas would be the normal velocity of flame propagation.

A combination of thermal and diffusion mechanisms explains the phenomenon of flame propagation. The region neighbouring a flame gets heated up by conduction and convection.

Chain carriers are also diffused from the flame to that region. As a result the pre-flame region becomes the zone of active reaction, and is converted into the new flame while the completion of reaction and relatively cooling effect extinguish the luminosity of the region where the flame existed earlier. The net effect is the advancement of the flame front.

Numerous attempts havse been made to predict flame velocity by purely thermal mechanism or purely diffusion mechanism or a combination of the two. There is no universally acceptable formula for calculating flame speed from the fundamental properties of the system. A simple derivation of v_n from flame propagation by heat conduction alone is given below.

Let us assume a stationary flame in which a fresh well-mixed combustible gaseous mixture moves continuously in one direction. The rate of heat flow by conduction per unit surface of gas layer is given by

$$q_1 = \frac{KT_m - T_0}{\delta}$$

where, K = average thermal conductivity.
δ = thickness of combustion zone, that is, width of flame front
T_m = temperature of the products of complete combustion
T_0 = initial temperature of the fresh gas

The rate of heat consumption by the incipient gas flow is

$$q_2 = v\rho_0 C_p (T_m - T_0)$$

where, v = gas flow rate = v_n for stationary flame
ρ_0 = density of fresh gas
$v\rho_0$ = mass flow rate per unit area
C_p = average specific heat

Equating q_1 and q_2 and substituting v_n for v, we get

$$v_n = \frac{K}{\rho_0 C_p \delta} = \frac{a}{\delta}$$

where, a = mean thermal diffusivity between T_0 and T_m.

Assuming $\delta = v_n t_r$

where, t_r = residence time in the reaction zone (reaction time), we finally get

$$v_n = \sqrt{\frac{a}{t_r}} \qquad (8.75)$$

Equation (8.75) gives a very important relationship that the flame velocity is directly proportional to the square root of thermal diffusivity and inversely proportional to the square root of the characteristic time of reaction.

Replacing t_r by $\frac{1}{K}$ or $\frac{1}{K_0 e^{-E/RT_m}}$

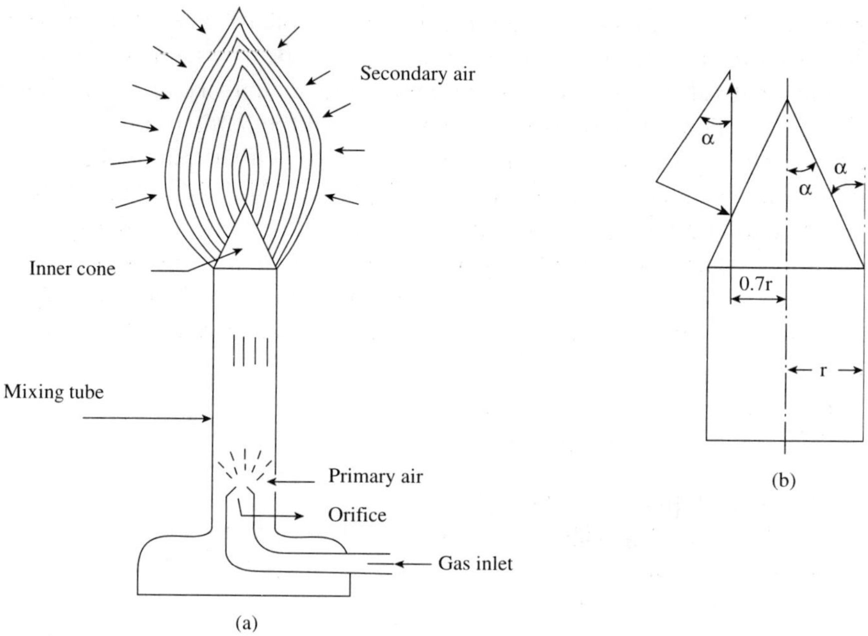

Figure 8.8 (a) Elements of a bunsen burner (b) Method for measuring V_n

where, K = specific reaction rate
K_0 = frequency factor
E = energy of activation
R = gas constant

we get

$$v_n = \sqrt{aK_0^{-E/RT_m}} \tag{8.76}$$

T_m is in fact the flame temperature. The conditions which increase the flame temperature will thus sharply raise the flame speed.

A more exact expression of V_n includes a factor f which is less than unity. This factor, in particular, is related to the difference between δ assumed above and the true width of the reaction zone.

$$v_n = f\sqrt{\frac{a}{t_r}} = f\sqrt{ake^{-E/RT_m}} \tag{8.77}$$

Many experimental techniques have been employed in determining normal velocity of flame propagation.

Nearly all of them fall under three groups:

1. Bunsen burner method or stationary flame method,
2. Spherical bomb method,
3. Glass tube method.

When proper care is taken and necessary corrections made, all these methods give comparable and reproducible results. The Bunsen burner method is briefly discussed here.

Figure 8.8a shows a Bunsen burner with a quiet flame. Fuel gas and primary air, 50% to 60% of the total air, are mixed in the mixing tube. The mixture burns with a flame anchored at the burner tip. The flame has an inner cone and an outer envelope. Combustion proceeds at the inner cone until the oxygen of the primary air is exhausted. The outer envelope is the zone of secondary combustion where the remaining combustibles burn in the secondary air derived from the atmosphere. The flame propagates in a direction normal to the surface of the inner cone. Since the flame is stationary, the normal velocity of flame propagation is vectorially equal and opposite to the velocity of the gas mixture issuing from the burner tip. It can be calculated from the knowledge of gas velocity and cone angle by using the vector triangle shown in Fig. 8.8b. The gas velocity varies along the cross-sectional area of the tube. It is maximum at the tube axis and almost zero at the wall. The local velocity at a distance of 0.7 times the radius from the axis represents the average velocity.

Let

$$v_{av} = \text{average velocity of gas–air mixture, m/s}$$

$$= \frac{V}{A}$$

A = cross-sectional area of the tube, m^2
V = flow-rate of the mixture. m^3/s
2α = cone angle, degree

Then the normal velocity of flame propagation

$$v_n = v_{av} \sin \alpha \qquad (8.78)$$

The angle α is determined from the angle of the flame projected on a screen, photographed or viewed through a telescope. The angle between the tangent to the inner cone surface and the tube axis gives α.

The factors affecting v_n are the composition of fuel–oxidant mixture, temperature, pressure and surface. The flame speed varies with composition and attains a maximum for a mixture which is slightly richer in fuel content than the stoichiometric composition (Fig. 8.9). This relationship is similar to the flame temperature–composition relationship and predictable from equation (8.76).

Flame speed depends upon the kinetics of combustion reactions and hence varies from fuel to fuel. The hydrogen flame is the fastest with v_n having a maximum value of about 2.5 m/s. Next comes acetylene, with v_n slightly less than 1.5 m/s. Carbon monoxide, paraffins and benzene have the maximum value of v_n in the range of 0.3–0.4 m/s. Propane is slightly above this range, while ethylene exceeds 0.6 m/s. All these values are for fuel–air systems. If oxygen is substituted for air, v_n is 5 to 10 times these values.

Flame speed rises if the initial temperature of the combustible mixture is increased or the pressure of the system decreased. By preheating the hydrogen–air mixture to 430°C and methane–air mixture to 680°C, the maximum value of v_n could be increased from 2.5 to 13.0 m/s,

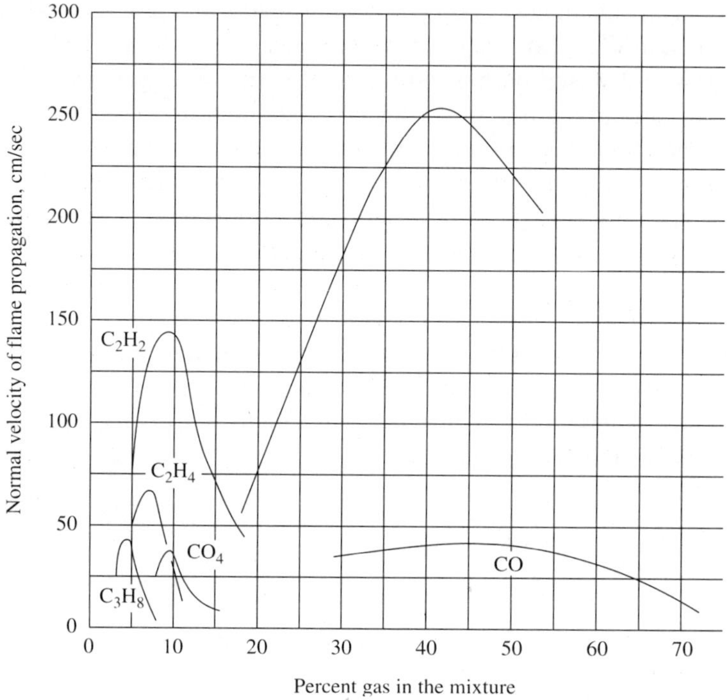

Figure 8.9 Flame velocity versus composition of gas–air mixture

and from 0.3 to 3.0 m/s, respectively. According to many investigators, v_n appears to vary as the square of the absolute temperature and inversely as the fourth root of the pressure.

The surface has a profound effect on flame propagation if the surface-to-volume ratio of the system is high. By exerting a cooling effect and adsorbing chain carriers, the container surface slows down the propagation of flame. In fact combustion can be completely stopped in very narrow tubes. This was first observed by Davis and formed the basis of his discovery of the miner's safety lamp bearing his name.

A knowledge of flame speed is of fundamental importance. It is one of the basic parameters that affect the rate of heat release in combustion. The solution of many practical problems in burners and furnaces needs a proper appreciation of the velocity of flame propagation.

8.6 LIMITS OF INFLAMMABILITY

A fuel–oxidant mixture is called inflammable when combustion liberates sufficient energy to cause a continued self-propagation of the flame throughout the unignited regions of the mixture. Like the limits of pressure (explosion limits) and temperature (ignition temperature), there also exist concentration limits for a system to be inflammable. These are known as the limits of inflammability. There are two limits of inflammability—the upper limit referring to the richest mixture and the lower limit to the leanest mixture—to propagate a flame when the surface effects are negligible.

An analysis of the mechanism of ignition temperature earlier indicated the existence of two limits of inflammability. A similar indication can be obtained from the velocity of flame propagation also. On both the sides of the optimum fuel–oxidant composition the flame speed declines. There should be a critical flame speed, actually a range of 10–12 cm/s, below which flame propagation is not possible owing to the over-whelming effect of heat loss. The flame should advance at a sufficiently high rate to compensate for the loss of heat. A mixture below the critical flame speed may be ignited at a point but the flame will disappear after travelling a short distance through the combustible mixture. Since the critical flame speed corresponds to the two compositions on the rich and lean sides, there are two limits of inflammability.

Inflammability limits are determined in an apparatus consisting of a 75 mm diameter vertical glass tube 1.5 m in height. The mixture is confined in the tube and then ignited by opening the lower end and applying a spark or a small flame. The two limiting concentrations are determined at which the flame just fails to propagate upwards throughout the whole length of the tube.

The limits of inflammability are influenced by many factors. A reduction in pressure narrows down the two limits, lowering the upper limit and raising the lower limit. The upper limit is much more affected than the lower. The limits tend to join at a critical pressure below which the mixture ceases to be inflammable at any concentration. The effect of pressure is explained by the increased diffusivity of chain carriers and other hot products towards the wall at a reduced pressure.

Ordinary variations of laboratory temperature have no appreciable influence on the limits. By raising the temperature of the mixture to an appreciable extent the limits may be widened. The effect is due to the enhanced flame temperature causing flame propagation in the mixtures of unfavourable composition.

The container size has negligible effect above 5 cm diameter. In the case of smaller sizes the limits are narrowed down by decreasing the diameter. This is the result of wall effect. The minimum tube sizes below which the flame does not propagate at room temperature and pressure depend upon the fuel, for H_2 it is 0.9 mm, for CO, 2.5 mm and for CH_4, 4.0 mm.

The direction of flame propagation affects the limits. The widest limits are obtained if the flame is to propagate upwards and the lowest limits result if flame propagates downwards. Intermediate results are provided by horizontal and inclined tubes. This variation is due to the buoyancy effect.

The limits of inflammability of combustible gases and vapours at ordinary temperatures and pressures in air are given in Table 8.2. Hydrogen, acetylene and ethylene oxide have an extremely wide range of inflammability (H_2: 4% to 75%). It is fortunate that petrol vapours and usual organic solvents (benzene, hexane, ethyl alcohol and acetone) have quite narrow limits of inflammability (benzene: 1.4% to 7.1%). Generally, the higher the maximum normal velocity of flame propagation of a combustible gas or vapour in air, the wider the limits of its inflammability in air. Carbon monoxide is the remarkable exception. Although its maximum flame velocity is less than 0.5 m/s, it has a very wide range of inflammability limits, 12.5% to 74%.

The inflammability limits of a mixture of combustibles in air may be approximately calculated by Le Chatelier's formula

$$L = \frac{100}{X_1/I_1 + X_2/I_2 + X_3/I_3 \cdots} \tag{8.79}$$

Table 8.2 Limits of inflammability of some gases and vapours at ordinary temperature and pressure in air

Gas or vapour	Per cent by volume		Gas or vapour	Per cent by volume	
	Lower limit L_1	Upper limit L_2		Lower limit L_1	Upper limit L_2
Inorganic substances			**Esters**		
Hydrogen	4.0	75.0	Methyl formate	5.9	20.0
Carbon disulphide	1.25	44.0	Ethyle formate	2.7	13.5
Carbon monoxide	12.5	74.0	Ethyl acetate	2.5	9.0
			Vinyl acetate	2.6	13.4
Hydrocarbons			Propyl acetate	2.0	–
Methane	5.3	14.0	Butyl acetate	1.7	–
Ethane	3.0	12.5			
Ethylene	3.1	32.0	**Bases**		
Acetylene	2.5	80.0	Methyl amine	4.9	20.7
Propane	2.2	9.5	Dimethyl amine	2.8	14.4
Propylene	2.4	10.3	Trimethyl amine	2.0	11.6
Cyclopropane	2.4	10.4	Ethyl amine	3.5	14.0
Butane	1.9	8.5	Diethyl amine	1.8	10.1
Butene-1	1.6	9.3	Propyl amine	2.0	10.4
Butene-2	1.8	9.7			
Butadiene	2.0	11.5	**Halogen compounds**		
Pentane	1.5	7.8	Methyl chloride	10.7	17.4
Hexane	1.2	7.5	Ethyl chloride	3.8	15.4
Cyclohexane	1.3	8.0			
Benzene	1.4	7.1	**Hydrocarbon fuels**		
Toluene	1.4	6.8	Petrol (gasoline)	1.4	7.6
Ethyl benzene	1.0	–	Pittsburgh natural gas	4.8	13.5
Alcohols			**Manufactured fuel gases**		
Methyl	7.3	36.5	Coke-oven gas	4.4	34.0
Ethyl	4.3	19.0	Oil gas	4.7	33.0
			Coal gas	5.3	32.0
Ethers			Carburetted water gas	5.5	36.0
Methyl	3.4	18.0	Water gas	7.0	72.0
Ethyl	1.9	48.0	Producer gas	17.0	70.0
Ethyl *n*-propyl	1.9	24.0	Blast furnace gas	35.0	74.0
Ethylene oxide	3.0	80.0			
			Bituminous coal g/m^2 t NTP x 10^4	0.0143	0.143
Aldehydes and ketones					
Acetaldehyde	4.1	55.0			
Butaldehyde	2.5	–			
Acetone	3.0	11.0			

Note: These figures are for upward propagation of flame in large vessels open at their lower ends.

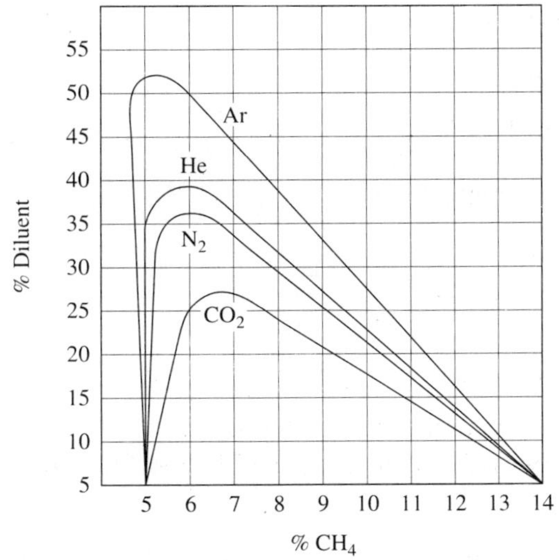

Figure 8.10 Effect of diluents on the limits of inflammability

where,

$x_1, x_2, x_3 \ldots$ = volume per cent of the combustibles in the mixture

$l_1, l_2, l_3 \ldots$ = limit (lower or upper, as the case may be) of inflammability of pure combustibles in air, per cent.

L = limit (lower or upper, as the case may be) of inflammability of the mixture in air, per cent.

It has been found that the lower limit of a mixture of combustibles can be accurately calculated but there is often appreciable divergence between the calculated and experimental values of the upper limit of a mixture. It clearly shows that the combustibles interact particularly when they are present in large quantities as in the upper limit of inflammability.

The influence of diluents is illustrated by Fig. 8.10. The very fact that the curves given by different diluents, namely, N_2, CO_2, He and Ar do not coincide shows that the diluents influence flame propagation to a greater or less extent. The small difference among N_2, He and Ar may be explained by their different thermal properties. However, the exceptional behaviour of CO_2 indicates that it is quite involved in the flame propagation.

Approximate calculation of the inflammability limits of combustibles containing diluents is based on the fact that the lower limit is actually the criterion of the combustible concentration while the upper limit is that of the combustible to oxygen ratio.

$$L'_1 = L_1 \times \frac{100}{100 - X_{dil}} \tag{8.80}$$

$$L'_2 = L_2 \times \frac{100}{100 - X_{dil}} \quad \text{if air is the diluent} \tag{8.81}$$

If there is no oxygen in the diluent,

$$L'_2 = \frac{10000L_2}{100x_{comb} + L_2(100 - x_{comb})} \tag{8.82}$$

If the diluent (other than air) contains O_2,

$$L'_2 = \frac{2100L_2}{21x_{comb} - L(x_{o_2} + 0.21x_{comb} - 21)} \tag{8.83}$$

L_1 and L_2 = lower and upper limits of the pure combustible

L'_1 and L'_2 are the limits of the combustible containing diluent

x_{dil}, x_{comb}, and x_{O_2} are per cent of diluent, combustible and O_2 respectively, in the combustible containing the diluent.

The limits of inflammability of gases and vapours in air are of great practical significance for precaution against the risks of fire and explosion. In using solvents and other inflammable substances it is necessary to keep the concentration of their vapours in the surroundings well outside the inflammability range. Sometimes nitrogen and carbon dioxide are used as diluents to bring the atmosphere below the lower limit of inflammability.

Example 1

Calculate the limits of inflammability of a gas mixture containing 25% CO, 40% H_2 and 35% CH_4.

Solution
From Table 8.2

		L_1	L_2
CO	–	12.5	74.0
H_2	–	4.0	75.0
CH_4	–	5.3	14.0

All the components are combustible. Applying Le Chatelier's formula (8.79)

$$(L_1)_{mixture} = \frac{100}{25/12.5 + 40/4.0 + 35/5.3} = 5.4\%$$

$$(L_1)\text{mixture} = \frac{100}{25/74.0 + 40/75.0 + 35/14.0} = 29.7\%$$

Example 2

Calculate the limits of inflammability of a gas containing 25.0% CH_4, 45.0% H_2, 10.0% CO, 10.0% CO_2 and 10.0% N_2.

Solutions

CO_2 and N_2 are the diluents.

Concentration of diluents = 10 + 10 = 20%

Composition of diluents in free combustible mixture

$$CH_4 = 25 \times \frac{100}{100-20} = \frac{25}{0.8} = 31.2\%$$

$$H_2 = \frac{45}{0.8} = 56.3\%$$

$$CO = \frac{10}{0.8} = 12.5\%$$

Limit of inflammability of the diluents-free mixture

$$L_1 = \frac{100}{31.3/5.3 + 56.3/4.0 + 12.5/12.5} = 4.8\%$$

$$L_2 = \frac{100}{31.3/14.0 + 56.3/75.0 + 12.5/74.0} = 31.7\%$$

Therefore, the limits of inflammability of the given mixture are

$$L_1 = 4.8 \times \frac{100}{100 - 20} = 6.0\%$$

$$L_2 = \frac{10000 \times 31.7}{100 \times 80 + 31.7(100 - 80)} = 36.7\%$$

8.7 STRUCTURE OF FLAME

The size and shape of a flame depend on its type, namely, premixed or diffusion, laminar or turbulent, and also on the burner dimensions. An elementary treatment of the subject is given here.

The premixed flame of a simple gas burner is shown in Fig 8.11. The burner has a nozzle diameter, d. The distance covered by the initial boundary of the flame is L, and may be called the zone of inflammation (height of inner cone). L_f is the flame length and indicates the zone of complete combustion. In the case of laminar flame, the inflammation zone is sharply pointed and has a thin surface (flame front). Under turbulent conditions, the front is distorted and has an appreciable width, δ. The distance between the visible front and the flame tip, L, represents the boundary of complete combustion. The most intense combustion takes place at the flame front while the process is much slower in the zone L, where diffusion of some air from the atmosphere is involved. The total flame length is

$$L_f = L_i + \delta + L_s$$

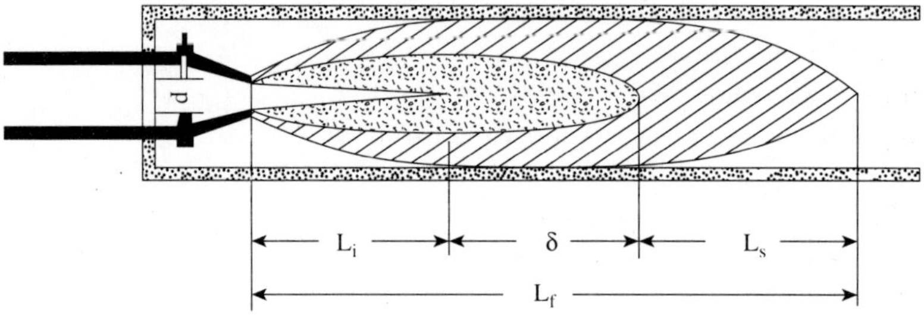

Figure 8.11 Premixed flame of a simple gas burner

For a laminar premixed flame, $\delta = o$ and

$$L_f = L_i + L_s$$

The length, L_i is related to the process of flame propagation from the periphery to the centre

$$L_i = \frac{v_{av} \times r}{v}$$

where,

v = velocity of flame propagation (laminar or turbulent, as the case may be)
r = tube radius
v_{av} = average linear velocity of the gas at the burner nozzle.

L_f increases with the increase in gas velocity and nozzle diameter and the decrease in velocity of flame propagation.

For a turbulent flame

$$v = v_{av}$$

that is, velocity of turbulent flame propagation is approximately equal to gas velocity.

Therefore, (L_i) turbulent $\approx r$ (8.84)

The height of the inner cone increases with the increase in gas velocity for a laminar flame but remains almost constant for a turbulent flame.

L_s is related mainly to the kinetic properties of the combustible mixture.

The width of flame front δ depends upon eddy diffusion.

$$\delta = \frac{\varepsilon}{V_n} = \frac{I' \times v'}{v_n} \quad (8.85)$$

where,

ε = eddy diffusivity
I′ = scale of turbulence (analogous to mean free path in molecular diffusion)

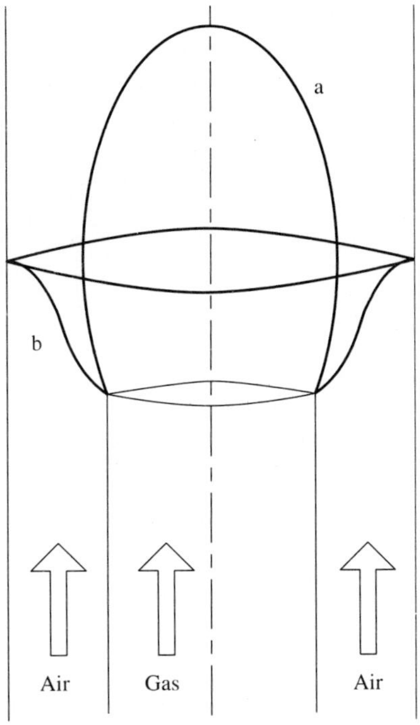

Figure 8.12 Simple laminar diffusion flame: (a) over-ventilated and (b) under-ventilated

v' = intensity of turbulence (root mean square value of instantaneous fluctuating velocity)
v_n = normal velocity of flame propagation of the mixture.

The front width increases with increase in eddy diffusivity and decreases in normal flame speed.

A simple laminar diffusion flame is illustrated in Fig. 8.12. Fuel gas and air flow separately but at the same linear velocity through two concentric tubes, the inner tube supplying the fuel jet. The structure of the diffusion flame is determined mainly by the process of mixing of gas and air, and not by the velocity of flame propagation. The mixing is achieved by either molecular diffusion or eddy diffusion, depending upon the laminar or turbulent conditions of the flame.

Laminar diffusion flame may be of two types. When air is in excess, a flame of the over-ventilated type is formed (Fig. 8.12a). With excess fuel, the flame is of the under ventilated type (Fig. 8.12b). The flame boundary is provided by the surface at which combustion is complete. Since combustion reactions are rapid it may be assumed that the fuel and air are essentially in the stoichiometric ratio at the boundary surface. When air is in excess, this surface closes in towards the axis as the diffusion of gas and air continues. The length of the flame corresponds to the condition of arrival of sufficient air at the tip for completing combustion. Beyond this point no fuel gas exists and the mixing of air and combustion products continues. A normal flame has the shape of the over-ventilated type. When fuel is in excess, the flame

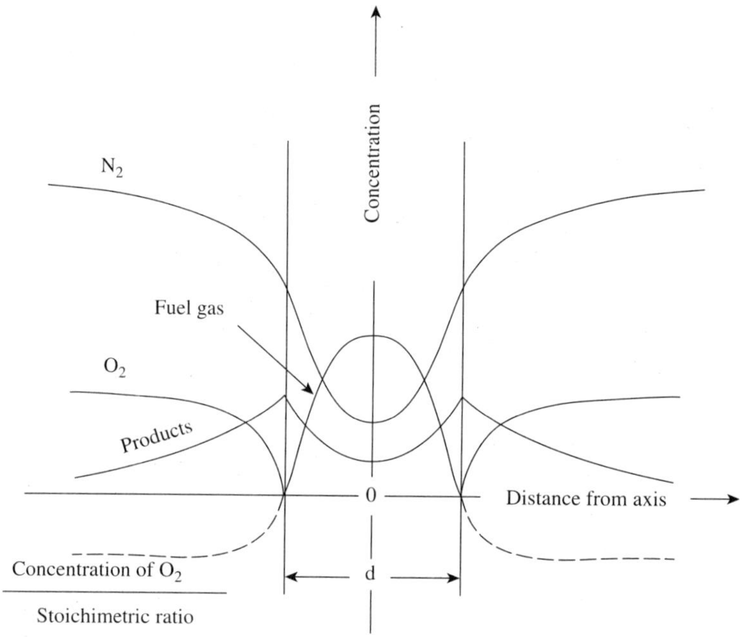

Figure 8.13 Concentration profile of a cross section of a laminar diffusion flame (schematic)

boundary moves upwards and the flame length corresponds to the arrival of fuel gas at the outer tube wall. The flame is bell-shaped.

The concentration profile of a cross section of the diffusion flame is shown in Fig. 8.13. The concentration of fuel gas is highest at the centre and falls to zero at the boundary. The oxygen is highest at the outer tube wall and disappears at the flame boundary. The combustion products are maximum at the boundary and slowly diffuse both towards the flame axis and the outer wall. The nitrogen is maximum at the wall and slowly drops to a minimum value at the centre.

A laminar diffusion flame is converted into the turbulent type by increasing the gas velocity beyond a critical value. However, the Reynolds number of the cold gas stream, that is the cold Reynolds number is no criterion for this conversion. Immediately beyond the burner rim, the temperature sharply rises and hence the gas viscosity also rises causing a sharp drop in the Reynolds number. Even when the cold Reynolds number exceeds 2,300, the value in the flame is much below the critical and the flame is still laminar. The critical cold Reynolds number for the conversion of laminar into turbulent flame may vary widely, say, 3,500 to 10,000 depending upon the fuel composition and the quantity of primary air.

Figure 8.14 depicts the transition from laminar to turbulent flame with the increase in jet velocity. The flame is quiet and possesses strong yellow luminosity in the laminar region. It is violent, flickery, noisy and devoid of yellow luminosity in the turbulent region. It may be seen that the flame length increases with the increase in jet velocity in the laminar region upto a maximum when flickering is first noted at the flame tip. With the further increase in gas velocity, a much larger portion of the flame is found to flicker and the total flame length is decreased.

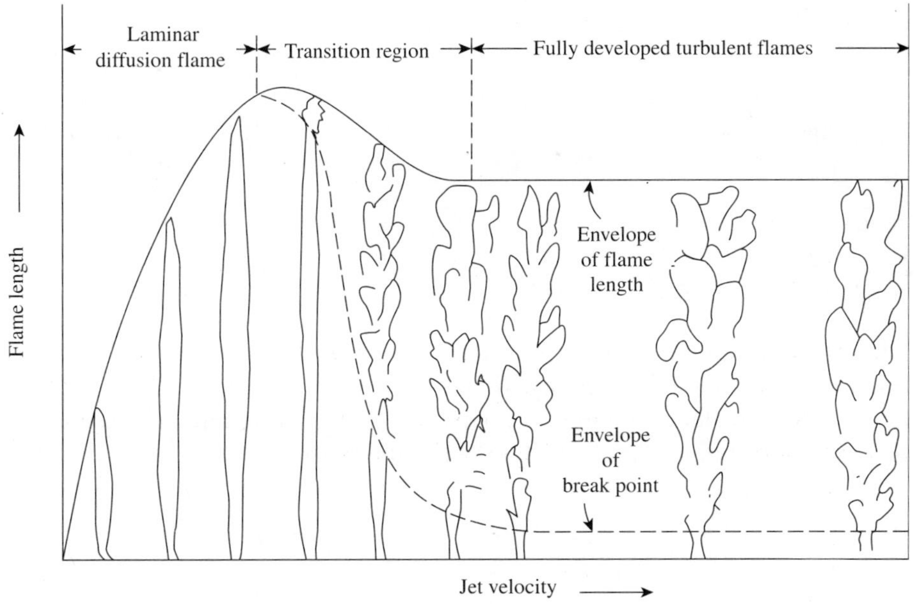

Figure 8.14 Progressive change in flame type with increase in jet velocity

This is the transition stage. In the turbulent region the flame length remains practically constant with increasing jet velocity. Above a critical velocity the flame is lifted and finally it is blown off when the jet velocity increases further.

The difference in the effect of jet velocity on flame length of laminar and turbulent diffusion flames may be explained in a simple manner.

From the concept of over-ventilated flame, it may be assumed that the flame length of a laminar diffusion flame corresponds to the condition at the flame tip on the axis of the jet, where the average depth of penetration of air into gas is approximately equal to the radius of the burner tube.

From the law of mean square of displacement

$$\bar{x}^2 = 2Dt$$

where,

$\bar{x}$ = root mean square of displacement
D = diffusion coefficient
t = time

Since, $\bar{x} = r$, the tube radius

$$r^2 = 2Dt$$

But

$$t = \frac{L}{v}$$

Therefore, $r^2 = \frac{2DL}{v}$

$$\text{or } L = \frac{vr^2}{2D} = \frac{V}{2\pi D} \tag{8.86}$$

where,
L = flame length
v = jet velocity
V = volume flow-rate of jet

For a given fuel and burner, r and D are constant. Hence, the length of a laminar diffusion flame is directly proportional to its jet velocity. If however, the volume flow-rate is kept constant by changing the tube diameter with the jet velocity, the flame length remains costant.

By analogy to the laminar flame, we may write the following relationship for the length of turbulent diffusion flame

$$L \propto \frac{vd^2}{\varepsilon}$$

where,
d = burner tube diameter
ε = eddy diffusivity
But, $\varepsilon = 1'\, v'$

where $1'$ = scale of turbulence
v' = intensity of turbulence

Experimental data show that

$$1' \propto d \text{ and } v' \propto v$$

Therefore, $\varepsilon \propto dv$

$$\text{Hence, } L \propto \frac{d^2 v}{dv}$$

$$\text{or } L \propto d$$

$$\text{or } \frac{L}{d} = \text{constant} \tag{8.87}$$

This simple relationship shows that the length of a turbulent diffusion flame is directly proportional to the nozzle diameter but independent of the velocity.

Since the flame represents the zone of combustion, its length is a measure of the reciprocal of the intensity of combustion and, therefore, of heat release. Flame length studies have great relevance in the rational design of a combustion chamber, be it for an internal engine or for a furnace.

8.8 FLAME STABILITY

A freely burning flame is considered stable when there is no flash-back or blow-off. The essential condition for flame stability is that the normal velocity of flame propagation is vectorially equal and opposite to the velocity of fuel–air mixture at the flame front. A consideration of all the factors affecting these two velocities is necessary for evaluating the conditions of flame stability.

The important factors to influence v_n are: nature of fuel, composition of fuel–air mixture and its temperature and tube diameter. The temperature of the mixture at the burner tip depends on heat transfer from the reaction zone, heat loss to the surroundings and also on the material of construction, size and shape of the burner wall. The gas velocity depends upon the desired volume flow-rate and burner nozzle diameter. It also varies along the cross-sectional area of the nozzle.

Condition for stable flame : $v_n = v_{av} \sin\alpha$
Condition for flash-back : $v_n > v_{av} \sin\alpha$
Condition for blow-off : $v_n < v_{av} \sin\alpha$

The angle α is not constant. Its value depends upon the location of the flame front. When $\alpha > 90°$, the flame will be pushed inside the burner tube and flash-back will occur. Therefore, the minimum gas velocity corresponds to the normal velocity of flame propagation for the system. On the other hand, if the flame front is developed at a large distance above the burner tip, that is, if $\alpha \rightarrow 0°$, there will be undue dilution of the combustible mixture by interdiffusion with the surronding atmosphere. It will cause undue lowering of the flame velocity and lead to blow-off of the flame. The range of gas velocity for a stable flame is the result of many variables operating on the system.

A stable flame may be obtained at high gas velocities either by holding a wire ring at a considerable distance from the burner tip or by inserting a wire partly into the burner tube. The gas velocity is slowed down in the vicinity of the wire in both cases. A conical flame is obtained on the wire ring while the upper point of the straight wire supports a fan-shaped flame.

An increase in the tube diameter enhances the tendency to flash-back while a decrease in the tube diameter tends to cause blow-off. Use of ceramic materials of low thermal conductivity in making the burner tip prevents the blow-off tendency of turbulent flames. The same, however, augments the flash-back tendency of laminar flames.

8.9 KINETICS OF LIQUID FUEL COMBUSTION

Liquid fuels do not burn in the liquid state. Vapours of highly volatile liquids are burnt on the principle of combustion of gaseous fuels. The other liquids may employ one of two basic principles.

1. In the simpler type the fuel is kept in a shallow, open vessel. Vapours of the fuel are formed at the surface on radiation by its own flame and these burn with the air forming a stable flame over the vessel.

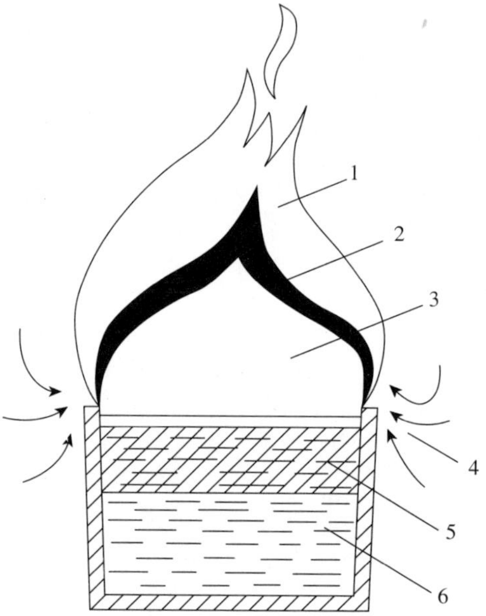

Figure 8.15 Combustion of a liquid fuel in an open vessel
1. Zone of complete combustion 2. Flame front 3. Vapour
4. Convection current 5. Heating zone 6. Liquid fuel

2. In the other method of liquid fuel combustion, the fuel is atomised, that is, divided into fine droplets. The latter are then burnt in air. Each droplet forms vapours of the fuel on heating and these vapours actually participate in the combustion reaction. Vaporisation of liquid fuels before combustion is possible owing to their boiling ranges falling below the spontaneous ignition temperatures. However, some low volatile liquid fuels are cracked before vaporisation and ignition. Their droplets undergo thermal decomposition and form combustible vapours and solid coke particles, both of which then burn in their respective ways.

The principle of combustion of liquid fuels in an open vessel is illustrated in Fig. 8.15. The liquid surface receives heat from the flame mainly by radiation. The vapours formed mix with air by diffusion and then burn at the flame front. Combustion is completed in the body of the flame. Once the fuel is ignited, stationary conditions are set up quite fast. Then the burning rate equals the rate of recession of the liquid, which is constant for the given conditions. The liquid temperature near the surface coincides with its average boiling point. The thickness of the heating zone is about 30 to 40 mm. The burning rate increases with the decrease in the latent heat of vaporisation and increase in the overall rate of heat transfer. The latter increases with the increase in the difference of flame temperature and boiling point. The burning rate is found to have a maximum value of 3.5 mm per minute.

The combustion of a liquid droplet is schematically shown in Fig. 8.16. The flame completely engulfs the droplet. The basic mechanism is the same as above. The droplet surface receives

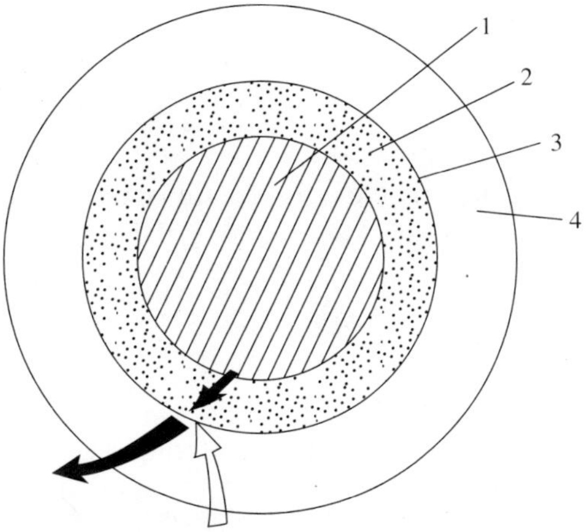

Figure 8.16 Combustion of a liquid fuel droplet
1. Liquid 2. Vapour layer 3. Combustion zone 4. Diffusion zone

heat from the flame. Vapours flow in a direction opposite to the heat flow and burn with air at the flame front. The droplet decreases in size with the progress of combustion until it is completely converted into products of combustion. A simple relation for the kinetics of the process is derived in the following.

Rate of heat transfer to the liquid per unit surface area is given by

$$q_l = h(T_g - T_l)$$

where,

h = heat transfer coefficient
T_g = temperature of the gaseous medium
T_l = temperature of the liquid, which may be considered uniform throughout the droplet in view of the small size

The chemical kinetics are very high at the high reaction temperature so that the partial pressure of the vapour at the surface is small and does not influence the rate of vaporisation. The latter will depend upon the liquid temperature

$$w = Ae^{-L/RT_l}$$

where,

w = rate of vaporisation
L = latent heat of vaporisation
A = constant governed by the nature of the liquid

Therefore, the rate of heat consumption per unit surface area is given by

$$q_2 = Lw = LAe^{-L/RT_l}$$

Under equilibrium conditions

$$q_l = q_2$$

$$\text{Therefore, } h(T_g - T_l) = LAe^{-L/RT_l} \tag{8.88}$$

This equation may be used to calculate the equilibrium temperature T_l and hence the rate of vaporisation. Under the conditions of very fast reactions, the latter gives the rate of combustion.

Therefore, rate of vaporisation

$$w = \frac{h}{L}(T_g - T_l)$$

For very small droplets and low values of Reynolds number, it is known that the Nusselt number Nu ≈ 2

Therefore,

$$h = \frac{k}{d}NU = \frac{k}{r}$$

$$\text{Therefore, } w = \frac{k}{Lr}(T_g - T_l) \tag{8.89}$$

where, k = thermal conductivity and d and r are the diameter and radius, respectively.

This is the simplified equation for the rate of vaporisation and combustion.

The time of complete combustion, t_b, of the droplet may be calculated from a relation between its radius and the quantity of heat received in a given time. This relation is

$$dQ = 4\pi r^2 h(T_g - T_l)\, dt$$

where, dq is the heat received by a droplet of radius r in time dt.

The mass vaporised in time dt

$$dG = \frac{dQ}{L}$$

Again, dG may be written as

$$dG = -4\pi r^2 \rho dr$$

where, ρ = liquid density.

$$\text{Therefore, } \frac{dr}{dt} = h\frac{T_g - T_l}{\rho L}$$

t_b may be obtained by assuming T_l = constant (boiling point) and integrating between $r = r_o$, $t = t_o$ and $r = r_o$, $t = t_b$ where r_o = initial radius of the droplet. Writing T_c (combustion temperature) for T_g, we thus get

$$t_b = \frac{L}{T_c - T_l}\int_o^{r_o} \frac{dr}{h}$$

At Nu = 2, $h = \frac{k}{r}$

$$\text{Therefore, } t_b = \frac{\rho L}{2K(T_c - T_l)} r_o^2 = k_o d_o^2 \tag{8.90}$$

where, K_o is a constant. This relation is very significant in stating that the time of burning of a droplet is directly proportional to the square of its initial diameter d_o. When t_b is in seconds and d_o in centimetres, the value of K_o is found to be of the order of 10^2.

8.10 KINETICS OF SOLID FUEL COMBUSTION

Solid fuels have been studied much more thoroughly than the liquid fuels. The ignition temperatures of the naturally occurring solid fuels, namely, wood, peat, ligmite and coal. are higher than the initial temperatures of active decomposition. These fuels also contain appreciable quantities of moisture. Therefore, they undergo drying and decomposition before actual combustion takes place. The volatile matter which contains combustible gases, vapours and also minute droplets of tarry substances, is thus evolved. The combustion of theses solid fuels consequently involves the combustion of the volatile matter and the combustion of the solid carbonaceous residue. The composition of the volatile matter, and hence its burning process, varies widely. The residue is essentially carbon. However, its reactivity depends on the type of fuel and hence its ignition characteristics also vary. The flame of a solid fuel is due to the combustion of the volatile matter and the carbon monoxide produced during the combustion of carbon. The visible indication of the combustion of carbon is a bright glow of the burning piece.

Carbon furnishes a typical case of heterogeneous combustion, that is, combustion between two different phases. The fuel is in the solid state and the oxidant in the gaseous state. This heterogeneous process has two basic components:

1. Delivery of the gaseous reactant to the surface of the solid by diffusion,
2. Chemical reaction of the solid and gas at the surface.

The overall kineties of the process depends in general on the rates of the two individual steps. When the latter differ considerably the overall kinetics are governed by the slower of the two. Thus we may have the diffusion-controlled combustion at high temperatures when the chemical reactions are extremely fast. Under the conditions of lower temperature, the chemical kinetics are slowed down and hence these govern the combustion process. In practical cases both the phenomena may be significant in controlling combustion.

The chemical kinetics are expressed by an equation of the Langmuir adsorption isotherm

$$R_c = \frac{k_1 k_2 p_s}{k_1 + k_2 p_s}$$

where,

R_c = specific reaction rate per unit surface area
p_s = oxygen concentration at the surface
k_1 = rate of formation of surface complex of oxygen with carbon
k_2 = rate of decomposition of the surface complex

This equation has two limiting cases

1. At high pressure, $k_1 p_s >> k_1$ (8.91)
   Therefore $R_c = k_1$, that is, the reaction is of zero order.
2. At low pressure, $k_1 >> k_2 p_s$, (8.92)
   Therefore $R_c = k_2 p_s$, that is, the reaction is of first order.

This variation in the order of the reaction can be achieved also by temperature variation when k_1 and k_2 have different temperature dependences.

The diffusional kinetics are given by the expression

$$R_d = k_o(p_o - p_s)$$

where,

R_d = rate of diffusion per unit surface area, which is equal to the rate of reaction per unit area where diffusion controls the process.
k_o = velocity constant for diffusion across the boundary layer.

$$= \frac{D}{\partial}$$

p_o = oxygen partial pressure in the main stream
D = diffusion coefficient.
δ = boundary layer thickness

If the carbon is porous, the kinetics of pore diffusion are given by

$$R_p = \sqrt{k_s SD \times p_s^{(m+1)}}$$

where,

R_p = rate of pore diffusion related to unit area of external surface, which is equal to the specific reaction rate at the pore surface related to unit area of external surface
k_s = velocity constant for the reaction at the surface, which may be k_1 or k_2 depending upon the conditions
S = internal surface area per unit volume
m = order of reaction with the limiting values of zero and one.

For a zero order reaction $k_s = k_1$ and $m = 0$

$$\text{Therefore, } R_p = \sqrt{k_1 S D p_s} \tag{8.93}$$

For a first order reaction, $k_s = k_2$ and $m = 1$

$$\text{Therefore } R_p = \sqrt{k_2 S D p_s} \tag{8.94}$$

The combined kinetics are given by the summation of rates at the surface and at the pores expressed per unit area of external surface.

$$R_t = R_s + R_p$$

where, R_t = total rate of reaction per unit area of external surface
R_s = rate of reaction at the external surface per unit area of external surface.

Under the conditions of low temperatures, $T < 800°C$, that is the ignition range

$$R_s = R_c \text{ of zero order} = k_1$$

$$R_p = \sqrt{k_1 S D p_s}$$

$$\text{Therefore } R_t = k_1 + \sqrt{k_1 S D p_s} = k_1\left[1 + \sqrt{\frac{SDp_s}{k_1}}\right] \tag{8.95}$$

If $R_s >> R_p$, $R_t = k_1$, the reaction is of zero order for non-porous carbon.

If $R_p >> R_s$, $R_t \propto P_s^{0.5}$, the reaction is of half order.

Under the conditions of high temperatures, $T > 1{,}200°C$, that is in the flame temperature region

$$R_s = R_c \text{ of first order} = k_2\, p_s$$

$$R_p = \sqrt{k_2 S D p_s}$$

$$\text{Therefore, } R_t = (k_2 + \sqrt{k_2 S D}) \times p_s \tag{8.96}$$

But $R_d = k_o\,(p_o - p_s)$

$$\text{Or } p_s = \frac{k_o p_o - R_d}{k_o}$$

The diffusion across the boundary layer may be assumed to control the overall kinetics at $T > 1200°C$, so that

$$R_d = R_t$$

$$\text{Therefore, } p_s = \frac{k_o p_o - R_t}{k_o}$$

We have also obtained,

$$R_t = (k_2 + \sqrt{k_2 SD}) \times p_s$$

Eliminating p_s and rearranging,we get

$$\frac{1}{R_t} = \frac{1}{k_o p_o} + \frac{1}{(k_2 + \sqrt{k_2 SD}) \times p_o} \qquad (8.97)$$

This shows that at high temperatures, the overall reaction is of first order with respect to the initial oxygen partial pressure; this is true for all cases whether pore diffusion is relatively important or not.

The above simple derivations confirm the experimental findings in the two temperature regions with lumps of carbon.

Practical fuel bed can be simultated by carbon channels. Experiments with single channels have shown that the burning rate is almost independent of air velocity in the laminar region but increases with it in the turbulent region; the rate of increase being higher at higher temperatures.

The combustion of pulverized fuel is simulated by the combustion of carbon particles. The overall kinetics of a single particle is given by

$$\frac{1}{R_t} = \frac{1}{k_o p_o} + \frac{1}{k_2 p_o} \qquad (8.98)$$

neglecting the pores in the small particles. Assuming relative velocity of the particles to air negligible, k_o = D/r. This bears out the enormous effect of size reduction in accelerating the process.

As in the case of liquid fuel droplets, the time of complete burning of carbon particles is found to be proportional to the square of the particle diameter.

$$t_b = k_d d_o^2$$

where,

t_b = time of complete burning of the particle
d_0 = initial particle diameter
k_d = constant of the order of 10^3 when t_b is in seconds and d_o in centimetres.

PROBLEMS

1. What percentages of stoichiometric air are present in the two limits of inflammability of the following combustible substances: hydrogen, carbon monoxide, acetylene, methane, propane, pentane, acetaldehyde, ethyl alcohol and acetane?

[Answer: 1008.4 and 14.0, 294.1 and 14.8, 327.7 and 2.1, 187.7 and 64.5, 186.8 and 40.0, 172.4 and 31.2, 196.6 and 6.9, 155.9 and 29.9, 169.8 and 42.5]

2. Compute the inflammability limits for a fuel of composition CH_4 90%, C_2H_6 4%, C_2H_4 2%, CO_2 2% and N_2 2%.

[Answer: 5.3% and 14.6%]

3. The atmosphere in a solvent handling plant contains 2% benzene and 1% toluene. How much nitrogen is to be added for avoiding fire risk?

[Answer: 115.1 litres N_2 per 100 litres polluted air]

Bibliography

1. Godsave GAE. Studies in the combustion of drops in a fuel spray– the burning of single drops in fuel. Fourth Symposium on Combustion, 1952, Cambridge, 1953.
2. Khitrin LN, *Fyzika Goreniya i Vzryva* (Physics of Combustion and Detonation), Moscow University Press, 1957. English translation, Jerusalem: IPST, 1962.
3. Lewis B and Von Elbe G. *Combustion, Flames and Explosions of Gases*, Third edition. Orlando: Academic Press, 1987.
4. Minkoff GJ and Tipper CFH. *Chemistry of Combustion Reactions*. London: Butterworths, 1962.
5. Thring MW and Essenthigh RH. Thermodynamics and kinetics of solid combustion. In Lowry HH, ed. *Chemistry of Coal Utilization*, Supp. Vol. 754–72, New York: Wiley, 1963.

9 Combustion Appliances

9.1 INTRODUCTION

The general requirements for smooth and efficient combustion may be identified as follows:

Fresh charge of fuel should be freely ignited as it enters the burning zone. Load fluctuations within the specified range should not cause any ignition problem. Radiation from the flame, hot fuel bed and hot refractory surface, and convection from hot gases should be adequate to ignite the fresh fuel. Special care is particularly required in burning solid fuels in lumps. Direct contact of glowing particles is sometimes used to achieve proper ignition. The composition of the fuel–primary air mixture must be well within the limits of inflammability.

Steady combustion is the basis for obtaining the desired rate of heat release. It is therefore necessary to maintain flame stability throughout the combustion process. Burners should not be operated below the stipulated capacity range. The flame should not suddenly come in contact with cold air or a cold surface. All factors causing flame extinction or flash-back are to be avoided. Some types of flame holders enhance flame stability.

High temperature combustion is no doubt a fast process but proceeds with a finite rate. Adequate combustion space should be provided for driving the process to its completion. In the absence of such space, there will be loss of combustibles with the exhaust gases, which will also accentuate the problem of smoke. The smoke may be of two types: brown or black. Brown smoke is due to unburnt combustible vapours, while black smoke is due to carbon black produced by the chilling of the flame when it impinges against a cold surface. Once formed, black smoke is difficult to burn. The shape of the flame should correspond to the geometry of the furnace and vice versa.

The quantity of air supply is important in achieving proper combustion. Excepting some premix systems, excess air is always required for complete combustion. If however, a reducing atmosphere is desired, excess air is to be avoided. Even otherwise, the quantity of air should be minimum, consistent with the requirement of a given system. Excess air leads to increased heat loss in the form of sensible heat of exhaust gases.

The method of air supply is another vital factor of efficient combustion. It is enough for the required oxygen to be present in the system; it is also necessary that there is intimate contact between oxygen and the combustibles. This is achieved by creating intense turbulence in the combustion space. The method of air supply is a powerful means for this.

Sufficient temperature of the combustion gases should be maintained in all parts of the combustion chamber. Suitable precautions are to be taken in the design and operation of the combustion system for this. Temperature is important for smooth ignition, stable combustion and smoke-free performance of the system. Theoretically, the most efficient combustion is that which leads to the maximum temperature possible. It is, however, advisable in practice to keep the furnace temperature below the maximum in many cases where other operation factors are important.

Besides the above general requirements, there are special ones for individual combustion systems. Some of these are discussed below.

9.2 GAS BURNERS

Numerous types of burners have been developed for buring gaseous fuels in domestic and industrial heating appliances. There are two basic types:

1. *Total or partial premix type*, in which a part or whole of the combustion air is mixed with the gas before it emerges out of the nozzle;
2. *Nozzle-mix type*, in which, the air is supplied to the burner tip after the gas leaves the nozzle.

Atmospheric or aerated gas burners

The kinetic energy of the gas is used to aspirate air from the atmosphere into a mixing tube which has the burner head at one end. A Bunsen burner is a typical example. Depending upon the gas pressure, there may be low pressure gas aspiration at 7.5 to 20 cm wg or high pressure gas aspiration at 0.25 kg/cm^2 gauge and above. A typical aerated gas burner is shown in Fig. 9.1. Its detailed description is given later in Section 9.2.1.

Air aspiration gas burners

Air is supplied from a compressor at a medium pressure of 0.15 to 0.35 kg/cm^2 gauge or above and its kinetic energy is used to aspirate gas at 7.5 to 20 cm wg into an aspirator body for mixing the two and then the mixture enters the combustion chamber. The gas is supplied at a constant pressure controlled by a governor (Fig. 9.2).

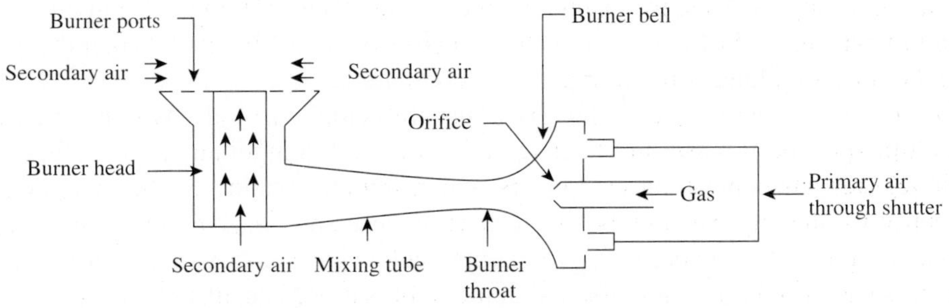

Figure 9.1 Atmospheric gas burner (low pressure gas aspiration)

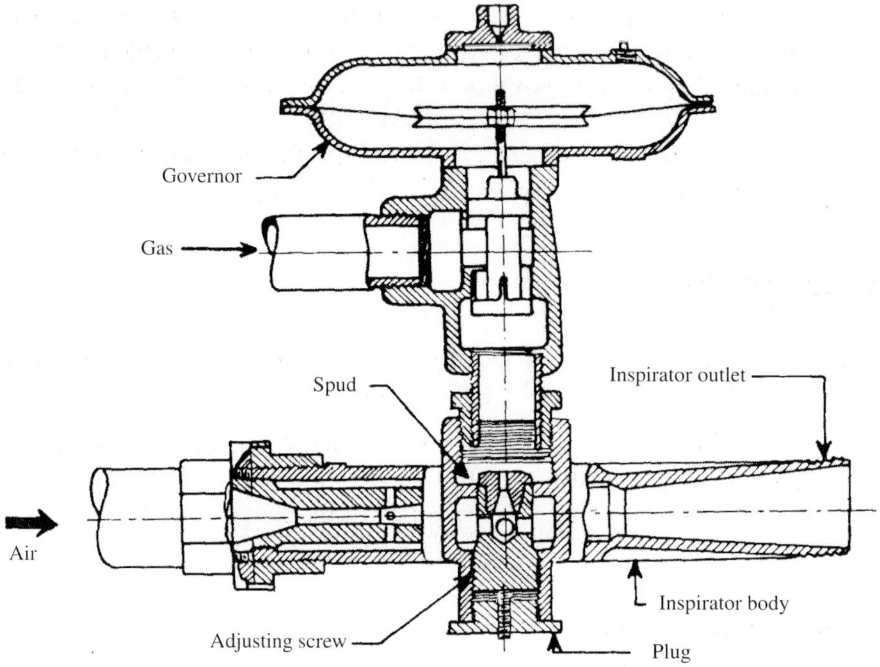

Figure 9.2 Air aspiration gas burner

Burners with an independent supply of gas and air to the mixer

In this system, the fuel and air are mixed in the desired proportion and the compressed mixture is then delivered to the combustion chamber under pressure. This system achieves total premix with high efficiency of mixing.

Typical methods of delivering gas and air in the nozzle-mix systems are given below.

1. *Separate air and gas ports* (Fig. 9.3): Streams of air and gas are separately led into the combustion chamber through ports of suitable size and shape. The introduction of almost parallel systems of gas and air are used to get a diffusion flame for uniform heating of large spaces, for example coke overns. By supplying gas and air at an angle and imparting a whirl to the air, a short intense flame is obtained. When the gas line is in the form of a long, straight pipe, it is called a *gun*.
2. *Gas ports and air register*: The gas is usually supplied by a gas gun to a single- or multiple-port head while air is delivered by a forced-draught air register (Fig. 9.4). The air register has controlling air louvers which initiate rotation of the incoming air and control its quantity. Such units give a short, intense and non-luminous flame. A diffuser plate is attached to the burner tip to maintain a low gas velocity in the central zone. This helps steady and prompt ignition and flame stabilisation. The diffuser also promotes rapid mixing of air and fuel. These burners are popularly known as blast burners.

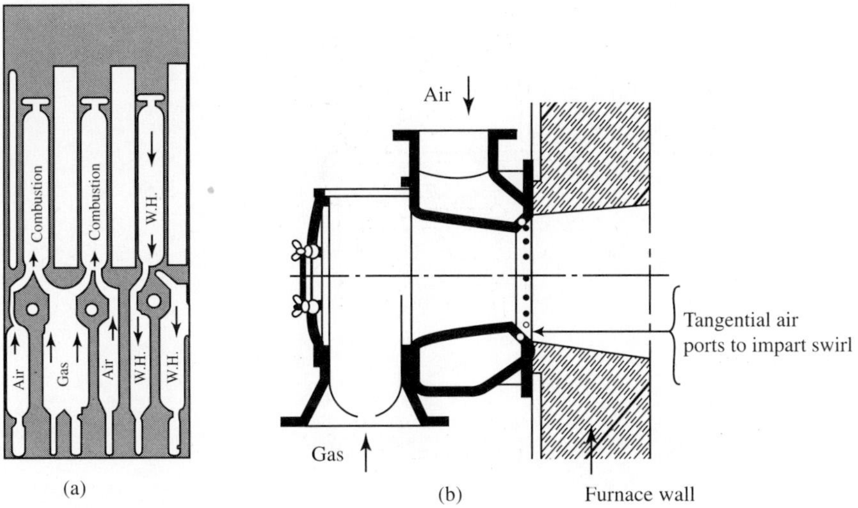

Figure 9.3 Combustion with separate gas and air ports: (a) Long flame (coke oven), and (b) Short flame

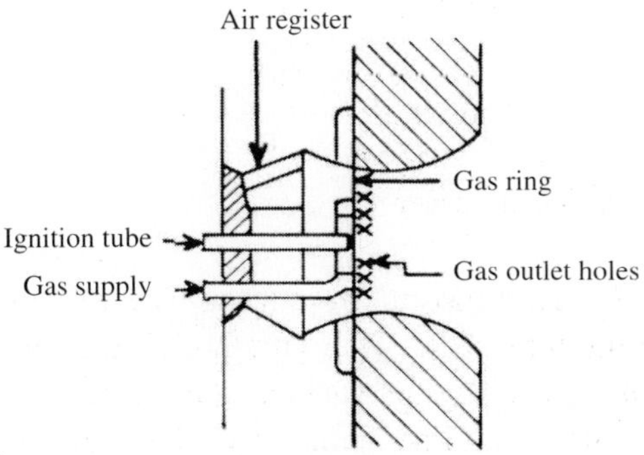

Figure 9.4 A blast burner with air register and gas gun

3. *Fanmix burner*: Unusually high burning rates are obtainable with this nozzle-mix system using as low as 5% excess air. The straight gas line ends in a turbine (spider) having holes drilled on one side of its blades. A fan is attached to the turbine (Fig. 9.5).

The emerging gas rotates the turbine and the fan. The latter delivers the combustion air at right angles to the gas streams. High turbulence results and leads to a thorough mixing of gas and air. The quantity of air is controlled by an air shutter.

Depending upon the structure of the flame, gas burners may be of the following types:

- *Torch burner*: Each burner gives one flame visibly projected from the burner tip.
- *Ring burner*: It is usually an atmospheric burner. The burner head is in the form of concentric rings having suitably drilled holes.

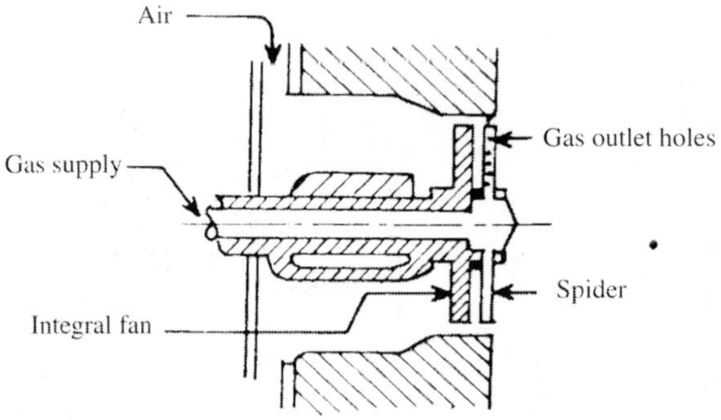

Figure 9.5 Fanmix burner

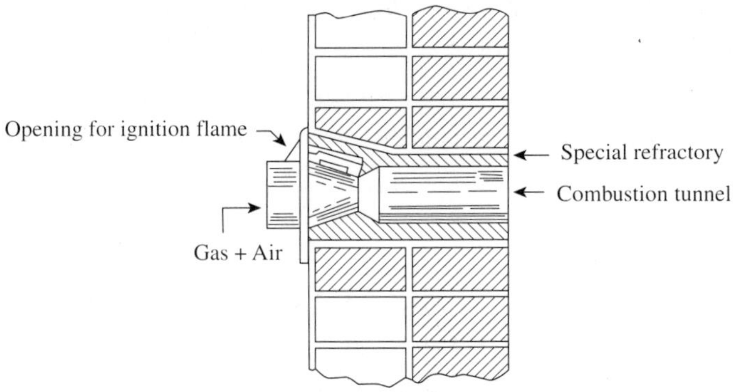

Figure 9.6 Surface-combustion tunnel burner

- *Pipe burner*: This is also an atmospheric burner. The mixing tube ends in a pipe with drilled holes or projecting tube ports. There may be single- or multiple-row drilling.
- *Flameless combustion types*: In the surface combustion system, the combustion is localised on the solid surface and no flame is visible. These systems are often referred to as flameless combustors. Surface combustion is used at both very high and very low gas velocity which will normally give rise to blow-off and flash-back problems, respectively. Figure 9.6 shows a tunnel burner using the surface combustion principle. A gas–air mixture is fed through a port into a circular or rectangular tunnel formed in a refractory block. The incandescent walls of the tunnel bring about surface combustion with very high intensity, say 15,000 kcal/h per square centimetre of port area. Heat transfer is by radiation from the incandescent refractory block. These burners suffer from the disadvantage of noisy operation brought about by the combustion progressing as a series of explosions in rapid succession inside the tunnel. Tunnel burners are manifold in groups so that one gas–air proportioner controls the group. The combustion is complete with the primary air supplied.
- *Pulsating combustors*: They are in a class by themselves. The periodic appearance of explosionflames characterise them (Chapter 8). Figure 9.7 shows a typical pulsating combustor.

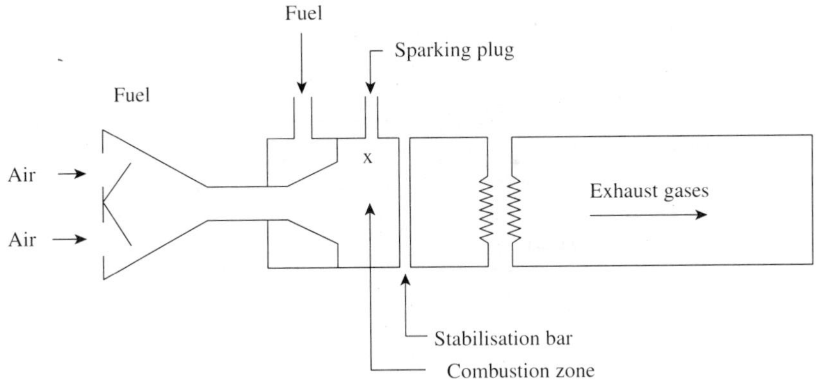

Figure 9.7 Pulsating combuster

The following burners are meant for special applications:

- *Immersion torch burners* (pipe in liquid): These are suitable for heating solutions, fused metals and salts. Heating tubes are immersed in the liquid. Torch burners fire salts at one end of the heating tubes while the flue gases exit from the other end. The liquid is heated indirectly by the heating tubes.
- *Radiant torch burners* (pipe in air or any other gas): The flame is confined within the pipe which heats the air or any other gas. Any contamination by the products of combustion is thus avoided. Radiant burners are often used in industrial drying.
- *Submerged combustion types*: Since tunnel burners do not need any secondary air for combustion; the refractory tunnel may be submerged in a liquid without disturbing the combustion. The combustion gases will bubble through the liquid which will receive heat from these hot gas bubbles. Figure 9.8 shows a simple burner using the submerged combustion principle. More intense and more uniform combustion is achieved by providing additional refractory surface inside the tunnel. The main application of the submerged combustion technique is in crystallising salts for which scaling of the heat-transfer surface is a serious problem. Burners are built with a rated capacity upto 1.3×10^6 kcal/h. Several of these burners may be used in one tank.

9.2.1 Atomospheric Gas Burners

These burners are based on the principle of the Bunsen burner. The variation occurs mainly in the design of the individual parts. The basic features of these burners are discussed below with reference to the burner shown in Fig. 9.1.

The burner consists of a fixed orifice, spud for gas inlet, a controllable shutter for air supply, a venturi-shaped mixing tube and a burner *head* with ports (holes) drilled in it. The narrow zone of the mixing tube is called its *throat* which diverges into the hind part called its *bell*. The gas rate is controlled by a valve in the gas line. When the gas is admitted through the orifice, its kinetic energy is increased owing to the high injection velocity and creates a low static

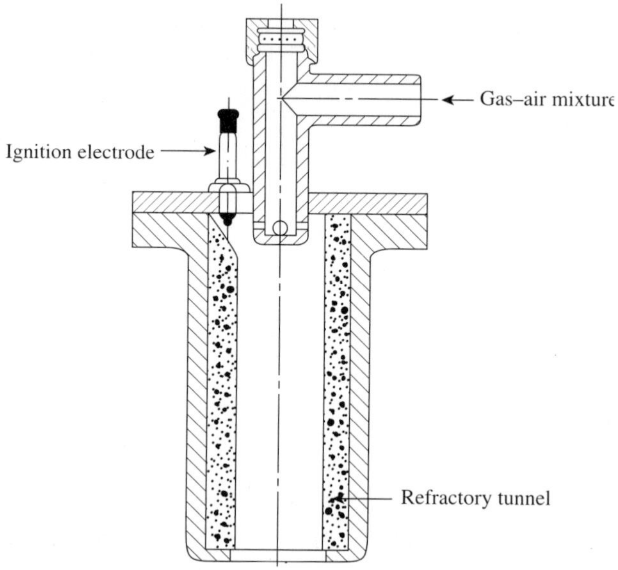

Figure 9.8 Gas burner of submerged combustion type

pressure at the throat. Primary air is sucked into the bell through the air shutter. During flow through the mixing tube from the bell to the burner head the gas and air mix uniformly and the static pressure is increased by the slowing down of the fluid stream. The combustible mixture burns in the form of tiny burner flames anchored on the ports of the burner head. The primary air is not enough to complete the combustion. Secondary air is entrained on the flame surface from the surrounding atmosphere by convection currents. The quantity of primary air depends upon the type of flame, burner design, input rate and type of gas.

The operation characteristics of an aerated gas burner is shown by a set of curves in Fig. 9.9; the coordinates are input rates and primary air supply. Stable burner operation is possible in the middle region confined by the curves 1, 2, 3 and 4.

Curve 1 represents the conditions of lifting and blow-off and sets the upper limit of input rate, or gas flow-rate and the primary air supply. This upper limit varies inversely with the quantity of primary air. A high input rate therefore requires an increased quantity of secondary air. *Curve 2* represents the lower limit of primary air. A low proportion of primary air gives a long, wavy and yellow flame in aerated burners and leads to the formation of soot and carbon monoxide by incomplete combustion. The supply of primary air becomes critical towards the range of high input rate as shown by the curves 1 and 2 tending to coverage on the right-hand side. Burner operation is not satisfactory in this region. The proportion of carbon monoxide in the products of combustion sharply increases for operation to the right of *Curve 4*. The conditions of flash-back prevail on the left of *Curve 3* which thus sets the lower limit of the input rate.

Figure 9.9 thus clearly shows that a given burner can operate within a range of gas flow-rates. The maximum gas rate is limited by the blow-off, incomplete combustion or inability

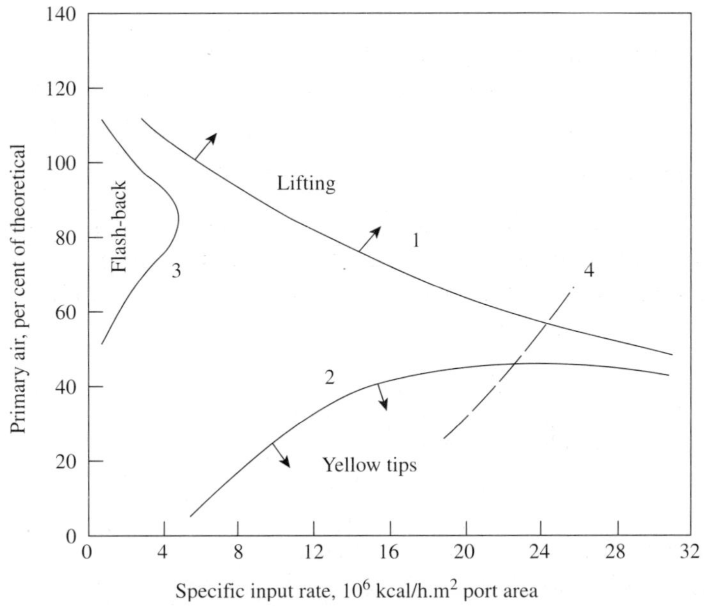

Figure 9.9 Operation characteristics of an aerated burner

to aspirate sufficient primary air. The minimum is limited by the flash-back. The ratio of maximum to minimum gas rate obtainable in a burner is called its *turndown ratio*. A value of 5:1 is desired in most gas-heated appliances. At a given gas rate, a long luminous flame is obtained with a low supply of primary air, while a short turbulent flame is obtained if the quantity of primary air approaches 100% of the theoretical.

All premix or partial premix burners give characteristic curves similar to Fig. 9.9. The relative shape and location of individual curves depend upon the burner design, the operating temperature and the type of gas used. In nozzle-mix burners there is no flash-back but the blow-off can be related to the corresponding conditions of Fig. 9.9. In industrial burners of very high input rate, either the flame stabiliser in the form of a physical obstruction, a diffuser, or the principle of surface combustion is used.

Equations and design data for a simple aerated burner

The gas flow-rate through the orifice is given by

$$V = 83{,}240KA\sqrt{\frac{\Delta pP}{TS}} \tag{9.1}$$

where,

V = rate of gas flow, Nm^3/h

K = Coefficient of discharge (usually 0.8–0.9 for fixed tubular orifice or spud).

A = orifice (port) area, m^2
Δp = pressure drop at the orifice, cm H_2O
P = initial gas pressure, cm Hg abs
T = gas temperature, K
S = specific gravity of gas (air = 1)

Knowing the input rate and the desired pressure drop at the orifice, it is possible to calculate the orifice diameter from equation (9.1). The desired pressure drop may be calculated from the primary air requirement with the following empirical equation for the aeration capacity of the burner.

$$P_A = \frac{\sqrt[4]{PS}}{\sqrt{C_N}} \times \left(\frac{5\times 10^6 \sqrt[4]{A_m A_p}}{\sqrt{Q}} \right) \times \frac{300}{T} \tag{9.2}$$

where,

P_A = primary air, per cent of theoretical air
P = gas pressure at orifice inlet, cm wg
C_N = net CV of gas, kcal/Nm^3
A_m = average area of mixing tube, $m^2 = \dfrac{A_1 + A_2}{2}$
A_1 = throat area, m^2
A_2 = area of mixing tube outlet, m^2
A_P = total port area, m^2
Q = heat input rate, kcal/h

The first term on the right-hand side of equation (9.2) is a function of the nature of the gas and its supply pressure. The second term is a function of the burner dimensions. The third term shows the effect of the temperature of the mixture.

The nature of application of the burner influences the primary air requirement. When a long flame is permitted in water heaters and furnaces, only 35%–40% primary air is recommended with natural gas and most manufactured gases. The figure is at least 55% with LPG. When a short flame is required as in cooking ranges, the primary air requirement is higher. Cooking ranges need 55%–60% and radiant heaters need at least 65%.

The total port area is computed from the design data of specific heat input rate for the burner, say, 6×10^6 kcal/m^2 per hour of port area.

Other design data:

Throat to total port area ratio = 0.20 to 1.0
Air shutter area = 1.25 to 2.25 times the total port area
Distance from gas orifice to throat = 0.5 to 2 times the throat diameter
Length of mixing tube = 6 times the throat diameter
Slope of mixing tube ~ 3°
Area of burner head = 1.5 to 2 times the total port area

It is clear from the above that the aerated burner design is sharply influenced by the nature of the gas. On the other hand the nature and quantity of available gases may widely vary. It is therefore necessary to know the basis of determination of interchangeability of gases.

The burner output rate is given by

$$Q = \eta V C_N$$

where, η = coefficient of combustion efficiency.

Combining equations (9.1) and (9.3), we get.

$$Q \propto \frac{C_N}{\sqrt{S}}$$

Furthermore, from equation (9.2), we have the primary air supply

$$P_A \propto \frac{1}{\left(\frac{C_N}{\sqrt{S}}\right)^{0.5}}$$

The ratio $C_N/\sqrt{S}$ is known as the Wobbe index. It is a useful parameter in assessing interchangability of gases with respect to aerated burners (see also Chapter 6). It is a dimensional quantity with the dimensions of calorific value of gases.

The Wobbe index is of great importance for gases supplied through the distribution system because aerated burners are widely used by a majority of customers. It should be realised that this parameter is of limited significance for other types of burners.

An *aeration test burner* (ATB) has been specified for evaluating variation in gas quantity. It is a Teclu burner of 30.5 cm height and 10.26 mm internal diameter and has a nipple of 0.99 mm diameter. The air shutter is rectangular and adjustable by a micrometre head. The given gas is burnt in it. It is aimed at determining the extent of the opening of the air shutter at which an exact cone of 6.35 cm (2.5 inches) height is obtained. The result is expressed as the ATB number. A variation of 5 units in the ATB number of a gas can seriously affect the performance of an aerated burner.

The maximum heat inputs obtainable with aerated burners of low gas pressure types are given in Table 9.1.

Table 9.1 Capacity of aerated gas burner (low gas pressure)

Type of gas	Max. input rate × 10^5 kcal/h.m^2 of port area			Wobbe Index $C_N/\sqrt{S}$ kcal/Nm^3
	Primary air per cent			
	40	55	70	
Coke oven gas	20,400	11,800	7,500	8,100
Natural gas	12,100	7,100	4,700	11,800
LPG	Yellow flame	8,200	5,100	21,200

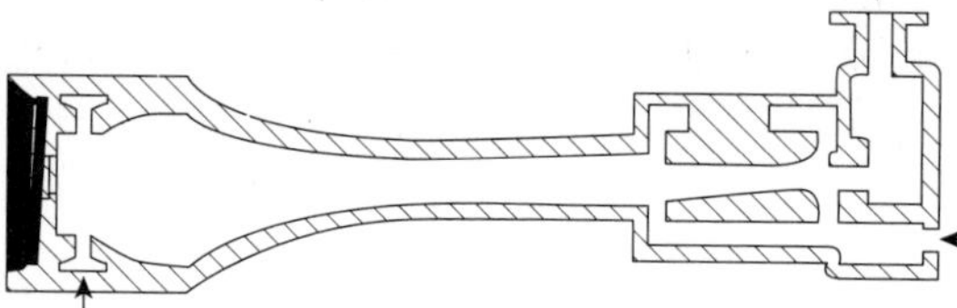

Figure 9.10 An aerated burner with two-stage aspiration and flame holder

Aerated burners of high pressure gas aspiration type work on the principle depicted in Fig. 9.1. However, better mixing and control of gas and air are obtained with a two-stage aspiration at the orifice (spud) and the burner throat (Fig. 9.10). This is also provided with a flame holder at the burner tip, which stabilises the flame by maintaining a zone of relatively low gas velocity when the gas pressure in the line is quite high.

9.3 OIL BURNERS

Basically there are two different classes of oil burners, namely, (i) vaporising type and (ii) atomising type. The vaporising burners are favoured for heating units of small size such as portable air heaters, small boilers and cooking stoves for domestic purposes. These are characterised by low cost and quiet operation. The atomising burners are used in industrial heating and to a much lesser extent in some of the big domestic heating appliances.

9.3.1 Vaporising Burners

Vaporising burners are of different designs and they are suitable for oils ranging from naphtha to light fuel oil. *Blow torches* and *portable gasoline stoves* burn straight-run gasoline. The volatile fuel is passed at a low pressure through a tube adjacent to the flame, where vaporisation takes place. The vapour stream issues out of an orifice at a high velocity and entrains primary air. The fuel–air mixture passes through a mixing tube and burns at the burner head similar to atmospheric gas burner principle.

Wick-type vaporising burners use superior kerosene. A wick raises the liquid fuel from a trough by capillary action. Radiation from the flame and adjacent hot surfaces vaporises the fuel from the wick head. The vapours are allowed to enter into an annular space between two vertical, perforated metallic cylinders and burn in the air drawn through the perforations by natural draught. The combustion takes place throughout the annular space with a blue flame. These burners are used as cooking stoves. A variation of the wick-type burner has a capacity upto 5 kg/h and is suitable in domestic boilers and central heating. Here an asbestos wick receives the fuel by a gravity-feed which may be regulated within a certain range. In ordinary

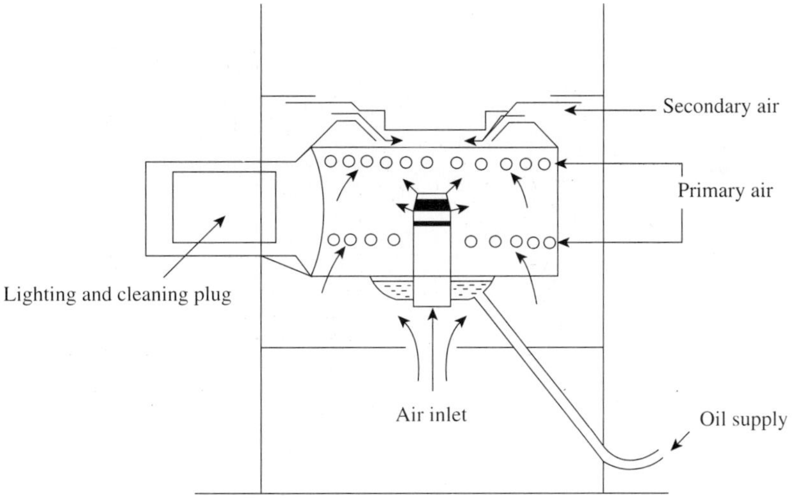

Figure 9.11 Pot-type burner

wick-type burners, the burning rate is controlled by varying the length of the exposed portion of the wick; the control is less satisfactory.

Figure 9.11 shows a ***pot-type burner***: another example of a common-type vaporising burner. There is a fuel reservoir at the bottom, supplied by an oil line. The oil surface is heated by radiation from the flame and the hot walls. The vapours mix with the primary air admitted through a central pillar and perforations in the pot wall. The fuel-air mixture is too rich at the oil surface to support combustion. Further up, it is diluted with air and becomes inflammable. The flame is supported at the top of the pot. These burners may work on natural or forced draught. Superior kerosene or light fuel oil is used. Some soot formation cannot be avoided and the pot burner should be periodically cleaned. The capacity may be upto 10 kg/h.

Wall-wiping flame rotary vaporising burners use coarse atomisation before the fuel is vaporised. A centrally located centrifugal spinner distributes the oil in the form of a spray of coarse droplets striking the hot stainless steel wall of the burner (Fig. 9.12). The vapours are formed and mixed with the air which is directed towards the wall by a fan. The fuel-air mixture rises upwards and burns on a stabilising grill with a blue flame. The flame burns around the inner circumference of the furnace only, and hence gets its characteristic name. These burners can burn up to 30 kg/h oil. High combustion efficacy is achieved by the good mixing of fuel and air. Power requirement is low and the operation noiseless. Kerosene and light fuel oil are used.

The ***petromax stoves*** burn kerosene. An upward moving fuel jet is allowed to strike against a hot metal surface at a high velocity. The vapours produced burn with a short and intense flame over the metal surface. The fuel is stored in a small reservoir below the fuel line and the necessary oil pressure is supplied by compressing air into the reservoir by a hand-driven piston. These stoves were quite popular in earlier times. Their use has declined owing to the risk of accidents and noisy operation, and to the advent of cheaper and safer stoves described in the previous paragraphs.

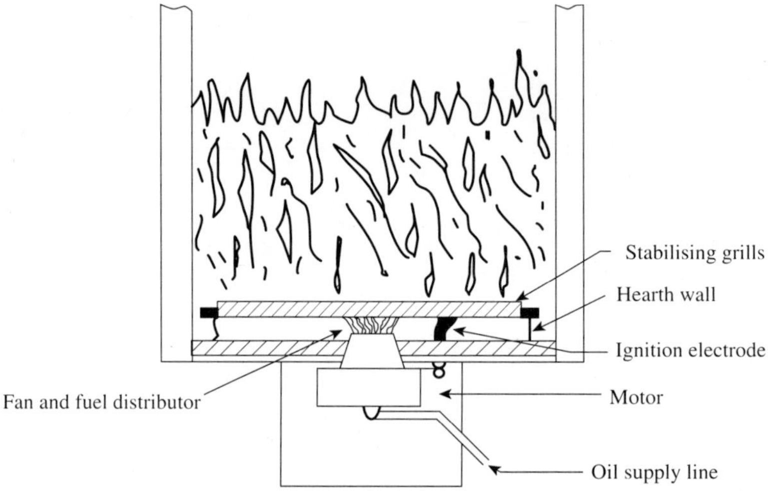

Figure 9.12 Wall-wiping flame rotary vaporising burner

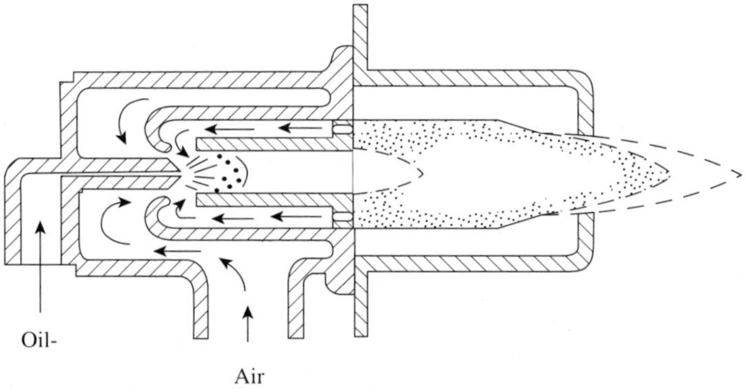

Figure 9.13 High speed vaporising burner

A ***high speed vaporising burner*** is shown in Fig. 9.13. A high rate of vaporisation is achieved by atomisation of the volatile liquid fuel and recirculation of hot combustion gases within the burner. This particular burner clearly bears out the difference between vaporising and atomising burners. The combustion chamber receives fuel vapours in the former and fuel droplets in the latter.

9.3.2 Atomising Burners

Atomisation prepares the liquid fuel for combustion by disintergrating it into droplets. Enormous surface area per unit weight of the fuel is created and this helps the heterogeneous combustion of the liquid fuel and the gaseous oxidant. It is well known that a liquid fuel like kerosene in a beaker cannot be ignited with a match but a spray of the same fuel easily ignites on the approach of a match flame and continues to burn by itself. Atomising burners are provided with an arrangement for the atomisation of liquid fuels before the actual combustion takes place.

There are basically three types of atomising burners differing from one another on the principle of atomisation. They are

1. Pressure jet atomising burner,
2. Twin-fluid or blast atomising burner,
3. Rotary atomising burner.

The mechanism of atomisation is more or less similar in them. Initially a jet or a thin film of liquid is obtained and allowed to emerge into the open atmosphere at a suitable velocity. Both the jet and unsupported film are unstable and break up into drops. Disturbances like swirling of the jet or rotation of the film hastens the disintegration process. Viscosity and surface tension of the liquid are of particular importance in atomisation. Other factors remaining the same, low viscosity and low surface tension are desirable for ease of dispersion and formation of small drops. The atomisation process is further assisted by the resistance, friction and turbulence of the air. The air resistance and friction become effective when any ruffling or distortion of the surface of the jet or film occurs. These forces are particularly significant at combustion chamber conditions owing to increased air viscosity at high temperatures. Local turbulance in the air causes deformation and disruption of the drops.

The velocity of the jet or film is one of the most important parameters influencing the process of atomisation. The higher the liquid velocity, the greater the disturbance in the liquid and also the resistance, friction and turbulence of air, and hence the greater the extent of atomisation.

A high speed jet disintegrates into a number of big droplets and a far greater number of *secondary droplets*. The big droplets are sub-divided into small ones during the flow of the spray. The atomisation is never complete in the immediate vicinity of the orifice, but it proceeds during the penetration of the spray into the surrounding medium. The resultant spray has a combination of droplets of various sizes. A considerable proportion of the droplets is in the extremely small size range, 25 μ or less. These assist in the initial ignition and promote the stability of the flame. The small droplets are quickly vaporised, or cracked and ignited. The resultant flame can support the combustion of the coarser ones. Very large droplets are to be avoided as far as possible because they may not get completely burnt in the time available, and they then create complications by depositing themselves on the walls of the combustion chamber.

The energy required to create a new surface by disintegrating the liquid into droplets is only a small fraction of the total energy consumed in the atomisation process. Energy losses (inertial, viscous and those due to mechanical inefficiency of the equipment) constitute the bulk of the total energy consumption.

Pressure jet atomisers are of two types. The plain orifice type is most widely used for fuel injection in diesel and other internal combustion engines. Very high oil pressures, upto 350 kg/cm^2 or even higher are used. The centrifugal swirl types are most popular in large industrial oil burners. Much lower oil pressures, 7 to 35 kg/cm^2 are required in these.

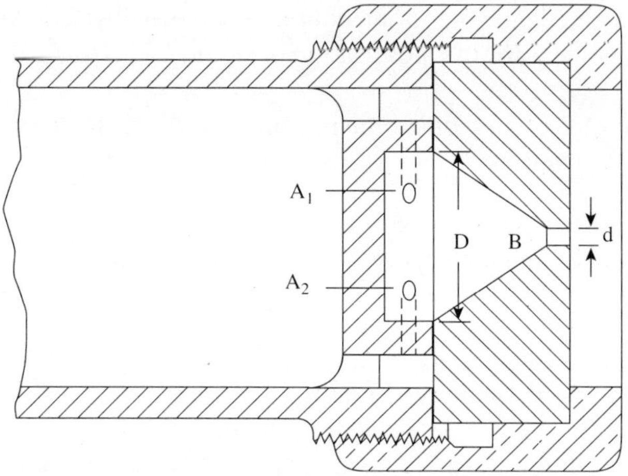

Figure 9.14 Swirl-type oil burner

A ***simple swirl-type atomiser*** is shown in Fig. 9.14. The oil enters the circular swirl chamber (or vortex chamber) through tangentially spaced ducts or slots (A_1 and A_2) and rotates in the chamber around an air core with an angular velocity varying inversely as the radius of the swirl. The rotating mass of liquid issues through an orifice, C, under the influence of two main forces. The translation force moves it forward while the centrifugal or spinning force makes it fly outwards tangentially. The result is the formation of a hollow, divergently conical and rapidly thinning film which finally provides a spray of droplets.

The oil viscosity should be about 70 Redwood 1 seconds for small nozzles and about 100 Redwood 1 seconds for larger nozzles. Heavier fuel oils should be preheated to this level of viscosity.

The output of a pressure atomiser, Q, increases with the increase in oil pressure, P, approximately following $Q \propto P^{0.5}$, and decreases with the decrease in oil viscosity. The mean droplet size d increases with viscosity, varies inversely as $P^{0.35}$ and increases with the orifice diameter, d_0, approximately following $d \propto d_0^{0.5}$. The droplet size varies between 10 μ and 200 μ.

The spray angle varies as the ratio of the total area of the tangential ducts or slots to the square root of the product of the orifice outlet diameter, d_0, and the swirl chamber diameter, D.

$$\text{Spray angle} \propto \frac{A_1 + A_2}{\sqrt{d_0 D}}$$

This angle is independent of the oil pressure and the slope of the internal surfaces of the swirl chamber. However, the slope of the outlet edge from which the spray develops influences the spray angle. The spray angle is widened by rounding the edge.

The simple pressure jet swirl-type atomisers are severely limited in flexibility owing to the dependence of output on oil pressure. A simple calculation from $Q \propto P^{0.5}$ shows that a wide range of oil pressure between 7 and 35 kg/cm^2 allows a ratio of only 1:2.2 between the lowest and the

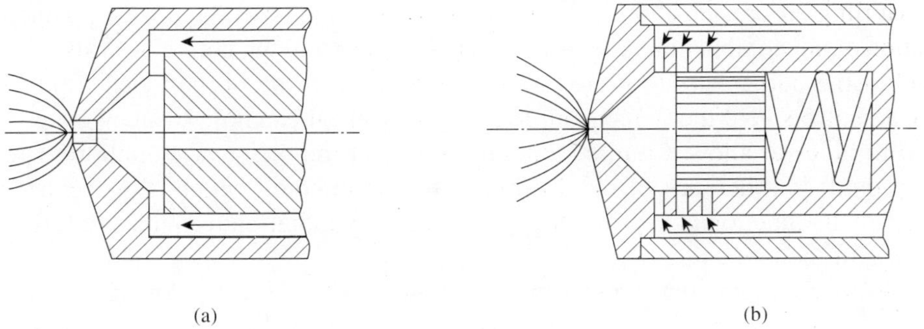

Figure 9.15 Wide range pressure jets: (a) Spill type and (b) Slot type

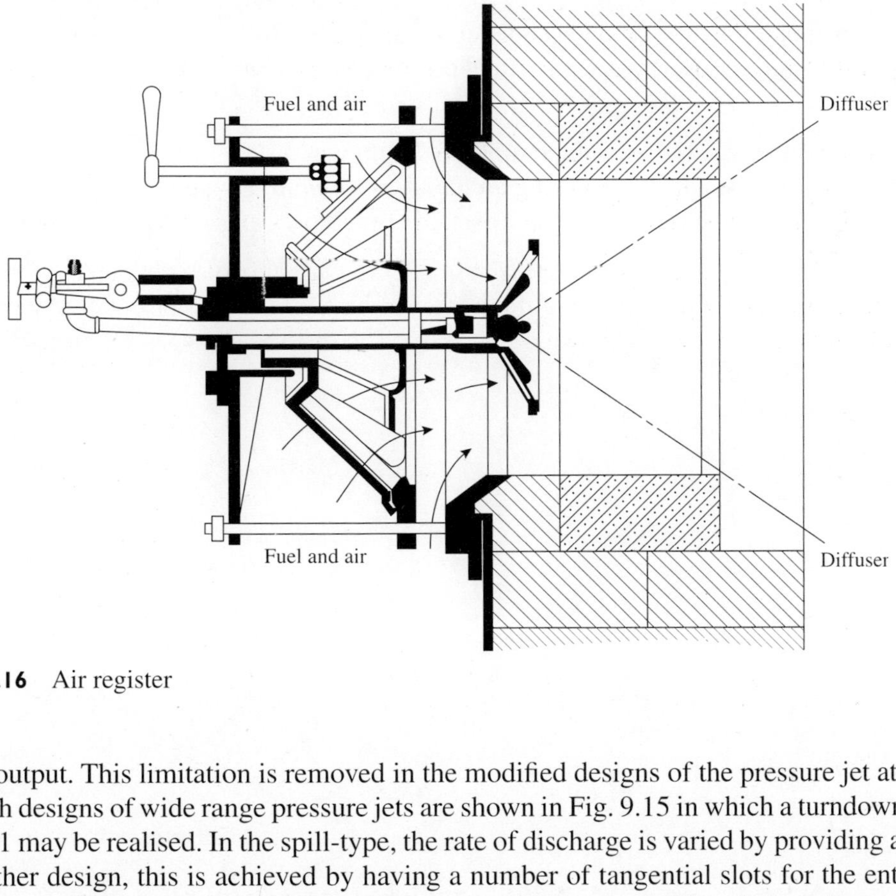

Figure 9.16 Air register

highest output. This limitation is removed in the modified designs of the pressure jet atomisers. Two such designs of wide range pressure jets are shown in Fig. 9.15 in which a turndown ratio of upto 10:1 may be realised. In the spill-type, the rate of discharge is varied by providing a bypass. In the other design, this is achieved by having a number of tangential slots for the entry of oil into the swirl chamber and varying the number of available slots with a sliding piston.

Combustion air flows from the windbox through an *air register* surrounding the burner (Fig. 9.16). The functions of the air register is to control the velocity and direction of the mixture of fuel and air, wrap the air around the flame and finally give a spinning motion to the air. The whirling air creates a low pressure zone in the vicinity of the nozzle tip and causes a recirculation of hot combustion gases which help in stabilising the flame. A diffuser is installed at the end of the burner *gun* to impart a flow of air around the oil particles, guide air along the flame and also shield the flame from high velocity air. It is a hollow metal cone, slotted or pierced with holes to allow a limited amount of air to promote steady ignition. *A refractory throat* is provided at the back of the diffuser. It has an opening which is concentric with the burner nozzle. Its function is to promote rapid ignition by radiation from the hot refractory and good mixing of the fuel and air.

The pressure jet atomising burners are very quiet and have a low operating cost. Power requirement is the lowest of all atomisers. These are perhaps the most widely used of all the atomising burners in existence. A turndown ratio of 6:1 is expected even in common installations of the wide-range types. Burners with a capacity upto 4,500 kg/h of oil are obtainable. Small domestic installations of the pressure jet-type are also available. The pressure jet atomiser is susceptible to choking by dirt in oil. The latter should be carefully strained.

Twin-fluid atomisers use an auxiliary fluid (air or steam) to atomise the oil and are of three main types:

1. Low pressure-type (LP) using air at 35 to 70 cm wg,
2. Medium pressure-type (MP) using air at 0.4 to 1 kg/cm^2 gauge
3. High pressure-type (HP) using air or steam at pressures exceeding 1 kg/cm^2 gauge.

Each of these types may again be either *inside-mix* or *outside-mix* type dedpending on whether the fuel and auxiliary fluid mix within the burner or beyond the burner outlet. A few inside-and-outside mixing systems are shown in Fig. 9.17. The inside-mix types are more commonly

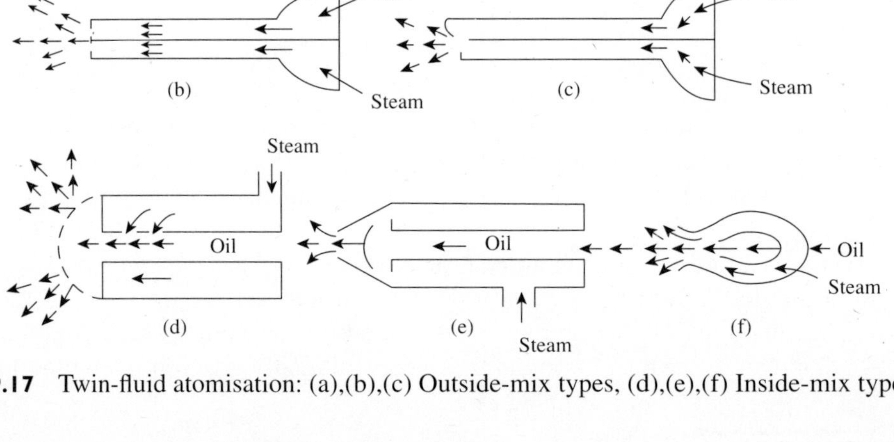

Figure 9.17 Twin-fluid atomisation: (a),(b),(c) Outside-mix types, (d),(e),(f) Inside-mix types

used. These provide high efficiency at the high rate of discharge, a high turndown ratio and a flexible flame structure.

The general principle in these atomisers is common. Oil flows through a central tube at a controlled rate and the auxiliary fluid meets it obliquely at a pre-determined angle as it emerges from the tube. The energy requirement is met by the auxiliary fluid and the oil pressure need not be more than that required to ensure a uniform flow through the nozzle.

Approximately the same kinetic energy of the auxiliary fluid is required for atomisation in the three types of twin fluid atomisers. Therefore, the higher the pressure of the auxiliary fluid, the lower the quantity required. High pressure atomisers need only 2% to 3% of the total combination air for atomisation while this figure is 3% to 15% for MP burners and as high as 20% to 100% for LP burners. On the weight basis, the steam requirement is similar to air requirement (0.3 kg/ kg oil) in the HP atomisers. Only 2.5% of the steam that could be generated in a boiler by a high pressure steam atomising burner is required for achieving the atomisation.

A high turndown ratio of upto 10:1 is obtained with HP and MP burners. The LP units normally have a low turndown ratio of 2:1 to 5:1, HP and MP burners use most of the combustion air as secondary air and thus permit high waste heat recovery by air preheating. The cost of air compression in HP burners is very high. Therefore, steam is more frequently used to atomise heavy residual fuels and coal tar fuels which require preheat for getting the correct viscosity.

LP units are used in domestic and industrial installations. They are quite versatile and built in wide size ranges. Industrial LP burners with a capacity of 2,000 kg/h are obtainable, and lighter grades of fuel oils are used.

MP burners are used in boilers and furnaces where greater flexibility and heat regeneration or recuperation are required.

When steam is substituted for compressed air, the same quality of atomisation can be obtained at a considerably lower pressure. High pressure steam burners are widely used in large installations where high pressure steam is available. These burners may have a capacity upto 3,500 kg/h. Heavy fuel oils are successfully burnt by HP steam atomising burners. Less preheating of the oil is required when steam is used to atomise it. Steam should be available at a pressure exceeding the oil pressure. The admission of the secondary air is through air registers. However, no diffuser is normally used. The flame shape is altered by changing the nozzle tips.

Rotary atomisers have a central stationary fuel line which delivers the oil to the inner surface of a rotating hollow tapered cup (Fig. 9.18). The cup is rotated at speeds of 3,600 to 10,000 rpm by either air-driven impellers or by an electric motor. Friction between the oil and the cup surface causes the former to rotate with the cup, and centrifugal force, assisted by the tapering of the cup causes it to flow towards the brim. Finally the oil is flung off in a thin film which readily disintegrates into fine droplets. The smoothness and uniformity of the cup surface results in the production of an even film with the same velocity at each point of the brim. The result is a higher degree of uniformity of droplet size than in the other atomisers.

About 15% of the combustion air is supplied as primary air by a fan attached to the rotating shaft of the motor driven cup. This air at a pressure of 10–25 cm wg gives a forward movement to the spray and also exerts additional atomising force on the droplets. Secondary air is supplied

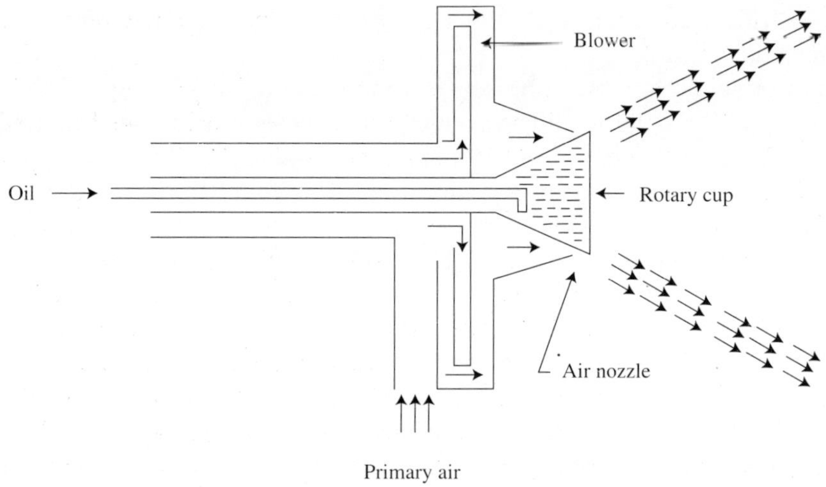

Figure 9.18 Rotary atomising burner

to the door underneath or through an opening in a chequered floor, the draught being provided by the stack. The initial spray angle is 180⁰ and the conical slope is given by the primary air steam. A fuel droplet size of 50 μ is claimed.

The oil pressure should be low, 5–8 cm wg and the viscosity should be in the range of 100–300 Redwood 1 seconds. Rotary atomisers can tolerate higher viscosity than the others. They may however lose atomising efficiency at low viscosities. The low viscosities used in a pressure jet system may cause the oil to 'slip' with the cup. The trouble due to choking by dirt is less. However, there is a greater chance of carbon deposition by cracking caused by the radiation from the surroundings after the burner is turned off.

Rotary atomising burners are extensively used in boilers and small installations. A high turndown ratio of 16:1 can be realised. Capacity up to 800 kg/h is available which is much lower than the maximum capacity of other atomiser types. Rotary atomising burners are compact (components combined in an integral unit), efficient and comparatively low in initial cost. The power requirements are also relatively low.

The rotary units are usually mounted horizontally in the furnace walls, although a few domestic units have a vertical shaft, producing a flame shaped like a sunflower.

9.3.3 Drip Feed System in Oil Burning

When the combustion chamber is above 600°C and the particular feeding system can be conveniently provided in the chamber, a very simple and cheap method of oil burning can be utilised. The fuel is gravity-fed into the hot environment of the furnace through a suitably shaped piece of metal tubing. The feed rate is controlled by a valve. The oil droplets fall on a current of combustion air which is drawn from the chambers undergoing cooling. The ignition is quite smooth and the combustion is stable. The drip feed system is particularly suitable for some kilns used in silicate industries.

9.4 COAL BURNING EQUIPMENT

Coal may be burnt by

1. Fixed or agitated beds on a grate,
2. Suspension in air in pulverised fuel burners or cyclone burners,
3. Fluidised bed combustors, in a slurry with water or in suspension in oil in atomisation burners.

Grate firing is employed in domestic and industrial units and also in thermal power stations. Pulverised fuel and cyclone burners are used in thermal power stations and large industrial units. Fluidised bed, coal–water mixture (CWM) and coal–oil mixture (COM) systems are in an advanced stage of development with limited number of units already in the market.

Coal may be burnt in a grate by hand firing or by using mechanical stokers. Domestic appliances, locomotive boilers and small industrial units use hand firing while medium to large industrial units are fired by mechanical stokers.

9.4.1 Hand Firing

The ordinary grates are usually made of iron bars with 6 to 10 mm gaps between them. There are two accepted methods of hand firing, namely *spreading* and coking.

Spreading method: A small quantity of coal is supplied at a time by spreading over a part of the fuel bed. Care is taken to maintain uniform bed thickness. Anthracite and free-burning coal are fired by this method.

Coking method: A considerable amount of coal is fed onto a dead plate just inside the door. The heap of fresh coal is slowly carbonised by the heat of the glowing bed. The volatile products pass over the bed and get burnt in the air rising through the grate. The coke is then distributed over the depleted bed and a fresh charge of coal supplied to the dead plate. Caking coals are successfully burnt by this method. In a variation of the coking method, firing is done in alternate sections—either over the front and back of the grate or to the right and left. This is known as the alternate method of hand firing.

The success of hand firing largely depends on the skill of the fireman. The ash is disposed of by shaking or dumping the grates periodically. However, the bed should be disturbed as little as possible. This precaution reduces clinker formation and prevents excessive loss of fuels with the refuse. The amount of air should be controlled with dampers. It should, however, be obvious that hand firing cannot safeguard the conditions of efficient combustion.

9.4.2 Mechanical Stokers

The mechanical stokers function on the principle of continuous coal feeding. The evolution of volatile matter is thus uniform and it becomes easier to control the air required for the primary and secondary combustion. There are five types of mechanical stokers, namely,

- under-feed stokers,
- chain- and travelling-grate stokers,
- vibrating-gate stokers,
- coking stokers,
- spreader stokers.

The last four types come under the general group of overfeed stokers. The spreader stokers are the correct example of overfeed stokers. The other three follow cross-feed pattern of fuel and air movement where the fuel moves horizontally and the air moves upwards at high angles to the fuel bed.

Under-feed stokers

As their name implies, they introduce the coal into a narrow horizontal retort or trough beneath the burning fuel bed. The incoming coal progressively forces the fuel out of the retort and onto the side grates, the rear end being closed. During its stay beneath the burning bed, the coal undergoes drying and carbonisation. The volatiles escape through the bed which ensures their smooth ignition. The coal is fed by means of a plunger or ram or a reciprocating retort bottom or a worm or screw. This feeding action places the bed under compression and closes any holes that may tend to form. Air is supplied through tuyeres (Fig. 9.19). The ash is usually removed by means of dumping grates at the sides and then discharged into ash pits.

The under-feed stokers are of single-retort and multiple-retort types. The single-retort stokers are widely used for firing medium to small installations with free-burning to moderately caking coals of high ash fusion point. Clinkering troubles are experienced if ash fusion point is not sufficiently high. The multiple retort stokers are meant for large installations.

The burning rate in under-feed stokers is about 100–250 kg of coal per square metre of the grate area per hour. The grate size in single-retort units may vary from 1 m^2 to 12 m^2. Owing

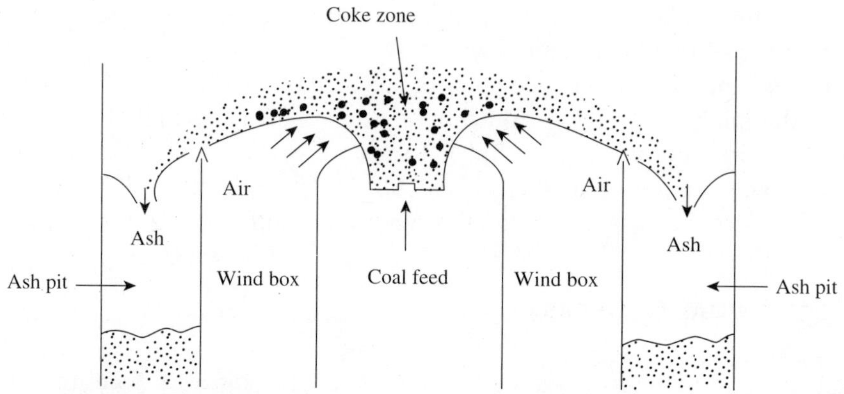

Figure 9.19 Single-retort under-feed stoker

to the absence of an ignition problem, the single-retort units are open or archless. The large multiple-retort units operate well with a long reach-arch for the avoidance of the drifting of fuel particles towards the dumping grates which are provided at the rear end of the multiple-retort units. This action of the rear arch will be discussed under chain-grate stokers.

Smokeless combustion is easiest to achieve in under-feed stokers. Such units constitute a major portion of coal firing applications to dwellings and small commercial enterprises.

Chain- and travelling-grate stokers

In the chain- and travelling-grate stokers the coal is fed on a forward moving endless grate from a hopper under an adjustable guillotine-type gate, which controls the thickness of the fuel bed. Ignition of the fresh coal is accompilished by a combination of three factors:

1. Radiation from a firebrick arch,
2. Hot gases blowing over the charge,
3. Spray of fine glowing fuel particles from the rear end onto the fresh coal.

Air supply is controlled at different positions on the grate by dividing the air space beneath into a number of compartments. A dead plate at the front blocks the air passage and thus prevents spreading of combustion towards the hopper. A complete combustion of the gases and vapours is ensured by supplying adequate secondary air and by destroying stratification with the help of air or steam jets. The moving grate carries ash over the rear end to an ash pit or other disposal means.

The chain-grate and travelling-grate stokers differ in the construction of the grate. An endless chain itself forms the grate in the former case, the individual links of the chain being sufficiently long. The edges of the links are usually serrated to provide passage for air. The travelling grate is made of bars which consist of small individual *castings* or *keys*. These keys are carried forward in endless belt fashion on chains, but do not themselves form a part of the chain. This arrangement is convenient in replacing broken or damaged keys without dismantling the chain.

The fuel bed in these stokers is not disturbed, and therefore the clinkering troubles are low. High ash coals with a relatively low ash fusion point can be successfully burnt. Low ash coals are actually unsuitable as these grates need a protecting ash layer. The chain- and travelling-grate stokers are best suited for free burning solid fuels, such as non-caking bituminous coals, lignite, anthracite and coke breeze. Strongly caking coals lead to an excessive fuel loss with the refuse.

Although furnaces with these moving grate stokers may be of an open type or archless, at least a short front arch and often a long rear arch are most widely used. The radiation from the front arch hinders the secondary combustion by promoting stratification. On the other hand, when a long rear arch is located low enough over the fuel bed, sufficient velocity is imparted to the furnace gas, which causes glowing fines to be carried over and deposited on the fresh charge. This raining of particles together with the radiation from the furnace nose maintains smooth ignition while turbulence created at the furnace throat goes a long way to break the stratification. Installation of jets at the front wall further improves the secondary combustion. Figure 9.20 shows a travelling-grate stoker with combination arch furnace.

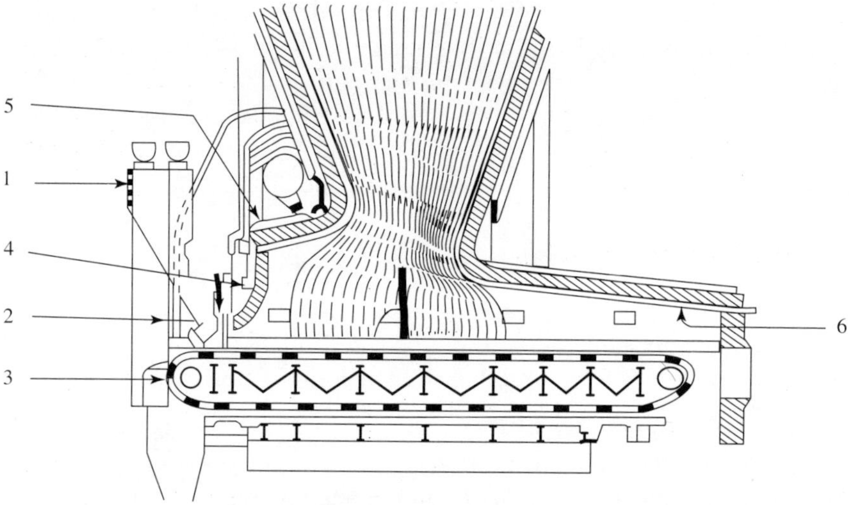

Figure 9.20 Travelling-grate stoker with combination-arch furnace (short front-arch and long rear-arch)
1. Hopper 2. Guillotine-gate 3. Grate 4. Furnace nose
5. Front-arch 6. Rear-arch 7. Furnace throat (behind the stoker structure and not visible.)

The burning rate in these stokers is 200–350 kg/m^2 per hour. The grate area may range from 1.5 to 70 m^2. The grate width corresponds to the furnace width and may vary from 1 m to as high as 7.5 m. In addition to being widely used for boilers of various sizes ranging from small to large, the moving-grate stokers are also popular for many special furnaces. The chain-grate stoker is one of the oldest forms of automatic coal firing appliance.

Vibrating-grate stoker

It consists of a flat inclined grate which is given a periodic forward and backward vibration in an essentially horizontal plane. This vibration affects the feeding of coal from the hopper, the movement of the fuel bed down the grate and finally the discharge of the refuse at the rear end. It also helps in maintaining a compact and uniform bed. The frequency of vibration is usually 600 cycles per minute. The stoker is normally vibrated every two minutes and the vibration lasts for a few seconds at a time. The burning rate is 150–200 kg/m^2 per hour. Vibrating stokers are quite popular in the new small to medium installations in some of the advanced countries. They are suitable for a wide range of coals with low-to-high ash and medium-to-high volatiles, low-to-high ash fusion point and weakly-to-strongly caking.

Coking stokers

The coal is fed by a reciprocating ram on to a reciprocating grate. The rate of firing is controlled by the speed of the to-and-fro movement of the grate and the ram, and by the length of travel of the ram. There is a dead plate at the front end for accomplishing coking of the fresh charge before the fuel is pushed along the grate bars. The stokers are suitable for medium-to-high volatile caking coals. Coking stokers and vibrating-grate stokers are similar in principle.

Spreader stokers

They have provision for continuously projecting coal into the furnace above the glowing fuel bed on a grate. The fines are partly burnt in suspension, while the large sizes form the fuel bed. These stokers thus combine the principles of combustion of coal on grate and in suspension. The coal is fed from the hopper by a throwing device which may be one of the following three:

1. Paddle-type underthrow spreader, in which the paddles revolve around a horizontal axis and throw the coal from their bottom periphery,
2. Paddle-type overthrow spreader, in which the paddles revolve in an opposite direction and throw the coal from the top of their periphery,
3. Steam or air jets.

The grate may be stationary, dumping or moving, thus determining the method of ash disposal. The moving grate is essentially an endless chair moving from the rear to the front end at a very slow speed, say 10 mm per minute. As the part of the coal is burnt in suspension, and as the caking properties of the remaining coal are reduced during the fall, the spreader stokers successfully burn the strongest caking coal. This is an important advantage over the chain-grate and travelling-grate stokers. In fact the spreader stokers have remarkable adaptability to a wide range of coals, ability to respond quickly to a change of load and relative freedom from slag and deposit problems. These advantages have contributed to the immense popularity of spreader stokers in small to large installations. The disadvantages of fly ash and smoke problems are largely obviated by using over-fire jets (Fig. 9.21). The spreader stokers are usually of the open- or archless-type.

High burning rates of 300 to 450 kg/m^2 per hour are quite typical in spreader stokers. By providing a rear arch in the continuous-discharge type spreader stoker, a very high rate of 750 kg/m^2 per hour can be achieved in locomotive boilers.

Sprinkler stokers

They are similar to spreader stokers. However, less attention is paid to the suspension combustion in these stokers. Their burning rates are also much lower, being about 150 kg/m^2 per hour. The sprinkler stokers have stationary grates.

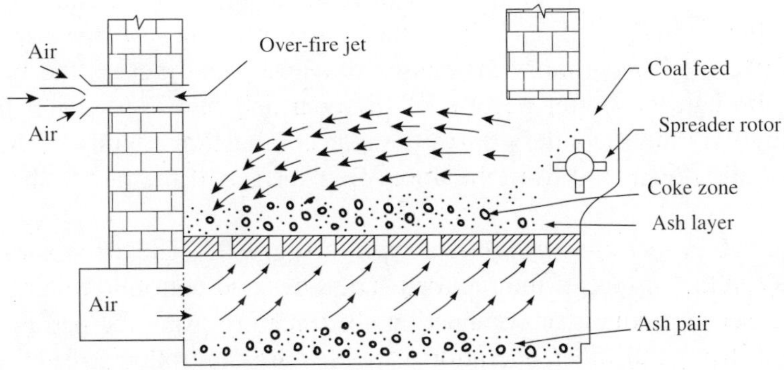

Figure 9.21 Spreader stoker

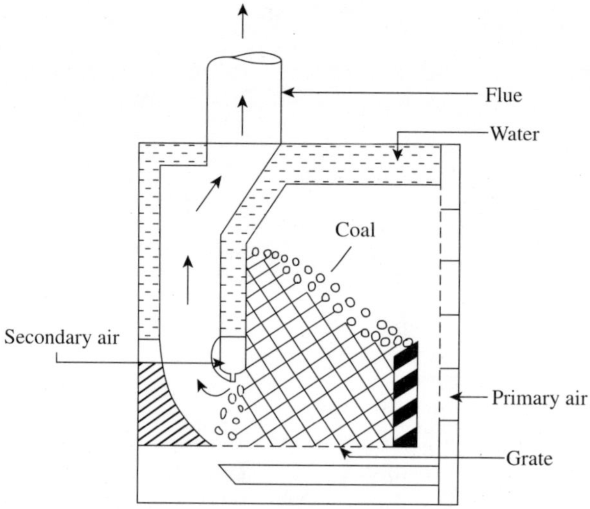

Figure 9.22 Downdraft heater

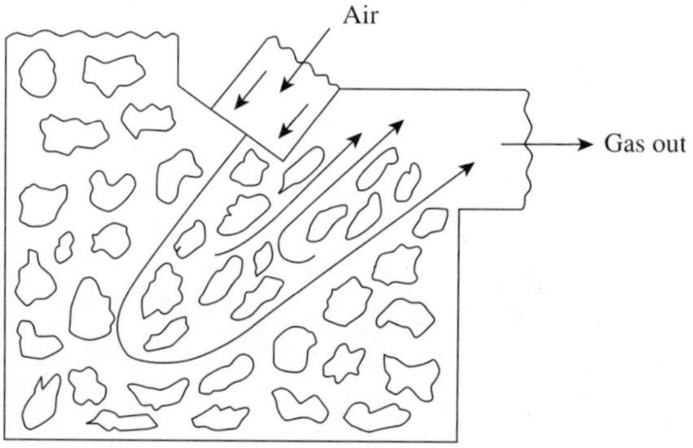

Figure 9.23 Principle of downjet firing

Downdraft heaters and *downjet* combustors have not been covered in the above discussions of lump-coal-firing appliances. Figure 9.22 shows the principles of downdraft combustion. Coal is fed by gravity. The primary air passes downward and across the bed. Any smoke formed passes through the hotter portion of the coal and burns, partly in this region and finally by mixing with the secondary air in the rear combustion chamber. The draft through the bed is essentially downward from the upper free surface of the bed to the bottom rear of the unit.

The principle of *downjet combustion* is shown in Fig. 9.23. The air is supplied through a jet that impinges on the surface of the fuel bed. The sides and bottom of the bed are enclosed. Therefore, the gaseous products of combustion are forced to leave the bed through its upper surface. The bed depth and other operation variables are controlled in such a way that the content of combustibles in the exit gas is negligible.

9.4.3 Pulverised Coal Firing

Most big installations like thermal power stations and cement rotary kilns fire coal in a powdered form. The combustion appliances used are called *pulverised fuel* (PF) burners. These burners can be designed for exceedingly high capacity. The capacity of such units is limited only by the size of the combustion space that can be provided. Boilers with more than 1×10^6 kg steam per hour are numerous and these have a huge cathedral-like size.

The fuel is pulverised in a ball mill or impact mill. The degree of fineness required depends upon the coal rank: 60% through 200 mesh (0.07 mm) for brown coals and sub-bituminous coals and upto 85% for anthracites. Not more than about 2% should be retained on a 50 mesh (0.3mm). Preheated air at 110–300°C is passed over the coal during grinding. This raises the capacity of the pulveriser by drying the fuel; the air also serves to transport the coal to the burner and acts as the primary air which constitutes about 15%–20% of the total combustion air. The temperature of the coal–air mixture fed to the burners is usually maintained at 70–80°C. The solid content of the coal–primary air mixture is about 0.5 kg/m^3. The old installations use the central storage or bin-and-feeder system wherein the prepared coals from several pulverisers were sent to a central bin for further delivery to individual burners. Current practice employs the unit or direct-fired system, in which the pulverisers directly feed the burners. The central storage systems suffer from the risk of explosion of coal–air mixtures and the tendency of the stored coal fines to granulate.

Depending upon the method of ash removal the PF furnaces may be *dry-bottom* or *slag-tap* type. Dry furnaces are more common of the two. In the dry-bottom furnace, the ash particles are cooled to below their softening point. The smaller particles—about 85% of the total ash—are entrained in the gas stream while the larger particles (15%) settle on the dry bottom of the combustion chamber. The settled refuse is removed periodically or continuously by a hydraulic system. Most of the fly ash is trapped in a collecting system while about 2% of the total ash escapes into the atmosphere. An electrostatic precipitator can ensure a total trapping of this ash.

In the slag-tap PF furnace, the ash flows molten from the combustion chamber into a tank of water below. It is converted into a suitable product for road filling, as cement aggregate and for brick making. The rest of the ash leaves the chamber as entrained particles. A collection system taps most of the fly ash while the loss to the atmosphere is about 2%. With the slag-tap type, coals with low ash fusion points present no difficulty while the furnace allows a higher combustion temperature with a high heat release rate. This type of furnace is not suitable for Indian coals with high ash fusion points.

There are three different types of PF burners. They are based on the method of firing: vertical, horizontal or tangential (Fig. 9.24).

In vertical firing, several burners are located in the furnace arch. The nozzle of each burner comprises a long narrow slot through which the coal–primary air mixture is introduced into the furnace vertically downward. After penetrating into a certain depth the flame reverses and then flows upward to attain a fantail shape with large width but small thickness. The long axis of the fantail is set at right angles to the front wall of the furnace. There are thus several

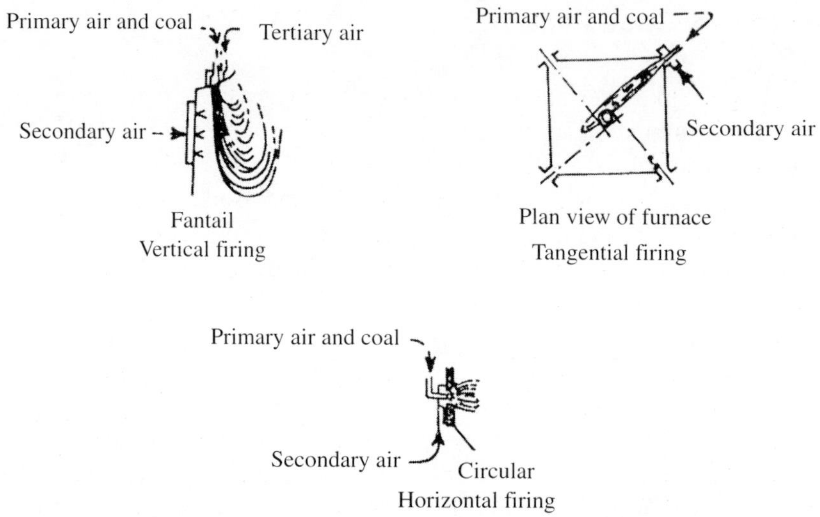

Figure 9.24 Pulverised coal firing

alternate lanes of flame and air space corresponding to the number of burners. The nozzle is riffled so as to ensure turbulence. Secondary air is supplied through ports in the front of the furnace. The length and intensity of the flame can be widely varied by controlling the points of secondary air supply. Tertiary air under pressure is admitted around the nozzles to help regulate the point of ignition. The depth of the flame travel may be adjusted by controlling the primary and tertiary air.

A horizontal PF burner essentially consists of a central coal nozzle which has a series of internal directing ribs. The coal nozzle is surrounded by a housing provided with adjustable vanes. The coal–primary air mixture enters the nozzle tangentially at the rear. The directing ribs spiral the mixture and assure its uniform composition over the periphery of the burner outlet. Secondary air is given a spinning action by the adjustable vanes as it flows through the housing around the coal nozzle. By controlling the angularity of these vanes it is possible to control the turbulence and hence the size and shape of the flame in the furnace. The burners are usually located in the front or the rear furnace wall. Secondary air is supplied under pressure.

In tangential firing, burners are located in each of the four corners of the furnace and tangentially aimed to a small imaginary circle lying at a horizontal plane at the centre. Intensive mixing occurs where the coal–air streams meet and promote rapid combustion. Ignition at each burner is aided by the flame from the preceding one. The whole furnace virtually acts as the burner. Therefore, it is not necessary to accurately proportion the coal–air mixture at individual burners. In some boiler designs, the burners can be tilted up or down, primarily to control the superheated steam temperature in case of variation in load.

The vertical firing leads to extended and progressive heat release and thus provide a relatively uniform temperature distribution throughout the furnace. It also furnishes a longer path of travel for fuel particles. Complete combustion of low volatile coal is therefore ensured.

Vertical firing is especially good for high, narrow furnaces. These burners are not easily adaptable for fuels other than coal.

Horizontal PF burners are capable of unusually high capacities—upto 400 kg coal per hour per burner. Therefore a lesser number of horizontal burners are required compared to other methods of PF firing. This considerably simplifies coal preparation and delivery system. The horizontal PF burners can be quickly changed over from one fuel to another including gas and oil. By installing the horizontal burners in both the front and back furnace walls, high turbulence can be created and extra-high firing rates achieved.

Tangential firing is the most effective method of achieving the high rate of heat release with solid fuels—upto 0.5 million kcal/m^3 per hour of the combustion space. These burners are also readily adaptable for oil or gas firing.

Although PF firing offers an efficient and intense method of combustion—85% to 90% efficiency—of solid fuels of even inferior quality, high ash and low rank coals and permits utilisation of coal slacks obtained during mining, transportation and washing, it suffers from certain limitations and undesirable features.

There are two serious disadvantages:

1. Large power requirement and maintenance cost for pulverisers,
2. Problems of fly ash such as erosion of boiler parts and induced draft fans, cost of collection of entrained particles and pollution of atmosphere.

These have generated increased interest for other methods of burning coal in large installations.

9.4.4 Cyclone Firing

This method of burning coal is similar to PF firing in that it employs the principle of combustion of coal particles in suspension. It is however, a step forward and incorporates a number of advantages over PF firing. The required degree of fineness is 95%, through 0.6 mm, which needs only simple grinders. The rate of heat release is very high, 5 million kcal/m^3 per hour. The problems of low ash fusion point coal is solved by maintaining more than 80% of the ash as a molten slag, and the problem of fly ash is significantly decreased to about 10% of total ash.

A cyclone burner consists of a water-cooled horizontal cylinder of about 2.5 m diameter and 3.5 m length (Fig. 9.25). Coal and primary air, that is, about 20% of total air, tangentially enter the burner at one end. Coal particles thus have a whirling motion. Secondary air, which is 75% of total air, is admitted in the same direction tangentially at the roof of the main chamber and imparts a further whirling motion to the solid. A very fast combustion ensures, and owing to the high rate of heat release gas temperatures exceeding 1,600°C are developed. The ash melts into a liquid slag and forms a layer on the walls. The incoming coal particles are thrown by the centrifugal force on to the slag layer and stick to it while the fast moving gases continue to apply a scrubbing action on them. This contributes to the achievement of very high rates of heterogeneous combustion of solid fuel. About 5% of total air is supplied as tertiary air at the burner axis to aid the burning of coal fines in the core of the vortex created by the whirling motion.

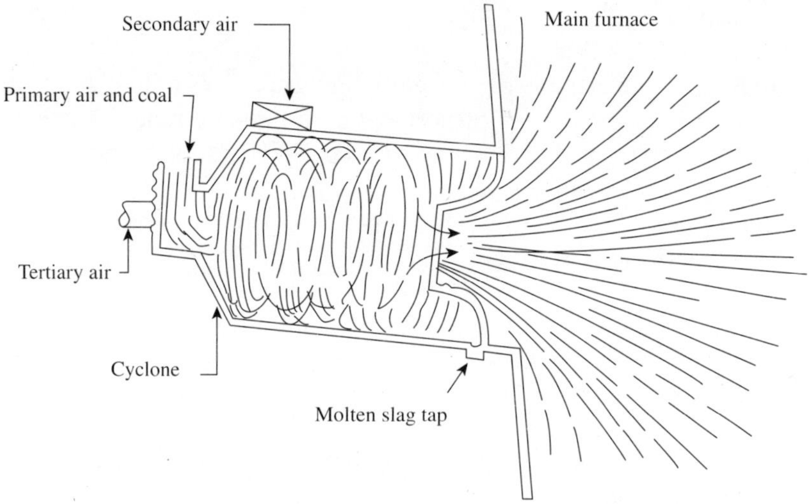

Figure 9.25 Cyclone firing

The exit end of the burner has a re-entrant throat which reverses the axial motion of the combustion gases and forces them radially inwards before they leave the cyclone. The molten slag slowly but continually flows down the wall toward the tap hole at the rear end. It is tapped into a slag tank and then disposed of. The exit gases entrain relatively small quantities of fly ash and enter into the main furnace chamber for delivering heat to the cold surfaces. The particle size of the fly ash is so small that erosion of heating surface (boiler tubes) is not experienced even at high gas velocities.

Cyclone burners are suitable for almost any grade of coal with minimum fuel preparation. They are popular in medium to large installations. Although they have a limited turndown ratio, large furnaces can be provided with a number of such units. The independent operation of these points offers the needed flexibility. Since Indian coals are characterised by high ash fusion points, cyclone furnaces are not used in the country.

9.4.5 Fluidised Bed Combustion (FBC)

This is another method of burning coal in suspension. The coal is crushed to an upper size of about 2–6 mm and fed to a refractory lined cylinder to form a bed 0.5–2.0 m in depth. Air is supplied through a perforated distributor plate at the bottom at a velocity of 0.4 to 4.0 m/s to form a fluidised bed. In this state the coal–air system behaves like a fluid; it finds its own level and has a hydrostatic head. A pressure drop of about 1 mm water gauge per millimetre depth of bed, equivalent to the weight of particles, is necessary to sustain fluidisation. Additional mass flow creates turbulence and bubbles. Accompanying the builing, particles are ejected from the bed. Some fall back to the bed and others are carried out of the vessel with the air stream. A clearance (freeboard) of about 3 m is provided at the top of the bed to avoid undue loss of particles by splashing. Coal is continuously fed to the bed via immersed nozzles fed by pneumatic pipelines.

The turbulence of a fluidised bed causes rapid mixing of the particles. The rapid relative movement between particles and fluidising medium is ideal for combustion of the particles. Temperatures range between 750 and 1,000°C, the lower level providing a reasonable margin above the minimum required for maintenance of combustion and the upper temperature should be such as to avoid sintering of ash particles. The burning rate of coal is about 40–400 kg/ m^2 per hour of bed area, giving a heat release of 0.25 to 2.5 million kcal/m^2 per hour. Ash may be removed from the bed by gravity flow and also as mechanical carry over with the exit gas. Some fine coal is also elutriated, measuring 2% to 10% of the input. This is recovered in a cyclone and recycled to the bed. A high combustion efficiency of 97% or more is achieved.

- A distinct advantage in fluidised bed combustion system is that cooling tubes may be immersed in the bed and approximately half of the heat can be removed and recovered by this means. A very high heat transfer coefficient, 125 to 430 kcal/m^2 C per hour prevails in the bed. The enthalpy of the exit gases can be extracted by usual waste heat boiler systems.
- There is a great flexibility in the choice of coal feed for FBC systems. Very high ash coals and even washery tailings are easily burnt.
- There are several favourable environmental considerations. The particulates from the fluidised combustor are coarser than from the PF burners and are therefore readily trapped in simple cyclones.
- The sulphur in the coal is converted to sulphate in the presence of limestone or dolomite fed into the bed. There is therefore no sulphur emissions in the exit gas.
- Since the bed temperature is below 1,000°C, little atmospheric nitrogen is converted to NO_x. This form of pollution is thus less in comparison with conventional combustion.

Fluidised bed combustors are now commercially available in India and abroad in applications like electricity generation and industrial boilers where steam is the main product. There are also FBC systems where hot flue gases are used to dry grass and other material. The FBC units discussed above operate at atmospheric pressure. Attention is also being given to the development of pressurised fluidised bed combustors at 6–12 atm for electricity generation in the combined cycle of gas turbines and steam turbines. Operation under pressurised conditions improves combustion efficiency and sulphur retention while lowering the NO_x emission; the whole unit also becomes more compact.

Bibliography

1. Bureau of Indian Standards, New Delhi.
   IS: 1342 (1988) Oil pressure stoves
   IS: 4246(2002), IS: 4473(2002), IS: 4760(1992) — Domestic gas stoves, ovens, for use with LPG
   IS: 5116(1996), IS: 5117(1993) — Commercial burners for use with LPG

2. *Gas Engineers Handbook*. New York: Industrial Press, 1966.
3. Giffen E and Muraszew A. *The Atomisation of Liquid* Fuels. London: Chapman and Hall, 1953.
4. Johnson AJ and Auth GH. *Fuels and Combustion Handbook.* New York: McGraw-Hill, 1951. Tokyo: Mcgraw-Hill and Kogakusha, 1973.
5. Perry IH. *Chemical Engineers' Handbook*, Fifth edition. Tokyo: McGraw-Hill and Kogakusha, 1973.
6. Sherman RA and Landry BA. Combustion Processes. In Lowry HH, ed. *Chemistry of Coal Utilization*, Supp. Vol. New York: Wiley, 1963.
7. Shindon L. *Gaseous Fuels*. Second edition. New York: American Gas Association, 1954.
8. Williams DA and Jones G. *Liquid Fuels.* London: Pergamon, 1963.

APPENDIX I

Mean Specific Heat of Gases at Constant Pressure Between 0°C and T°C (Mass basis, kcal/kg, °C)

t°C	Oxygen	Nitrogen	Air	Carbon monoxide	Hydrogen	Carbon dioxide	Water vapour	Methane	Ethylene	Ethane	Benzene vapour	Propane	Butane
0	0.218	0.248	0.240	0.249	0.40	0.195	0.443	0.496	0.342	0.397	0.224	0.359	0.342
100	0.220	0.248	0.240	0.249	3.44	0.208	0.445	0.545	0.387	0.450	0.273	0.407	0.390
200	0.223	0.249	0.242	0.250	3.45	0.219	0.450	0.590	0.431	0.501	0.316	0.456	0.438
300	0.237	0.250	0.243	0.252	3.46	0.228	0.456	0.634	0.470	0.548	0.356	0.504	0.486
400	0.230	0.252	0.246	0.254	3.46	0.236	0.462	0.677	0.508	0.594	0.390	0.553	0.534
500	0.234	0.254	0.248	0.257	3.47	0.244	0.470	0.719	0.542	0.636	0.422	0.601	0.582
600	0.237	0.256	0.250	0.259	3.48	0.250	0.477	0.758	0.574	0.676	0.448	0.650	0.630
700	0.240	0.258	0.253	0.262	3.49	0.256	0.485	0.794	0.602	0.713	0.474	0.698	0.678
800	0.243	0.261	0.256	0.265	3.50	0.261	0.494	0.830	0.628	0.748	0.494	0.747	0.726
900	0.245	0.264	0.258	0.268	3.51	0.266	0.501	0.864	0.651	0.780	0.514	0.795	0.774
1000	0.247	0.266	0.260	0.270	3.53	0.270	0.510	0.896	0.671	0.810	0.534	0.844	0.822
1100	0.249	0.268	0.263	0.273	3.55	0.274	0.517						
1200	0.251	0.271	0.265	0.275	3.57	0.277	0.525						
1300	0.253	0.272	0.267	0.277	3.59	0.280	0 531						
1400	0.255	0.275	0.269	0.278	3.61	0.283	0.540						
1500	0.256	0.276	0.271	0.280	3.63	0.85	0.546						
1600	0.258	0.278	0.272	0.282	3.65	0.288	0.552						
1700	0.258	0.280	0.273	0.283	3.67	0.290	0.560						
1800	0.260	0.281	0.274	0.285	3.69	0.292	0.566						
1900	0.262	0.282	0.276	0.286	3.71	0.293	0.572						
2000	0.263	0.284	0.277	0.287	3.74	0.295	0.577						
2100	0.264	0.285	0.278	0.289	3.76	0.297	0.583						
2200	0.265	0.286	0.279	0.290	3.78	0.298	0.588						
2300	0.266	0.288	0.280	0.291	3.80	0.300	0.595						
2400	0.267	0.288	0.281	0.292	3.82	0.301	0.598						
2500	0.268	0.289	0.282	0.293	3.84	0.302	0.603						
2600	0.269	0.291	0.283	0.294	3.86	0.303	0.608						
2700	0.270	0.291	0.284	0.295	3.87	0.304	0.612						
2800	0.271	0.292	0.284	0.295	3.89	0.305	0.617						
2900	0.272	0.293	0.285	0.296	3.91	0.306	0.621						
3000	0.274	0.294	0.286	0.297	3.93	0.307	0.623						

Note: Values for the undissociated gas. At temperatures above 1,500°C dissociation must be taken into account.

APPENDIX 2

Mean Specific Heat of Gases at Constant Pressure Between 0°C and T°C (Mass basis, kcal/kg, °C)

t°C	Oxygen	Nitrogen	Air	Carbon monoxide	Hydrogen	Carbon dioxide	Water vapour	Methane	Ethylene	Ethane	Benzene vapour	Propane	Butane
0	0.312	0.311	0.310	0.311	0.306	0.384	0.356	0.355	0.428	0.532	0.78	0.707	0.886
100	0.315	0.311	0.311	0.311	0.309	0.409	0.358	0.390	0.485	0.603	0.95	0.802	1.011
200	0.319	0.312	0.313	0.312	0.310	0.431	0.362	0.422	0.539	0.671	1.10	0.898	1.135
300	0.324	0.314	0.315	0.315	0.311	0.449	0.367	0.454	0.589	0.735	1.24	0.993	1.260
400	0.329	0.316	0.318	0.318	0.311	0.464	0.372	0.484	0.635	0.796	1.36	1.089	1.384
500	0.334	0.319	0.321	0.321	0.321	0.480	0.378	0.514	0.678	0.853	1.47	1.184	1.509
600	0.339	0.322	0.324	0.324	0.313	0.492	0.384	0.542	0.718	0.906	1.56	1.280	1.633
700	0.343	0.324	0.328	0.328	0.314	0.503	0.390	0.568	0.754	0.956	1.65	1.375	1.758
800	0.347	0.328	0.331	0.332	0.315	0.513	0.397	0.594	0.786	1.003	1.72	1.471	1.882
900	0.350	0.331	0.334	0.335	0.316	0.522	0.403	0.618	0.815	1.046	1.79	1.566	2.007
1000	0.353	0.334	0.337	0.338	0.318	0.530	0.410	0.641	0.840	0.085	1.86	1.662	2.131
1100	0.356	0.337	0.340	0.341	0.319	0.538	0.416						
1200	0.359	0.340	0.343	0.344	0.321	0.544	0.422						
1300	0.362	0.342	0.345	0.346	0.323	0.550	0.427						
1400	0.364	0.345	0.347	0.348	0.325	0.556	0.434						
1500	0.366	0.347	0.350	0.350	0.327	0.560	0.439						
1600	0.368	0.349	0.352	0.352	0.328	0.565	0.144						
1700	0.369	0.351	0.353	0.354	0.330	0.569	0.450						
1800	0.372	0.353	0.355	0.356	0.332	0.573	0.455						
1900	0.374	0.354	0.357	0.358	0.334	0.576	0.460						
2000	0.376	0.356	0.358	0.359	0.336	0.580	0.464						
2100	0.377	0.357	0.360	0.361	0.338	0.583	0.469						
2200	0.378	0.359	0.361	0.362	0.340	0.586	0.473						
2300	0.380	0.361	0.362	0.364	0.342	0.589	0.478						
2400	0.382	0.361	0.363	0.365	0.344	0.591	0.481						
2500	0.383	0.363	0.365	0.366	0.345	0.593	0.485						
2600	0.385	0.365	0.366	0.368	0.347	0.595	0.489						
2700	0.386	0.365	0.367	0.369	0.348	0.597	0.492						
2800	0.387	0.367	0.368	0.369	0.350	0.599	0.496						
2900	0.389	0.368	0.369	0.370	0.352	0.601	0.499						
3000	0.391	0.369	0.370	0.371	0.353	0.603	0.501						

Note: Values for the undissociated gas. At temperatures above 1,500°C dissociation must be taken into account.

APPENDIX 3

Enthalpy of Gases Above 0°C (Mass basis, kcal/kg)

t°C	Oxygen	Nitrogen	Air	Carbon monoxide	Hydrogen	Carbon dioxide	Water vapour	Methane	Ethylene	Ethane	Benzene vapour	Propane	Butane
100	22.0	24.8	24.0	24.9	344	20.8	44.8	54.5	38.8	45.0	27.3	40.7	39.0
200	44.7	49.8	48.4	50.0	689	43.8	90.2	118.0	86.1	100.1	63.1	91.2	87.6
300	68.0	75.0	73.1	75.6	1037	68.6	136.8	190.4	141.2	164.5	106.5	151.3	145.8
400	92.2	100.7	98.4	101.7	1386	94.5	185.1	270.7	202.9	237.5	156.1	221.2	213.6
500	117.0	127.0	124.1	128.3	1734	122.1	235.1	359.3	270.9	318.2	210.6	300.7	291.0
600	142.2	153.6	150.4	155.7	2087	150.3	286.4	454.6	344.2	405.5	269.2	390.0	378.0
700	167.9	180.8	177.2	183.4	2442	179.3	339.0	555.9	421.7	499.2	331.2	488.9	474.6
800	194.2	208.9	204.8	212.2	2802	209.9	394.7	664.3	502.4	598.6	396.1	597.6	580.8
900	220.6	237.2	232.5	241.0	3166	239.1	451.5	777.6	585.7	702.3	463.5	715.9	696.6
1000	247.4	266.1	260.7	270.0	3533	269.8	509.5	896.1	671.2	809.4	533.0	844.0	822.0
1100	274.2	295.4	289.1	300.0	3902	301.0	569.0						
1200	30.13	324.8	317.8	329.8	4287	332.2	629.5						
1300	329.3	354.1	346.6	360.1	4670	363.7	691.3						
1400	356.8	384.4	375.7	390.3	5077	395.8	755.4						
1500	384.2	414.5	405.6	420.3	5448	427.9	819.3						
1600	411.8	444.3	434.9	451.2	5842	460.0	884.5						
1700	439.6	474.6	464.4	481.8	6242	492.2	952.2						
1800	468.3	505.7	494.2	512.7	6653	524.9	1019.1						
1900	497.3	535.8	523.6	543.9	7061	557.5	1086.3						
2000	526.0	566.8	554.0	574.7	7473	590.0	1155.7						
2100	554.3	597.4	583.8	606.4	7887	622.9	1226.2						
2200	582.7	629.0	613.1	637.7	8318	656.0	1295.6						
2300	612.0	660.0	644.2	669.1	8741	689.0	1367.3						
2400	641.7	690.5	673.9	700.8	9170	721.7	1437.4						
2500	670.7	722.8	705.3	732.6	9601	755.1	1508.4						
2600	700.7	754.5	735.4	764.7	10037	787.7	1581.7						
2700	729.5	785.4	766.4	796.1	10463	821.1	1653.0						
2800	759.1	817.4	796.7	827.5	10905	854.0	1726.7						
2900	789.8	848.7	827.2	859.2	11352	887.8	1799.6						
3000	820.8	880.1	857.8	891.0	11788	921.2	1870.0						

Note: Values for the undissociated gas. At temperatures above 1,500°C dissociation must be taken into account.

APPENDIX 4

Enthalpy of Gases Above 0°C (Volume basis, kcal/Nm^3)

t°C	Oxygen	Nitrogen	Air	Carbon monoxide	Hydrogen	Carbon dioxide	Water vapour	Methane	Ethylene	Ethane	Benzene vapour	Propane	Butane
100	31.5	31.1	31.1	31.1	30.9	40.9	35.8	39.0	48.5	60.3	95	80.2	101.1
200	63.8	62.5	62.6	62.5	62.0	86.1	72.5	84.4	107.8	134.2	200	179.6	227.0
300	97.2	94.2	94.5	94.5	93.3	134.7	110.0	136.2	176.7	220.5	371	297.9	378.0
400	131.7	126.5	127.3	127.1	124.6	185.6	148.8	193.6	254.0	318.4	544	435.6	553.6
500	167.1	159.5	160.6	160.4	155.9	239.8	189.0	257.0	339.0	426.5	734	592.0	754.5
600	203.2	193.0	194.6	194.6	187.7	295.3	230.2	325.2	430.8	543.6	938	768.0	979.8
700	239.9	227.1	229.2	229.2	219.6	352.3	273.0	397.6	527.8	669.2	1154	962.5	1230.6
800	277.4	262.4	264.9	265.2	252.0	410.5	317.3	475.2	628.8	802.4	1380	1176.8	1505.6
900	315.2	298.0	300.8	301.2	287.7	469.8	363.0	556.2	733.5	941.1	1615	1409.4	1806.3
1000	353.4	334.2	337.3	337.8	317.7	530.1	409.6	641.0	840.0	1085.0	1857	1662.0	2131.0
1100	391.7	371.1	374.0	375.0	350.9	591.4	457.4						
1200	430.5	408.0	411.2	412.3	385.5	652.7	506.0						
1300	470.4	444.9	448.4	450.1	420.0	714.6	555.7						
1400	509.7	482.9	486.0	487.9	454.8	777.7	607.2						
1500	548.8	520.7	524.7	525.4	489.9	440.6	658.6						
1600	588.3	558.3	562.6	564.0	525.4	903.8	711.0						
1700	628.0	596.2	600.7	602.3	561.3	967.1	765.4						
1800	669.0	635.3	639.3	640.9	598.3	1031.2	819.2						
1900	710.4	673.1	6774	679.9	635.0	1095.3	873.2						
2000	751.4	712.1	716.6	718.4	672.0	1159.2	929.0						
2100	791.8	750.5	755.2	758.0	709.3	1223.7	985.7						
2200	832.4	790.2	793.1	797.1	748.0	1288.9	1041.5						
2300	874.3	829.2	833.3	836.4	786.1	1353.6	1099.1						
2400	916.7	867.4	871.7	876.0	824.6	1417.8	1155.5						
2500	958.2	908.0	912.4	915.8	863.4	1483.6	1212.5						
2600	1001.2	947.8	95.13	955.9	902.6	1547.6	1271.5						
2700	1042.1	986.7	991.5	995.1	940.9	1613.1	1328.8						
2800	1084.4	1026.9	1030.7	1034.4	980.7	1677.7	1388.0						
2900	1128.3	1066.2	1070.1	1074.0	1020.9	1744.2	1446.6						
3000	1172.6	1105.6	1109.7	1113.7	1060.1	1809.8	1503.2						

Note: Values for the undissociated gas. At temperatures above 1,500°C dissociation must be taken into account.

APPENDIX 5

Equilibrium Constants, K_p, of Some Combustion Reactions

Temperature K	Dissociation of water vapour	Dissociation of carbon dioxide	Producer gas equilibrium	Formation of carbon monoxide	Water gas equilibrium	Dissociation of hydrogen	Dissociation of oxygen	Dissociation of water vapour into H_2 and OH	Formation of nitric oxide
	$\frac{P_{H_2O}}{P_{H_2} \times P_{O_2}}$	$\frac{P_{CO_2}}{P_{CO} \times P_{O_2}^{0.5}}$	$\frac{P_{CO}^2}{P_{CO_2}}$	$\frac{P_{CO}}{P_{O_2}^{0.5}}$	$\frac{P_{CC} \times P_{H_2O}}{P_{CC_2} \times P_{H_2}}$	$\frac{P_{H_2}}{P_H^2}$	$\frac{P_{O_2}}{P_O^2}$	$\frac{P_{H_2O}}{P_{OH} \times P_{H_2}^{0.5}}$	$\frac{P_{NO}}{P_{O_2}^{0.5} \times P_{NO_2}^{0.5}}$
(1)	(2)	(3)	(4)	(5)	(6)	(7)	(8)	(9)	(10)
298.16	1.11×10^{40}	1.11×10^{45}	1.01×10^{-21}	1.12×10^{24}	1.01×10^{-5}	–	–	–	–
300	6.12×10^{39}	5.49×10^{44}	1.55×10^{-21}	8.48×10^{23}	1.11×10^{-5}	5.50×10^{70}	1.23×10^{50}	1.95×10	8.14×10^{-16}
400	1.74×10^{29}	2.57×10^{32}	5.21×10^{-14}	1.34×10^{19}	6.76×10^{-4}	5.50×10^{51}	3.24×10^{58}	1.13×10^{33}	6.99×10^{-43}
500	7.68×10^{22}	1.01×10^{35}	1.77×10^{-9}	1.79×10^{16}	7.60×10^{-3}	–	–	–	–
600	4.29×10^{18}	1.16×10^{30}	1.87×10^{-6}	2.17×10^{14}	3.69×10^{-2}	4.27×10^{33}	7.24×10^{36}	2.95×10^{21}	6.04×10^{-8}
700	3.83×10^{15}	3.45×10^{16}	2.67×10^{-4}	9.22×10^{12}	1.11×10^{-1}	–	–	–	–
800	1.94×10^{13}	7.85×10^{13}	1.09×10^{-2}	8.55×10^{11}	2.48×10^{-1}	1.18×10^{23}	9.77×10^{25}	1.66×10^{15}	5.62×10^{-4}
900	3.14×10^{11}	6.93×10^{11}	1.92×10^{-1}	1.33×10^{11}	4.53×10^{-1}	–	–	–	–
1000	1.15×10^{10}	1.58×10^{10}	1.90	3.00×10^{10}	7.29×10^{-1}	1.94×10^{17}	2.75×10^{19}	2.75×10^{21}	8.55×10^{-4}
1100	7.63×10^{8}	7.21×10^{8}	1.22×10^{1}	8.80×10^{9}	1.06	–	–	–	–
1200	7.92×10^{7}	5.52×10^{7}	5.71×10^{1}	3.15×10^{9}	1.44	2.57×10^{13}	1.15×10^{15}	8.36×10^{8}	5.25×10^{-4}
1300	1.16×10^{7}	6.29×10^{8}	2.08×10^{2}	1.31×10^{9}	1.84	–	–	–	–
1400	2.23×10^{5}	9.81×10^{5}	6.29×10^{2}	6.17×10^{8}	2.27	4.24×10^{10}	8.55×10^{11}	1.31×10^{7}	1.92×10^{-3}
1500	5.32×10^{5}	1.97×10^{5}	1.62×10^{3}	3.19×10^{8}	2.70	–	–	–	–
1600	1.50×10^{5}	4.83×10^{4}		1.80×10^{8}	3.09	3.39×10^{8}	3.76×10^{9}	5.73×10^{5}	5.08×10^{-3}
1800	1.83×10^{4}	4.70×10^{3}		6.75×10^{7}	3.90	7.82×10^{6}	5.50×10^{7}	5.01×10^{4}	1.08×10^{-2}
2000	3.40×10^{3}	7.30×10^{2}		3.06×10^{7}	4.66	3.79×10^{5}	1.86×10^{6}	7.11×10^{3}	1.98×10^{-2}
2200	8.53×10^{2}	1.61×10^{2}		1.59×10^{7}	5.31	3.16×10^{4}	1.16×10^{5}	1.44×10^{3}	3.25×10^{-2}
2400	2.69×10^{2}	4.59×10^{1}		9.14×10^{6}	5.85	3.96×10^{3}	1.14×10^{4}	3.78×10^{2}	4.91×10^{-2}
2600	1.01×10^{1}	1.60×10^{1}		5.74×10^{6}	6.31	6.81×10^{2}	1.61×10^{3}	1.22×10^{1}	6.97×10^{-2}
2800	4.35×10^{1}	6.41		3.78×10^{6}	6.78	1.50×10^{2}	2.99×10^{2}	4.68×10^{1}	9.38×10^{-2}
2000	2.10×10^{1}	2.94		2.62×10^{6}	7.13	4.02×10^{1}	6.92×10^{1}	2.00×10^{1}	1.22×10^{-4}
3200	1.11×10^{1}	1.50		1.88×10^{6}	7.43	1.27×10^{1}	1.93×10^{1}	9.62	1.52×10^{-4}
3500	4.93	6.30×10^{-1}		1.24×10^{6}	7.83	2.87	3.72	3.78	2.03×10^{-4}
4000	1.66	2.00×10^{-4}		6.93×10^{5}	8.32	3.93×10^{-1}	1.15×10^{-4}	1.08	2.98×10^{-4}

Note: Carbon, where involved, is in the form of solid ß-graphite.

INDEX